Understanding the Infinite

Understanding the Infinite

Shaughan Lavine

Harvard University Press

Cambridge, Massachusetts

London, England

1994

Library of Congress Cataloging-in-Publication Data
Lavine, Shaughan.
Understanding the infinite / Shaughan Lavine.
p.   cm.
Includes bibliographical references and index.
ISBN 0-674-92096-1 (alk. paper)
1. Mathematics—Philosophy.   I. Title.
QA8.4.L38   1994
511.3'22—dc20
93-49697
CIP

Preface

In writing this book I have tried to keep mathematical prerequisites to a minimum. The reader who is essentially innocent of mathematical knowledge beyond that taught in high school should be able to read at least halfway through Chapter VIII plus parts of the rest of the book, though such a reader will need to skip the occasional formula. That is enough of the book for all of the major ideas to be presented. The Introduction may seem daunting since it refers to ideas that are not explained until later—trust me, they *are* explained. A reader who learned freshman calculus once, but perhaps does not remember it very well, and who has had a logic course that included a proof of the completeness theorem will be in fine shape throughout the book, except for various "technical remarks," an appendix to Chapter VI, and a few parts of Chapter IX. Those few technical discussions require varying degrees of mathematical sophistication and knowledge of general mathematical logic plus occasional knowledge of elementary recursion theory, model theory, or modal logic.

Thanks are due Bonnie Kent, Vann McGee, Sidney Morgenbesser, and Sarah Stebbins for their infinite patience in listening to my many half-baked ideas and for their substantial help in culling and completing them while I was writing this book. As they learned, I cannot think without the give and take of conversation. Thanks also to Ti-Grace Atkinson, Jeff Barrett, William Boos, Hartry Field, Alan Gabbey, Haim Gaifman, Alexander George, Allen Hazen, Gregory Landini, Penelope Maddy, Robert Miller, Edward Nelson, Ahmet Omurtag, David Owen, Charles Parsons, Thomas Pogge, Vincent Renzi, Scott Shapiro, Mark Steiner, and Robert Vaught for their thoughtful comments on an early version of the book. Those comments have led to significant improvements. And thanks to Thomas Pogge for his substantial help in correcting my translations from German. Any remaining mistakes are, of course, my own.

Thanks are also due my parents, Dorothy and Leroy Lavine, not only for their moral support, which I very much appreciated, but also for their generous financial support, without which the preparation of this book would not have been possible. My wife, Caroline, and daughter, Caila, deserve the most special thanks of all, for tolerating with such understanding my absences and the stresses on our family life that the writing of a book inevitably required. This book is dedicated to them.

Contents

Understanding the Infinite

I

Introduction

In the latter half of the nineteenth century Georg Cantor introduced the infinite into mathematics. The Cantorian infinite has been one of the main nutrients for the spectacular flowering of mathematics in the twentieth century, and yet it remains mysterious and ill understood.

At some point during the 1870s Cantor realized that sets—that is, collections in a familiar sense that had always been a part of mathematics—were worthy of study in their own right. He developed a theory of the sizes of infinite collections and an infinite arithmetic to serve as a generalization of ordinary arithmetic. He generalized his theory of sets so that it could encompass all of mathematics. The theory has become crucial for both mathematics and the philosophy of mathematics as a result. Unfortunately, Cantor had been naive, as Cantor himself and Cesare Burali-Forti realized late in the nineteenth century and as Bertrand Russell realized early in the twentieth. His simple and elegant set theory was inconsistent—it was subject to paradoxes.

The history of set theory ever since the discovery of the paradoxes has been one of attempting to salvage as much as possible of Cantor's naive theory. Formal axiom systems have been developed in order to codify a somewhat arbitrarily restricted part of Cantor's simple theory, formal systems that have two virtues: they permit a reconstruction of much of Cantor's positive work, and they are, we hope, consistent. At least the axiomatic theories have been formulated to avoid all of the known pitfalls. Nonetheless, they involve certain undesirable features: First, the Axiom of Choice is a part of the theories not so much because it seems true—it is at best controversial—but because it seems to be required to get the desired results. Second, since present-day set theory is *ad hoc*, the result of retreat from disaster, we cannot expect it to correspond in any very simple way to our uneducated intuitions about collections. Those are what got Cantor into trouble in the first place.

1

We can never rely on our intuitions again. The fundamental axioms of mathematics—those of the set theory that is its modern basis—are to a large extent arbitrary and historically determined. They are the remote and imperfectly inferred remnants of Cantor's beautiful but tragically flawed paradise.

The story I have just told is a common one, widely believed. Not one word of it is true. That is important, not just for the history of mathematics but for the philosophy of mathematics and many other parts of philosophy as well. The story has influenced our ideas about the mathematical infinite, and hence our ideas about mathematics and about abstract knowledge in general, in many deep ways.

Both elementary number theory and the geometry of the Greeks, for all that they are abstract, have clear ties to experience. They are, in fact, often thought to result from idealizing that experience. Modern mathematics, including much of the mathematics of physics, is frequently thought to be abstract in a much more thoroughgoing sense. As I shall put it, modern mathematics is not only abstract but also remote, because it is set-theoretic:[1] The story tells us that modern axiomatic set theory is the product not of idealization but of the failure of an attempted idealization.

Since science and often mathematics are thought of as quintessential examples of human knowledge, modern epistemology tries to come to grips with scientific and mathematical knowledge, to see it as knowledge of a typical or core kind. That poses a serious problem for epistemology, since mathematical knowledge and the scientific knowledge that incorporates it is thought to be so remote.

The whole picture of mathematical knowledge that drives the epistemology is wrong. As this book will demonstrate, set theory, as Cantor and Ernst Zermelo developed it, is connected to a kind of idealization from human experience much like that connected to the numbers or to Euclidean geometry.

Cantor studied the theory of trigonometric series during the 1870s. He became interested in arbitrary sets of real numbers in the process of making

---

1. When I say that modern mathematics is set-theoretic, I am not referring to the so-called set-theoretic foundations of mathematics, which play little role in this book. What I have in mind is the ubiquitous use of set-theoretic concepts in mathematics, concepts like open set, closed set, countable set, abstract structure, and so on and on. The concepts mentioned were, as we shall see in Chapter III, introduced by Cantor in the course of the same investigations in which he introduced his theory of infinite numbers and their arithmetic.

that theory apply to more general classes of functions. His work was part of a long historical development that had in his day culminated in the idea that a function from the real numbers to the real numbers is just any association—however arbitrary—from each real number to a single other real number, the value of the function. The term *arbitrary* is to make it clear that no rule or method of computation need be involved. That notion of a function is the one we use today.

Cantor's study of the theory of trigonometric series led him to this progression of transfinite "indexes":

$$0, 1, \ldots, \infty, \infty + 1, \infty + 2, \ldots, \infty \cdot 2, \ldots, \infty \cdot 3, \ldots,$$
$$\infty^2, \ldots, \infty^3, \ldots, \infty^\infty, \ldots, \infty^{\infty^\infty}, \ldots.$$

Cantor's set theory began as, and always remained, an attempt to work out the consequences of the progression, especially the consequences for sets of real numbers. Despite the usual story, Cantor's set theory was a theory not of collections in some familiar sense but of collections that can be counted using the indexes—the finite and transfinite ordinal numbers, as he came to call them. Though Cantor came to realize the general utility of his theory for codifying a large part of mathematics, that was never his main goal.

Cantor's original set theory was neither naive nor subject to paradoxes. It grew seamlessly out of a single coherent idea: sets are collections *that can be counted*. He treated infinite collections as if they were finite to such an extent that the most sensitive historian of Cantor's work, Michael Hallett, wrote of Cantor's "finitism." Cantor's theory is a part of the one we use today.

Russell was the inventor of the naive set theory so often attributed to Cantor. Russell was building on work of Giuseppe Peano. Russell was also the one to discover paradoxes in the naive set theory he had invented. Cantor, when he learned of the paradoxes, simply observed that they did not apply to his own theory. He never worried about them, since they had nothing to do with him. Burali-Forti didn't discover any paradoxes either, though his work suggested a paradox to Russell.

Cantor's theory had other problems. It did not, in its original form, include the real numbers as a set. Cantor had, for good reason, believed until the 1890s—very late in his career—that it would include them. (Most everything else I am saying here is known to one or another historian or mathematician, but the claim that Cantor had a smooth theory that broke down in the 1890s is new here. It is argued in detail in §IV.2.)[2] Cantor grafted a new assumption on to his theory as soon as he realized he needed it, an assumption that allowed him to incorporate the real numbers, but the assumption caused big trouble.

The new assumption was his version of what is today the Power Set Axiom. The trouble it caused was that his theory was supposed to be a theory of collections that can be counted, but he did not know how to count the new collections to which the Power Set Axiom gave rise. The whole theory was therefore thrown into doubt, but not, let me emphasize, into contradiction and paradox. It seemed that counting could no longer serve as the key idea. Cantor did not know how to replace it.

Zermelo came to the rescue of Cantor's theory of sets in 1904. He isolated a principle inherent in the notion of an arbitrary function, a principle that had been used without special note by many mathematicians, including Cantor, in the study of functions and that had also been used by Cantor in his study of the ordinal numbers. Zermelo named that principle the Axiom of Choice. Though the principle had been used before Zermelo without special notice, no oversight had been involved: the principle really is inherent in the notion of an arbitrary function. What Zermelo noted was that the principle could be used to "count," in the Cantorian sense, those collections that had given Cantor so much trouble, which restored a certain unity to set theory.

The Axiom of Choice was never, despite the usual story, a source of controversy. Everyone agreed that it is a part of the notion of an arbitrary function. The brouhaha that attended Zermelo's introduction of Choice was a dispute about whether the notion of an arbitrary function was the appropriate one to use in mathematics (and indeed about whether it was a coherent notion). The rival idea was that functions should be taken to be given only by rules, an idea that would put Choice in doubt. The controversy was between advocates of taking mathematics to be about arbitrary functions and advocates of taking mathematics to be about functions given by rules—not about Choice *per se,* but about the correct notion of function. Arbitrary functions have won, and Choice comes with them. There is, therefore, no longer any reason to think of the Axiom of Choice as in any way questionable.

Zermelo's work was widely criticized. One important criticism was that he had used principles that, like Russell's, led to known contradictions. He hadn't. In order to defend his theorem that the real numbers can be "counted,"

---

2. The reference is to Chapter IV, Section 2. A reference to §2 would be a reference to Section 2 of the present chapter.

Zermelo gave an axiomatic presentation of set theory and a new proof of the theorem on the basis of his axioms. The axioms were to help make it clear that he had been working on the basis of a straightforwardly consistent picture all along. That is a far cry from the common view that he axiomatized set theory to provide a consistent theory in the absence of any apparent way out of the paradoxes.

There *was* a theory developed as a retreat from the disastrous Russellian theory and its precursor in Gottlob Frege, namely, the theory of types. But it never had much to do with Cantorian set theory. I discuss it only in so far as that is necessary to distinguish it from Cantorian set theory. In the process of discussing it, I introduce a distinctive use Russell suggested for something like schemas,[3] a use that shows that schemas have useful properties deserving of more serious study. Such a study is a running subtheme of this book.

It did not take long for Thoralf Skolem and Abraham Fraenkel to note that Zermelo's axioms, while they served Zermelo's purpose of defending his theorem, were missing an important principle of Cantorian set theory—what is now the Replacement Axiom. The universal agreement about the truth of the Replacement Axiom that followed is remarkable, since the axiom wasn't good for anything. That is, at a time when Replacement was not known to have any consequences about anything except the properties of the higher reaches of the Cantorian infinite, it was nonetheless immediately and universally accepted as a correct principle about Cantorian sets.

Chapters II–V establish in considerable detail that it is the historical sketch just given that is correct, not the usual one I parodied above, and they include other details of the development of set theory. Just one more sample—the iterative conception of set, which is today often taken to be the conception that motivated the development of set theory and to be the one that justifies the axioms, was not so much as suggested, let alone advocated by anyone, until 1947.

There are three main philosophical purposes for telling the story just sketched. The first is to counteract the baneful influence of the standard account, which seems to have convinced many philosophers of mathematics that our intuitions are seriously defective and not to be relied on and that the axioms of mathematics are therefore to a large extent arbitrary, historically

---

3. A *schema* is a statement form used to suggest a list of statements. For example, $X = X$, where the substitution class for X is numerals, is a schema that has as instances, among others, $0 = 0$, $1 = 1$, and $2 = 2$.

determined, conventional, and so forth. The details vary, but the pejoratives multiply.

On the contrary, set theory is not riddled with paradoxes. It was never in such dire straits. It developed in a fairly direct way as the unfolding of a more or less coherent conception. (Actually, I think there have been two main strands in the development of the theory, symbolized above by the notion of counting and by Power Set. As I discuss in §V.5, it could be clearer how they fit together. One symptom of our lack of clarity on the issue is the independence of the Continuum Hypothesis. But that is a far cry from the usual tale of woe.)

The second purpose is to show what as a matter of historical fact we know about the Cantorian infinite on the basis of clear and universal intuitions that distinctively concern the infinite. The two most striking cases of things we know about the Cantorian infinite on the basis of intuition are codified as Choice and Replacement. How we could know such things? It seems completely mysterious. The verdict has often been that we do not—our use of Choice and Replacement is to a large extent arbitrary, historically determined, conventional, and so forth. But that is not true to the historical facts of mathematical practice, facts that any adequate philosophy of mathematics must confront. (Allow me to take the liberty of ignoring constructivist skepticism about such matters in the Introduction. I shall confront it in the text.)

The third purpose is to make clearer the nature of intuition—the basis on which we know what we do. I have been using the term *intuition* because it is so familiar, but I do not mean the sort of armchair contemplation of a Platonic heaven or the occult form of perception that the term conjures up for many. Whatever intuition is, it is very important to mathematics:

> In mathematics, as in any scientific research, we find two tendencies present. On the one hand, the tendency toward *abstraction* . . . On the other hand, the tendency toward *intuitive understanding* fosters a more immediate grasp of the objects one studies, a live *rapport* with them, so to speak, which stresses the concrete meaning of their relations.
>
> . . . It is still as true today as it ever was that *intuitive* understanding plays a major role in geometry. And such concrete intuition is of great value not only for the research worker, but also for anyone who wishes to study and appreciate the results of research in geometry.   (Page iii of David Hilbert's preface to [HCV52].)

The quotation is from a book about geometry, but the point is far more general.

Just as one scientific theory can displace another because of its superior ability to systematize, one mathematical theory can displace another. Unexpected developments can spawn new theories, which can in turn lead to fruitful developments in old theories and become so intertwined with them that the new and the old become indistinguishable. We shall see examples of those things: The modern notion of a function evolved gradually out of the desire to see what curves can be represented as trigonometric series. The study of arbitrary functions, in the modern sense, led Cantor to the ordinal numbers, which led to set theory. And set theory became so intertwined with the theories of functions and of the real numbers as to transform them completely. That is all a part of the story told in Chapters II and III. Mathematics does not have the same ties to experiment as science, but the way mathematics evolves is nonetheless very similar to the way that science evolves.

The view of mathematics just outlined is usually thought to be antithetical to the possibility of any distinctive sort of mathematical intuition. New mathematics has been thought to evolve out of old without any further constraint than what can be proved. But that cannot have been right for most of the history of modern mathematics: from, say, the first half of the seventeenth century until the second half of the nineteenth there was no coherent systematization or axiomatization for much of mathematics and certainly no adequate notion of proof.

Mathematicians necessarily saw themselves as working on the basis of an intuitive conception, relying to some extent on what was obvious, to some extent on connections with physics, and to some extent—but only to some extent, since proof was not a completely reliable procedure—on proof. (See Chapter II.) I believe that most mathematicians today still see themselves as working in much the same conceptually based and quasi-intuitive way, though that is much harder to show, since rigorous standards of proof and precise axiomatizations are now available. The intuitive conceptions that underlie mathematical theories evolve, as do the theories, but the intuitions both constrain the theories and suggest new developments in them in unexpected ways.

The development of set theory is an excellent example of the positive and necessary role intuition plays in mathematics. Because set theory is in so many respects unlike the mathematics that had gone before, it is clear that prior training was far from an adequate guide for Cantor. Besides, the progression that he found does, in some sense, have clear intuitive content. There is a great and mysterious puzzle in the suggestiveness of Cantor's progression that can hardly be overstated. The progression is infinite, and we have absolutely no experience of any kind of the infinite. So what method are we using—what method did Cantor use—to make sense of the progression? The question is another version of the one raised above about Choice and Replacement.

It is difficult to understand how we can know any mathematical truths at all, since the subject matter of mathematics is so abstract. But the problem is particularly acute for truths about the infinite. There is no doubt that we know that $2 + 2 = 4$ in some sense or other, and that that knowledge is somehow connected to our experience that disjoint pairs combine to form a quadruple. The facts are indisputable and have multifarious connections to human experience. But there *is* genuine doubt about the truth of, say, $\aleph_2 + \aleph_2 = \aleph_2$, because, for example, there is doubt about whether there could be $\aleph_2$ things.[4] Everyone agrees we must in some sense accept that $2 + 2 = 4$, but it is reasonable to be altogether skeptical about the infinite. Worse still, it is not clear what connections to human experience truths about the infinite might have. A modern philosopher of mathematics put it this way:

> The human mind is finite and the set theoretic hierarchy is infinite. Presumably any contact between my mind and the iterative hierarchy can involve at most finitely much of the latter structure. But in that case, I might just as well be related to any one of a host of other structures that agree with the standard hierarchy only on the minuscule finite portion I've managed to grasp.   [Mad90, p. 79]

There is a general philosophical problem about knowledge of abstract objects, mathematical objects in particular. But the special case of knowledge of infinite mathematical objects is a distinctive problem for which distinctive solutions have been suggested. Chapters VI and VII are concerned with that problem of the infinite. In Chapter VI, I survey various accounts of mathematical knowledge of the infinite that attempt to show how it can come out of experience. They begin with a theory of knowledge and try to fit mathematics to it. Intuitionism, various forms of formalism, and one version of David Hilbert's program are discussed. I use a Russellian picture of schemas to clarify how Hilbert's finitary mathematics could avoid any commitment to the infinite. It is a consequence of each of the philosophies surveyed that we could not know what we in fact do.

---

4. The symbol is a capital Hebrew aleph. $\aleph_2$ (pronounced "aleph two") stands for one of Cantor's infinite numbers.

In Chapter VII, I survey various accounts of mathematical knowledge of the infinite that go in the opposite direction. They begin with mathematics and try to fit a theory of knowledge to it. Kurt Gödel's views and those of Willard Van Orman Quine and Hilary Putnam are discussed. Each fails to account for the higher reaches of set theory. I also discuss Skolem's skeptical challenge to mathematical knowledge of the infinite—a history of which is a part of Chapter V—and the attempt to use second-order logic to block it. While I conclude that the Skolemite criticism of second-order logic has merit, I propose a related solution to the skeptical problem, one dependent on the use of schemas, that I believe succeeds.

None of the philosophies discussed in Chapters VI and VII could solve the problem of the infinite because none of them faced up to the main issue— What is the source of our intuitions concerning the Cantorian infinite? In more prosaic and somewhat over-simple terms, what do the ellipses, the triples of dots, in the written form of Cantor's transfinite progression suggest to us? Whatever that is is a large part of what led Cantor to his theory.

Finding an answer is important for many reasons. Our set theory is incomplete—it is inadequate for resolving many of the problems to which it gives rise. Anything that helps to clarify the sources of our axioms may help to suggest more axioms or help to adjudicate between the additional ones that have already been proposed. That is important both for mathematical reasons and because the apparent hopelessness of finding new axioms has itself become a source of skepticism about the mathematical theory of the infinite.

The apparent problem in accounting for the mathematical infinite led to the split between the philosophers discussed in Chapter VI and those discussed in Chapter VII. Each side seems today to be a council of despair. The resulting impasse has had repercussions far beyond the philosophy of mathematics. It has affected all modern epistemological theories.

In Chapter VIII, I propose that the source of our intuitions concerning the Cantorian infinite is experience of the indefinitely large. That is, our image of what the ellipses represent arises from our idea of going on for much longer than we have so far—going on indefinitely long. The proposal may gain some plausibility from the fact that children go through a stage at which they think the infinite literally is nothing more than the indefinitely large.

The proposal is nothing new, but I give a substantial new argument for it, making use of a mathematical theory of the indefinitely large developed by Jan Mycielski. In order to show that the theory can serve as a codification of the actual historical and psychological source of our intuitions concerning the infinite, it is necessary to show four things: (1) that the theory does not presuppose the infinite and is therefore suited in principle to be a source of intuitions concerning the infinite in that it does not presuppose what it is to explain; (2) that the theory formalizes ordinary experience of the indefinitely large and is therefore a reconstruction of intuitions that we have, as a matter of actual psychological fact; (3) that it does lead to set theory, and that it is therefore rich enough to explain what we have set out to explain; and (4) that it coheres well with the actual development of set theory, and thus that it can be taken to capture the intuitions that played an actual historical role.

To show the first, that the infinite is not presupposed, it is necessary to present the theory in such a way that it involves no commitment to the infinite. That is done using schemas. As a bonus this presentation shows, using mathematical work of Mycielski, that the theory enables us to provide a counterpart for ordinary set-theoretic mathematics that involves no commitment to the infinite.

To argue for the second, that the theory is a reasonable codification of our experience of the indefinitely large, I show how it can be applied to make some parts of the calculus more obvious—connected with daily experience—than they are when given the usual presentation involving limits. That—in addition to the plausibility of the theory in itself—shows how natural and intuitive the theory is, and, as you will see for yourself, how close to your pre-theoretic intuitions.

I show the third, that the theory does lead to set theory, by showing that set theory, including Choice and Replacement, arises by extrapolation, in a precise mathematical sense, from the theory of the indefinitely large.

The chief argument for the fourth, that the theory coheres well with the actual development of set theory, is that the theory of the indefinitely large helps us to make sense of Cantor's "finitism." Cantor saw himself as making an analogy between the finite and the infinite. We can now make precise sense of that: his procedure, analyzed and reconstructed, was that of extrapolating from the indefinitely large to the infinitely large.

The process of idealization that connects the finite to the infinite will be shown not to be very different in principle from the one that connects pencil dots to geometrical points. Points are, more or less, idealized dots, while infinite sets are, more or less, idealized indefinitely large collections. Thus, set theory is of a piece with arithmetic and geometry: all three have a close association with familiar types of experience. The apparently mysterious character of knowledge of the infinite is dissolved.

Infinity, Mathematics' Persistent Suitor

. . . But, from the very nature of an irrational number, it would seem to be necessary to understand the mathematical infinite thoroughly before an adequate theory of irrationals is possible. The appeal to infinite classes is obvious in Dedekind's definition of a cut. Such classes lead to serious logical difficulties.

It depends upon the individual mathematician's level of sophistication whether he regards these difficulties as relevant or of no consequence for the consistent development of mathematics. The courageous analyst goes boldly ahead, piling one Babel on top of another and trusting that no outraged god of reason will confound him and all his works, while the critical logician, peering cynically at the foundations of his brother's imposing skyscraper, makes a rapid mental calculation predicting the date of collapse. In the meantime all are busy and all seem to be enjoying themselves. But one conclusion appears to be inescapable: without a consistent theory of the mathematical infinite there is no theory of irrationals; without a theory of irrationals there is no mathematical analysis in any form even remotely resembling what we now have; and finally, without analysis the major part of mathematics—including geometry and most of applied mathematics—as it now exists would cease to exist.

The most important task confronting mathematicians would therefore seem to be the construction of a satisfactory theory of the infinite . . . If the reader will glance back at Eudoxus' definition of "same ratio" . . . he will see that "infinite difficulties" occur there too . . . Nevertheless some progress has been made since Eudoxus wrote; we are at least beginning to understand the nature of our difficulties.

[Bel37, pp. 521–522]

11

With this chapter, I hope to make better known a few aspects of the history of the mathematical infinite that are known at least in outline to many mathematicians. The chapter is a work of exposition, not of scholarship. Little of what I shall say is controversial.[1] If I succeed in making the story accessible without introducing detailed knowledge of Fourier series or of the distinction between convergence and uniform convergence, the chapter will have served its purpose.

The modern-day theory of the infinite did not begin with an effort to produce a theory of the infinite, and it did not build on a long history of attempts at mathematical theories of the infinite. It began instead with an attempt to clarify the foundations of analysis and specifically of the calculus—that is, it grew out of the development of our theory of rates of change and of areas under curves. The infinite has entered present-day mathematics in large part as the result of attempts to make sense of the notion of an arbitrary curve or function.

The story of the hugely successful application of analysis to physics is one that is too well known to bear retelling here. Let me simply note that analysis could not in Newton's time and cannot today be regarded as just one among many branches of mathematics: it is the one whose application, especially to physics, has been the most fruitful. It is therefore the branch of mathematics through which mathematics makes its most intimate contact with physics, the sciences, and the natural world.

§1. Incommensurable Lengths, Irrational Numbers

Most of us have been taught at one time or another that Pythagoras discovered that the square root of two is irrational. That is very likely not true, though our historical information concerning the Pythagoreans is sparse. First of all, many of the discoveries of the Pythagoreans are attributed to Pythagoras himself, and it is very likely that some other member of the Pythagorean school made the discovery. Indeed, the discovery is attributed to Hippasus of

---

1. I have relied heavily on Morris Kline's *Mathematical Thought from Ancient to Modern Times* [Kli72] and the articles in *From the Calculus to Set Theory* [GG80b], edited by I. Grattan-Guiness. My analysis of the development of the calculus has been heavily influenced by Philip Kitcher's *The Nature of Mathematical Knowledge* [Kit83]. I have also made some use of Florian Cajori's *History of Mathematics* [Caj85] and Dirk J. Struik's *A Concise History of Mathematics* [Str87]. Various more specialized historical works, cited in the text when necessary, have served as useful correctives.

Metapontium (fifth century B.C.E.) among others. Legend has it that he made the discovery while at sea with the other Pythagoreans and that he was tossed overboard for his trouble. (See [Hea81, vol. 1, pp. 154–157] and [Hea56, vol. 1, pp. 411–414].)

Second, and much more important, the only numbers the Pythagoreans had anything to do with were whole numbers—no rational numbers, and certainly no irrational ones. They knew many things about geometrical proportions between *geometrical* magnitudes. For example, they knew that two strings of the same type and tension whose lengths were in the ratio of three to two would, when plucked, produce notes a musical interval of a fifth apart. The ratio of three to two meant approximately that the two lengths could be measured by a common unit so that one was three times the length of that unit, while the other was twice that length. That was in no way associated with the fractions or rational numbers 3/2 or 2/3.

The lengths of the two strings in our example were *commensurable*—measurable by whole-number multiples of a common unit. What the Pythagoreans had discovered was not that the square root of two is irrational but that the side and the diagonal of a square are not commensurable. That made it impossible to continue the Pythagorean program of identifying geometry with the theory of the numbers, which were, for the Greeks, just the whole numbers.

Sometime in the century following the work of Hippasus of Metapontium, Eudoxus gave an ingenious theory of incommensurable ratios, a theory that remains the basis of our understanding today. Incommensurable ratios arose within geometry, and his theory was entirely geometric. Indeed, Eudoxus contrasted geometric magnitudes with numbers, which increase a unit at a time. The main idea of his theory of incommensurable ratios is more or less this: a is in the same ratio to b that c is to d if for any whole numbers n and m, na is less than, equal to, or greater than mb if and only if nc is, respectively, less than, equal to, or greater than md.

Less than a century later, the Eudoxian theory was codified in Book V of Euclid's *Elements*. Book II showed how to do what algebra there was geometrically: Numbers are represented or, probably more accurately, replaced by lengths, angles, areas, and volumes. The product of two lengths is an area; the product of three, a volume. One can add and subtract lengths from lengths, areas from areas, and so forth. Numbers and algebra have in effect been eliminated in favor of geometry, and the foundations of the geometrical theory of ratios or proportions are those of Eudoxus.

The ratios of magnitudes, commensurable and incommensurable, are not stand-ins for numbers, rational and irrational. No procedure is given, for example, for adding or multiplying ratios of magnitudes.

Neither are the magnitudes themselves—lengths and the like—stand-ins for rational and irrational numbers. One can add them, but the product of lengths, for example, is an area. Euclid was careful to state (Definition 3) that a ratio can only relate magnitudes of the same kind. That is, in particular, one cannot relate lengths and areas in a ratio. Unlike the product of numbers, a product of lengths is an entity of a different kind.

In Book X, Euclid investigated and classified ratios between lines that we would represent as having lengths of the form $\sqrt{\sqrt{a} \pm \sqrt{b}}$ for commensurable a and b. Ratios between lines that cannot be expressed in that form were not discussed in the *Elements*.

Leonardo of Pisa (Fibonacci) was educated in Africa, and he traveled widely. He reintroduced Euclid's *Elements* and other Greek mathematical works to Europe. He also disseminated Arabic numerals and methods of calculation. In 1220, Leonardo published his discovery that the roots of $x^3 + 2x^2 + 10x = 20$ are not expressible in the form $\sqrt{\sqrt{a} \pm \sqrt{b}}$. The Arabs worked freely with irrational numbers, and Leonardo's discovery showed that not every number could be constructed within the Euclidean strictures of compass and straightedge. But no adequate foundation had been provided for the use of irrational numbers.

In succeeding centuries the use of irrational numbers became increasingly common among European mathematicians, but it was not clear in what sense they were numbers. In his *Arithmetica Integra* (1544) Michael Stifel wrote,

Since, in proving geometrical figures, . . . irrational numbers . . . prove exactly those things which rational numbers could not prove . . . we are moved and compelled to assert that they truly are numbers . . . On the other hand, other considerations compel us to deny that irrational numbers are numbers at all. To wit, when we seek [to give them a decimal representation] . . . we find that they flee away perpetually, so that not one of them can be apprehended precisely in itself . . . Now that cannot be called a true number which is of such a nature that it lacks precision . . . Therefore, just as an infinite number is not a number, so an irrational number is not a true number, but lies hidden in a kind of cloud of infinity.   [Kli72, p. 251]

As we shall see, Stifel's remarks were prescient: the basis of the irrational numbers was not adequately clarified until *infinite* numbers were allowed into mathematics.

The ties to geometry remained strong. Stifel said that "going beyond the cube just as if there were more than three dimensions . . . is against nature" [Kli72, p. 279]. René Descartes, around 1628 (in *Regulae ad Directionem Ingenii*), explicitly allowed irrational numbers for continuous magnitudes. In 1637 Descartes took the product of lengths to be a length, not an area, and viewed polynomials as determining curves [Des54]. (See also [Gro80] and [Mah73].) Newton introduced number as "the abstracted ratio of any quantity, to another quantity of the same kind," including incommensurable ratios, and introduced multiplication, division, and roots in terms of ratios in his university lectures, published in 1707 as *Arithmetica universalis sive de compositione et resolutione arithmetica liber* [Whi67, vol. 2, p. 7].

Until now we have been considering the geometry of straight lines (and rectangles, and so forth) and their magnitudes. We shall now turn to the geometry of curves and the areas they bound. Once more, Eudoxus did basic work that Euclid incorporated in the *Elements*, in Book XII. Archimedes went even further in developing what is called the method of exhaustion. The method remained the only fully worked out and thoroughly justified one for computing areas and volumes until the nineteenth century, but the details are not central to our story.

§2. Newton and Leibniz

In the first half of the seventeenth century various curves were introduced or described by means of motion. That was not new, but this method of description came to play an increasingly central role. In 1615 Marin Mersenne defined the cycloid as the path traced out by a point on the edge of a rolling circle. The cycloid was not new; the definition was. Galileo Galilei showed in *Discorsi e dimostrazione matematiche intorno a due nuove scienze* (1638) that the path of a cannonball was a parabola, and he regarded the curve as the path of a moving point.

Many techniques were devised for computing various properties of curves, in part building on the method of exhaustion: techniques for computing maxima and minima, locating tangent lines, and computing areas and volumes. The mathematicians involved included Pierre Fermat, Descartes, Isaac Barrow, Johann Kepler, Bonaventura Cavalieri, Gilles Personne de Roberval,

Evangelista Torricelli, Blaise Pascal, John Wallis, Sir Christopher Wren, William Neile, Gregory of St. Vincent, James Gregory, and Christiaan Huygens. But Isaac Newton and Gottfried Wilhelm Leibniz soon systematized the techniques into the calculus, and so we shall only briefly look at the work of the others.

The new study of curves and motion led to a new definition of the line tangent to a curve (Roberval, *Brieves Observation sur la composition des mouvemens et sur le moyen de trouver les Touchantes des Ligne Courbes*, ca. 1636, published 1693). The Greek definition of a line tangent to a curve is a line touching the curve at a point. Roberval defined a tangent to a curve as the direction of the velocity of a moving point tracing the curve.

In his *Arithmetica Infinitorum* (1655), Wallis studied infinite sums and products. He also gave a correct general definition of the limit of an infinite sequence of numbers, a definition that did not surface again until around 1820. (For example, the limit of the sequence $1, \frac{1}{2}, \frac{1}{4}, \dots$ is 0. See §5.) Newton studied the *Arithmetica Infinitorum* and used its techniques to convince himself that the binomial theorem—which gives the coefficients of the expansion of $(a + b)^n$ for arbitrary n—also held when n was negative or fractional. In those cases, there are infinitely many coefficients—one obtains an expansion of $(a + b)^{m/n}$ as an infinite sum or *series*. (As an example of a series—though not one derived from the binomial theorem—the limit of the series $1 + \frac{1}{2} + \frac{1}{4} + \dots$ is 2.) Such series were crucial for Newton's development of the calculus, to which we now turn.

In *De Analysi per Aequationes Numero Terminorum Infinitas* (circulated in 1669, published 1711), Newton gave a considerably more general version of the following derivation: Suppose that the area z under a curve is given by $z = x^2$. (See Figure 1, which is not drawn to scale.) Suppose x increases by a "moment" o, that is, in our present-day Leibnizian terminology, by an infinitesimal.[2] (The term *moment* was presumably suggested by thinking of x as time.) Then the area under the curve increases by ov, and so $z + ov = (x + o)^2$, where the right-hand side is obtained by using $z = x^2$, which we have assumed true, at the point at which the x coordinate has value $(x + o)$. Multiplying out, $z + ov = x^2 + 2ox + o^2$, and since $z = x^2$, it follows that

---

2. The history of analysis from this point on depends heavily on present-day ideas about infinitesimals, on which see §VIII.3. Those ideas are used to adjudicate what arguments have a reasonable reconstruction in modern terms, and hence are to be viewed as correct, and which do not.

$ov = 2ox + o^2$. We now divide through by o to obtain $v = 2x + o$. At this point, Newton took o "infinitely small" to obtain $y = 2x$, since (from the figure) v is equal to y when o is infinitely small.

As Newton himself admitted, the method is "shortly explained rather than accurately demonstrated." The derivation accomplishes two things at once: First, it shows that the rate of change of x^2 is $2x$ (on the right-hand side we computed the change $(x + o)^2 - x^2$ divided by the "time" o in which the change occurs to obtain the rate of change). Second, it shows that the rate of change of the area z is the curve y bounding that area (on the left-hand side we computed the rate of change of z and obtained y). The equation $y = 2x$ thus asserts that the rate of change $(2x)$ of the area $(z = x^2)$ bounded by a curve (y) is the curve itself. That is Newton's version of the fundamental theorem of the calculus[3]—for $z = x^2$. Newton did not use that example. He made

---

3. Here is all you need to know about the fundamental theorem of the calculus. I have omitted how to handle negative values since the details don't matter for our story. I have also omitted important restrictions on the applicability of the theorem. They were far from worked out in the days of Newton and Leibniz. Differentiation is pretty much the operation that takes a function f to the function g that graphs the slope or, equivalently, the rate of change of f (that is, $g(x)$ is the slope of f at x or the rate of change of f at x if we think of x as representing time). Integration is pretty much the operation that takes a function f to the function g that graphs the area under f (that is, $g(x)$ is the area under the graph of f between 0 and x). The fundamental theorem of the calculus states that integration and

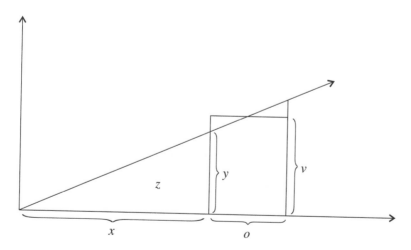

Figure 1. Newton's derivation.

it clear that one could use $z = ax^m$, where m could be negative or fractional, expanding the right-hand side not by multiplying it out but by using the binomial theorem. He thus obtained the result that the rate of change of ax^m is max^{m-1}. He then expanded other equations involving x as infinite series of terms of the form ax^m and applied the result term by term to compute other rates of change.

In a subsequent work (*Methodus Fluxionum et Serierum Infinitarum*, written in 1671, published 1736), Newton called a variable quantity a "fluent" and its rate of change a "fluxion." He computed rates of change by computing the fluxion of a fluent, and he found areas by finding the fluent of a fluxion. He now regarded fluents as generated by continuous motions instead of as being built up as static assemblages of moments. The moment o is now conveniently thought of as "an infinitely small interval of time." The idea of taking a curve to be the path of a moving point thereby became fundamental. Newton had introduced an early form of the idea of functional dependence—with time as an auxiliary independent variable.

In a third paper (*Tractatus de Quadratura*, written 1676, published 1704), Newton attempted to eliminate the moments, or infinitesimals. He said, "Lines are described . . . not by the apposition of parts, but by the continued motion of points," and "Fluxions are, as near as we please, as the increments of fluents generated in times, equal and as small as possible, and to speak accurately, they are in the prime ratio of nascent increments." His computations were much as before, but the new excuse for dropping terms involving o at the end was "Let now the increments vanish and their last proportion will be . . . " To the modern ear, that phrase suggests the beginnings of the theory of limits that eventually became a crucial part of the foundations of analysis. In contrast to his concern about the increment o, Newton did little to provide a basis for his use of series—infinite sums [Kit83, p. 234].

Let us now turn to Leibniz's independent discovery of the calculus. Whereas Newton relied heavily on temporal ideas and infinite series, Leibniz assimilated curves to sequences of numbers. In 1666, while Newton was completing the main part of his development of the calculus, Leibniz published a work, *De Arte Combinatoria*, on what seems a different subject. Consider the

---

differentiation are inverse operations, which means that if g is the integral of f, then f is the derivative of g.

following sequences of numbers. Each sequence on a line below another sequence consists of the differences between the terms in the sequence above it:

$$0, \quad 1, \quad 2, \quad 3, \quad \ldots$$
$$1, \quad 1, \quad 1, \quad \ldots$$
$$0, \quad 0, \quad \ldots$$

Now, we start with squares:

$$0, \quad 1, \quad 4, \quad 9, \quad 16, \quad \ldots$$
$$1, \quad 3, \quad 5, \quad 7, \quad \ldots$$
$$2, \quad 2, \quad 2, \quad \ldots$$
$$0, \quad 0 \quad \ldots$$

Finally, with cubes:

$$0, \quad 1, \quad 8, \quad 27, \quad 64, \quad 125, \quad \ldots$$
$$1, \quad 7, \quad 19, \quad 37, \quad 61, \quad \ldots$$
$$6, \quad 12, \quad 18, \quad 24, \quad \ldots$$
$$6, \quad 6, \quad 6, \quad \ldots$$
$$0, \quad 0, \quad \ldots$$

Leibniz noted that the second differences for the sequence of natural numbers, the third differences for the sequence of squares, and so forth, all vanish. He also recognized that each sequence could be recovered as the successive sums of its first member and the members of the sequence below it—that is, by putting the differences back together. In 1673, during the time between Newton's second paper and his third, Leibniz connected those facts to the study of curves by thinking of a curve as a sequence of successive points. He later came to think of the successive points as differing by infinitesimals. When the succession of points is such that their x coordinates differ by a constant amount, the successive, infinitesimally close x values are thought of as giving the order of the terms in the sequence, while the y values may constitute the terms themselves. Thus, a curve is conceived of in terms of a sequence of values much like the sequences Leibniz had investigated earlier. At this stage, dx (a notation Leibniz introduced a couple of years later) is 1 since the terms are the first, second, third, and so forth, while dy is the actual difference between adjacent terms. He thus saw that if the unit is infinitely small, then the sum of the ys gives the area under the curve and the differences dy ($dy = dy/dx$, since $dx = 1$) are the slopes of the tangent lines. He recognized that (in a now familiar notation he introduced later) $\int dy = y$: the sum of the differences is the original series. That is the beginning of his version of the fundamental theorem of the calculus—the integral ($\int$) of the differential (dy) of y is y. The integral is Leibniz's—and our—term for pretty much what Newton had called a fluent, and the differential played pretty much the role of Newton's fluxion.

Leibniz also made use of the "characteristic triangle," which he adopted from Pascal. (See Figure 2.) The characteristic triangle is abc, where ac is simultaneously a straight line and part of the curve. The curve was in effect viewed as a polygon with infinitesimal sides. The triangle abc is similar to the triangle ABa, which has sides of ordinary finite length. The line Aa is tangent to the curve. Those facts exemplify the main reasons why the characteristic triangle was useful.

Using the above ideas, Leibniz had most of the essential features of his calculus by 1675. The details took a couple of years more. Unlike Newton, Leibniz preferred to avoid the use of infinite series.

At first Leibniz had little to say about the nature of the dxs and dys. In 1680 he said that "these dx and dy are taken to be infinitely small, or the two points on the curve are understood to be a distance apart that is less than any given length." The differential dy is a "momentaneous increment." In 1684, Leibniz defined a tangent as a line joining two points that are infinitely close. In 1690 he said (in a letter to Wallis):

> It is useful to consider quantities infinitely small such that when their ratio is sought, they may not be considered zero but which are rejected

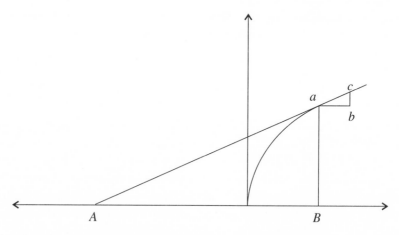

Figure 2. The characteristic triangle.

as often as they occur with quantities incomparably greater. Thus if we have $x + dx$, dx is rejected. But it is different if we seek the difference between $x + dx$ and x.   [Kli72, p. 385]

The infinitesimals are on occasion taken to be vanishing or incipient quantities, or indefinitely small quantities smaller than any finite quantity.

While Newton and Leibniz were struggling with infinitely small and nascent quantities, Wallis was fairly clear about the nature of the number line. He accepted irrationals as numbers, and he thought of the Eudoxian theory of ratios of magnitudes, as presented in Book V of Euclid's *Elements,* as arithmetical. He identified rational numbers with repeating decimals. But the calculus became such a central part of mathematics that the unclarity of its basic concepts infected virtually all of the work of mathematicians. Proof was almost completely abandoned.

In 1673 Leibniz had taken a curve to be given by an equation, but he called any quantity varying along the curve—for example, the length of the tangent line from the curve to the x axis—a function. The function is not, however, a function of a variable but of the curve [Bos74, p. 9]. Newton, at least in principle, did not give a curve any special status: he took fluxions of fluents and fluents of fluxions equally. What Newton considered were quantities obtained from others by means of (possibly infinite) algebraic combinations, primarily infinite sums of finite combinations—what today would be called series.

When one considers infinite series it is necessary to consider convergence if one wishes to avoid absurd results. For example, the series

$$x + x^2 + x^3 + \cdots$$

diverges when $x = 2$, and so does not have a value, but for $x = \frac{1}{2}$, it becomes

$$\frac{1}{2} + \frac{1}{4} + \frac{1}{8} + \cdots,$$

which converges to 1. The terms *convergent* and *divergent* appeared in the work of James Gregory while Newton was developing the calculus. In that period, Lord Brouncker showed some series to be convergent.

Newton did not show as much facility with the distinction between convergent and divergent series as Gregory and Brouncker. He noted that some series (like the one of our example) should be used only for small values of x, while others should be used only for large values of x. He noted that some series become infinite at some values of x and that they are useless for those values.

In 1713 Leibniz devised a test for the convergence of some series, but by and large Newton and Leibniz and their successors simply treated series as on a par with finite sums. Indeed, at around 1800 Joseph-Louis Lagrange tried to provide an algebraic basis for analysis that used infinite series without regard for whether or not they converged.

Leibniz and Newton showed at least occasional concern for the convergence of series. In contrast, the Bernoulli brothers, who studied Leibniz's work and corresponded with him often, did extensive work concerning series, and they showed almost no awareness of any need for caution. Wrong results were described as paradoxes. The Bernoullis also made substantial positive contributions, but those are irrelevant to our concerns here, with a few exceptions. From 1697 on, Johann Bernoulli employed the notion of any "quantity" formed from a variable and constants using algebraic and transcendental expressions. He called such a quantity a "function" beginning in 1698, adopting the term that Leibniz had used earlier. I shall refer to such functions as "analytic expressions," a term used by Leonhard Euler [Bos74, p. 10], to emphasize the difference between such expressions and functions in the modern sense.[4] Johann's work marked the beginning of a transition from a focus on the study of geometrical curves to a focus on the study of analytic expressions. Johann also deemphasized the geometrical basis of the notion of an integral as an area by simply defining the integral to be the inverse of the differential. The fundamental theorem of the calculus was thus absorbed into the definition. That style of defining the integral was dominant into the nineteenth century. Euler solidified the change away from geometry in a series of influential textbooks published from 1748 to 1770. He adopted and generalized Johann Bernoulli's definition of a function. (See [Bos80, pp. 73–79].) Johann also did some work on the vibrating-string problem, a problem that we shall be discussing in detail later.

§3. Go Forward, and Faith Will Come to You

Euler investigated infinite sums in the 1730s. I shall give two examples he used to illustrate the difficulties that such series cause. Formal long division

---

4. Analytic expressions bear only the most remote of relations to "analytic functions" in the modern technical sense, which are certainly not what I mean.

of polynomials—the process you were taught in high school—yields

$$\frac{1}{(1+x)^2} = 1 - 2x + 3x^2 - 4x^3 + \cdots$$

and

$$\frac{1}{1-x} = 1 + x + x^2 + x^3 + \cdots.$$

Plugging $x = -1$ into the first series, one obtains (because $1/0$ is ∞)[5]

$$\infty = 1 + 2 + 3 + 4 + \cdots.$$

Plugging $x = 2$ into the second series, one obtains

$$-1 = 1 + 2 + 4 + 8 + \cdots.$$

The series for -1 is term by term greater than or equal to the series just above it for ∞, and it is term by term greater after the first two terms. Hence, Euler noted, one might conclude that $-1 > \infty$ and infer that ∞ is in some sense between the positive and negative numbers. He also plugged $x = -1$ into the series for $1/(1-x)$ to obtain

$$\frac{1}{2} = 1 - 1 + 1 - 1 + \cdots,$$

as Leibniz had done. He considered and rejected the idea that one should restrict attention to the sums of convergent series [Kit83, pp. 242–244].

Euler was no idiot. He was very interested in computing the sums of infinite series, and he developed many techniques for computing them that involved divergent series. The problematic series did not get him into trouble because he could always check his results, at least approximately, by summing a few terms of whatever series he was considering. The divergent series that could cause problems also led to too many successes to simply be

---

5. Dividing by 0, as in the equation under discussion, can lead to trouble: Since $0 \cdot 0 = 0 \cdot 1$, one might conclude that $0 = 1$. But using ∞ as a notation for $n/0$, for $n > 0$, is not problematic. The mistake here lies in the use of divergent series, not in the division by 0.

dropped [Kit83, p. 250], though he did things with them that later mathematicians would view with horror [Kit83, p. 323n].

Besides his work on series, Euler solidified the separation of analysis from geometry, introduced functions (analytic expressions) of more than one variable, and gave differential coefficients, essentially derivatives, a crucial role. In 1734 Euler considered a notion of function considerably broader than the analytic expressions we have seen before: he allowed a function to be formed by putting together parts of curves, and even allowed curves freely drawn. He also introduced the now familiar notation $f(x)$ for a function of x.

In that same year George Berkeley, the Anglican bishop of Cloyne, published *The Analyst,* a devastating critique of the foundations of analysis.[6] Like Newton, Berkeley had doubts about matters related to infinitesimals, not infinite series. His criticisms in large part sound exactly right to the modern ear. He understood the value of the methods: "The Method of Fluxions is the general key by help whereof the modern mathematicians unlock the secrets of Geometry, and consequently of Nature" [Ber34, p. 66]. Nonetheless, he said that the mathematicians of his age took more pains to apply their principles than to understand them [Ber34, p. 99]. He pointed out that derivations like the one by Newton described above are incoherent, since one divides through by o and later assumes o equal to zero [Ber34, p. 72]. That criticism may be a bit unfair to Newton, who can, as we have seen, be read as having some idea of using something like limits to replace the procedure of setting o equal to zero [Kit83, p. 239n].

Berkeley criticized Newton's theory of fluxions as ultimate ratios of evanescent increments in a familiar passage:

> And what are these fluxions? The velocities of evanescent increments? And what are these evanescent increments? They are neither finite quantities, nor quantities infinitely small, nor yet nothing. May we not call them the ghosts of departed quantities?   [Ber34, p. 88]

Leibniz's infinitesimals fared no better:

> [Our modern analysts] consider quantities infinitely less than the least discernible quantity; and others infinitely less than those infinitely small ones; and still others infinitely less than the preceding infinitesimals, and so on without end or limit.   [Ber34, p. 68]

---

6. Bernard Nieuwentijdt made similar criticisms of the calculus forty years earlier. He was widely read and provoked a reply from Leibniz. See §VIII.3 for a discussion. See [Man89] for information on other early criticisms of the calculus.

He continued:

> Nothing is easier than to devise expressions or notations, for fluxions
> and infinitesimals of the first, second, third, fourth, and subsequent or-
> ders . . . But if we remove the veil and look underneath, if, laying aside
> the expressions, we set ourselves attentively to consider the things them-
> selves which are supposed to be expressed or marked thereby, we shall
> discover much emptiness, darkness, and confusion; nay, if I mistake not,
> direct impossibilities and contradictions.   [Ber34, p. 69]

And finally,

> In all this the ultimate drift of the author [Newton] is very clear, but his
> principles are obscure.   [Ber34, p. 94]

In *A Defence of Free-thinking in Mathematics* (1735, a reply to a reply
to *The Analyst*), Berkeley summarized the various contemporary attempts at
foundations of analysis:

> Some fly to proportions between nothings. Some reject quantities be-
> cause infinitesimal. Others allow only finite quantities and reject them
> because inconsiderable. Others place the method of fluxions on a foot
> with that of *exhaustions,* and admit nothing new therein. Some maintain
> the clear conception of fluxions. Others hold they can demonstrate about
> things incomprehensible. Some would prove the algorithm of fluxions
> by *reductio ad absurdum;* others *a priori.* Some hold the evanescent in-
> crements to be real quantities, some to be nothings, some to be limits. As
> many men, so many minds . . . Some insist the conclusions are true, and
> therefore the principles . . . Lastly several . . . frankly owned the objec-
> tions to be unanswerable.   [Ber35a, p. 133]

That seems a pretty fair summary of the state of affairs. Euler, by the way,
was one who endorsed "proportions between nothings":

> There is no doubt that every quantity can be diminished to such an extent
> that it vanishes completely and disappears. But an infinitely small quan-
> tity is nothing other than a vanishing quantity and therefore the thing
> itself equals 0.   (*Institutiones*, 1755; see [Kli72, p. 429].)

He then went on to explain how dy/dx, which was $0/0$, could have a definite
value. Such was the state of the art. The above quote is from one of Euler's
textbooks, which overall had a tremendous positive influence in organizing
analysis into a coherent study of analytical expressions [Bos80, pp. 53, 76].
    The complete lack of rigor came to be considered a virtue:

But the tables have turned. All reasoning concerned with what common sense knows in advance, serves only to conceal the truth and to weary the reader and is today disregarded.   (Alexis Claude Clairaut, *Eléments de géométrie*, 1741; see [Kli72, p. 619].)

The attitude of Clairaut was widespread. Nonetheless, there were a number of attempts to give a more adequate development of the calculus. Those attempts were motivated by particular mathematical problems—not by any desire to prove the obvious more carefully. Euler's texts emphasized "differential coefficients," which are the derivatives that ultimately replaced differentials in the foundations of analysis [Bos80, p. 74]. Jean Le Rond d'Alembert and Benjamin Robins, like Wallis about a century earlier, emphasized limits [Bos80, p. 91]. D'Alembert said, "The theory of limits is the true metaphysics of the calculus." But he never actually worked out a presentation of the calculus on that basis; indeed it is likely that he could not have done so, since he, like Robins, considered the limits of *variables* (variable quantities) not of *functions* with a specified independent variable [Bos80, p. 92]. In the absence of a correct presentation, he also is said to have said, "Allez en avant, et la foi vous viendra" (Go forward, and faith will come to you).

§4. Vibrating Strings

D'Alembert made significant progress on *the vibrating-string problem:* given the tension in a string and its initial position, figure out how the string moves when it is released. The problem had been studied by Brook Taylor in 1713 [Kli72, p. 478] and by Johann Bernoulli in 1727 [Kli72, p. 479]. Johann's son, Daniel Bernoulli, did related work in 1732 [Kli72, p. 489]. But in 1746 (work published 1747 and 1749) d'Alembert wrote down essentially the modern partial differential equation involved and gave a general solution.

I am discussing the vibrating-string problem for good reason. The vibrations of a string can always be represented as an infinite sum or series of sine waves.[7] Since the initial position of a string can in a sense be arbitrary—we can stretch the string into any shape—it apparently follows that *any* curve, that is, any function, can be represented as a sum or series of sine waves, briefly, a *trigonometric series*. The modern definition of function—of an arbitrary function—evolved as part of the attempt to formulate that conclusion

---

7. We add functions or curves by adding them at each point. Thus, the sum $F = f + g$ reduces to an ordinary sum of values at each point: for example, $F(3) = f(3) + g(3)$. Similarly, an infinite sum or series of functions reduces to an ordinary series at each point.

as a theorem, and set theory evolved more or less as part of the attempt to prove the theorem. The details are an important part of the history I wish to present. In the end it has turned out that the "theorem" is not quite true— there are functions that cannot be represented using trigonometric series—but the ones that can be so represented include functions far stranger than anyone had thought possible.

D'Alembert's general solution to the vibrating-string problem required that the initial curve of the string be periodic, that is, that it repeat the same shape it had on one length of string over and over. He therefore required that the initial position of the string be a periodic analytic expression. That was a substantial restriction on the allowed initial positions of the string: one couldn't start with a string in just any arbitrary configuration, but only one given by a periodic analytic expression.

Shortly after seeing d'Alembert's work, Euler wrote a paper, published in 1748 [Kli72, p. 505], in which he allowed the initial function describing the position of the string to be any function (meeting some other constraints that I omit here) on an interval. He ensured that the function was periodic by simply duplicating its values outside that interval. The function had to be zero at its endpoints since the string was fixed at its endpoints. Thus the duplications matched up at the ends. The function also had to be free of jumps since the string was in one piece. But, and this is the key point, there was no requirement that the function be given by a single (periodic) analytic expression. Euler's broad notion of function from 1734 was now being put to an important mathematical use. In the paper of 1748 he also saw that the motion of the string is periodic in time (the string resumes its initial shape at regular intervals) and that at least some solutions to the problem can be written as sums of sines and cosines.

By 1755 Euler defined a function thus [Kli72, p. 506]: "If some quantities depend on others in such a way as to undergo variation when the latter are varied, then the former are called functions of the latter." He specifically intended to allow functions that are not given by a single equation on the entire domain. In 1763 he wrote to d'Alembert that allowing this more general notion of a function "opens to us a wholly new range of analysis" [Kli72, p. 507].

Daniel Bernoulli, in his work of 1732–1733, was the first to recognize that a string could vibrate at many frequencies—the fundamental frequency that had been studied by Taylor and Daniel's father Johann, and the harmonics (multiples) of that frequency [Kli72, p. 480]. In the early 1740s, Daniel Bernoulli said that a vibrating bar can vibrate at two harmonic frequencies at once. The statement is based on physical understanding, not mathematical derivation. In 1753 he went further: after seeing the work of d'Alembert and Euler, he said that "all the new curves given by d'Alembert and Euler are only combinations of . . . [sinusoidal] vibrations" [Kli72, p. 509]. He was still relying on physics, not mathematics. His claim was that all the new curves can be represented by trigonometric series.

Euler and d'Alembert objected at once to Daniel Bernoulli's claim [Kli72, pp. 509–510]. Euler believed that the functions he had allowed, functions that are defined by different equations in different intervals, could not be a sum of sine functions. A function could not simultaneously be "discontinuous" (given by different expressions in different intervals) and "continuous" (given by a single expression—a sum of sine functions). Moreover, despite his liberal notion of function, a notion that made it possible to piece together a periodic function from nonperiodic ones, Euler argued that since every trigonometric series must be periodic, no nonperiodic function (analytic expression) could be equal to a trigonometric series [GGR72, pp. 245–247]. Euler's attention was now on the analytic expressions themselves. Bernoulli held his ground, and the three continued to disagree with each other through the 1770s [Kli72, p. 513]. Lagrange and the Marquis Pierre Simon de Laplace eventually entered the fray. That all seems a bit bizarre when one sees that Euler (in 1750–1751), d'Alembert (in 1754), and Clairaut (in 1757) had all discovered general methods of representing arbitrary functions by trigonometric series [Kli72, pp. 456–459]. They applied those methods only when they had some (usually physical) reason to believe that a trigonometric-series representation ought to exist. The mathematics did not stand on its own. Indeed, several of the derivations were not correct by our standards. Since nothing could be proved in a reliable way, one tried to confirm the results of a mathematical derivation on some independent grounds. If a result was contrary to expectations, it was often dismissed.

The disagreement about trigonometric series, the paradoxes arising from the use of infinite series, and other disputes created a real internal mathematical need for clarification of the foundations of analysis. The fundamental concepts of function, derivative, and integral had no adequate definition. They had been used in a manner suggested by their applications to simple functions—especially polynomials. As the notion of function was broadened, mainly as a result of work on the vibrating-string problem, that analogical procedure became less and less adequate.

The problem became even more acute with the work of the Baron Jean Baptiste Joseph de Fourier on heat conduction, which involved him in the

problem of trigonometric series. The reception of his work gives some indication of the controversy that surrounded it. In 1807 he submitted a paper to
the Academy of Science of Paris. It was rejected by Adrien Marie Legendre,
Laplace, and Lagrange. But, to encourage Fourier, the problems he studied
were made the subject of an 1812 prize. Fourier's 1811 paper won the prize
but was not published. In 1822 Fourier published his great *Théorie analytique
de la chaleur.* It included part of the 1811 paper. Fourier became secretary of
the Academy two years later, and he had the Academy publish the paper of
1811. (See [Kli72, p. 672].)

Fourier came to see, by a complicated process that we shall not consider,
that if for some coefficients b_ν

$$f(x) = \sum_{\nu=1}^{\infty} b_\nu \sin \nu x \quad \text{for } 0 < x < \pi,$$

that is, if the function f could be represented by a trigonometric series[8] in the
interval from 0 to π, then for every value of ν

$$b_\nu = \frac{2}{\pi} \int_0^\pi f(s) \sin \nu s \, ds.$$

He took that to mean that the coefficients b_ν of the sum would have to be
$2/\pi$ times the area under the curve $f(s) \sin \nu s$ between $s = 0$ and $s = \pi$.[9] As
we have seen, formally similar results had been obtained by others, including
Euler, d'Alembert, and Clairaut. But Fourier departed from the practice of his
day and did not interpret the integral as an inverse differential but geometrically, as an area [GG80a, p. 107]. He observed that the area involved is well
defined for an extremely wide variety of fs. Fourier had not been studying vibrating strings, but heat flow, though the mathematics is much the same. His
function f represented not the initial position of a vibrating string but the initial distribution of heat in a metal bar. A string must be in one piece, but the
temperature in a bar can jump: to produce a bar with a jump in temperature,
take a hot bar and a cold bar and join them together. (See [Haw80, p. 152].)

---

8. I shall use the term *trigonometric series* to mean a sum of sines, as indicated in the
text. Though a general trigonometric series would allow cosines as well, I shall just ignore
that fact. The details of what is gained by allowing cosines are irrelevant for our purposes.

9. Actually, Fourier wasn't quite right. Cantor straightened out the problems in the
1870s, and that work led him to set theory—as we shall see below.

Certainly no single analytic expression need be available for Fourier's temperature function f. The function f might be "freely drawn," and it could even jump about in value. The existence of an inverse differential was of no particular concern, and so Fourier was freed from a restriction to analytic expressions. He concluded that *every* function was representable by a special trigonometric series on the interval from 0 to π, the trigonometric series determined by the formula for the b_νs. Such a series is called a *Fourier series*. Fourier tested his opinion by computing the first few b_ν for many fs and plotting the sums of the first few terms of the corresponding sequences. The results looked good.

Summing the trigonometric series above yields a function that is odd and periodic: the values between $-\pi$ and 0 are just the opposites of the values between 0 and π, and the values between $-\pi$ and π are repeated over and over. Thus, the trigonometric series for an *arbitrary* function f, while it may yield the same values as the function between 0 and π, will not yield the same values as the function elsewhere if the function is not odd and periodic. For example, the trigonometric series for the absolute value function (see Figure 3a) has for its sum the "sawtoothed" function in Figure 3b. Fourier stated without fanfare that functions that agree on an interval need not agree elsewhere. That is obviously true on his conception of functions, and it shows how big a change that conception was. For Fourier, in contrast to Euler and Lagrange, functions consist of their values, not the expression used to compute them, and hence may be considered on arbitrarily restricted intervals.

Fourier used his techniques to greatly advance the art of solving partial differential equations. They were too successful to be ignored. Indeed, Siméon-Denis Poisson thought the techniques could be extended to yield general methods to solve all partial differential equations. That did not turn out to be the case, but Poisson did greatly expand their domain of application, and they remain essentially the only available techniques for obtaining exact solutions to partial differential equations subject to boundary conditions [Str87, p. 150]. Since such equations are at the heart of mathematical physics, Fourier's peculiar functions—functions that need not be defined by closed expressions, that could jump about, that did not have the same analytic expression everywhere—became part of the repertoire of every mathematician. In his book, Fourier said [Str87, p. 150], "In general the function $f(x)$ represents a succession of values or ordinates each of which is arbitrary . . . We do not suppose these ordinates to be subject to a common law; they succeed each other in any manner whatsoever." The functions he actually employed were considerably less general. The old basis for analysis had relied on analogies between polynomials and other analytic expressions. It was no longer adequate.

The trigonometric series with coefficients b_ν developed by Fourier—the Fourier series—are infinite discrete series of sine waves. It was natural for the analysts of the day to seek a corresponding integral—to represent a function not as a sum of sines of discrete harmonic frequencies but as a sum of sines of continuously varying frequencies. Results concerning such Fourier integrals were obtained by Fourier (1811), Poisson (1816), and Augustin Louis Cauchy (1816), each of whom was aware of the work of the other two [Kli72, pp. 679–681].

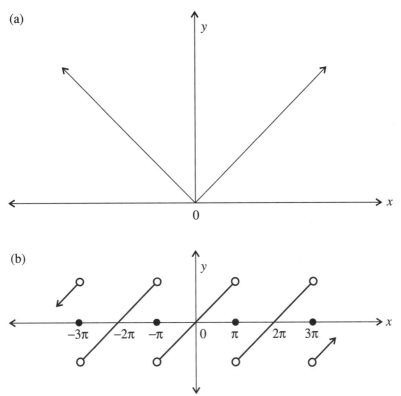

Figure 3. The functions graphed in (a) and (b) agree between $x = 0$ and $x = \pi$. The graph in (b) is obtained by reflecting that piece to obtain the values between $x = 0$ and $x = -\pi$, and then duplicating the graph between $x = -\pi$ and $x = \pi$ for the other values of x. I have given the correct value of the trigonometric series, 0, at the jumps, but Fourier drew vertical lines. (See [GG80a, p. 107] and [Bot86, p. 70].)

§5. Infinity Spurned

It was Cauchy, surely in part stimulated by his work on Fourier integrals, who brought rigor into analysis. His rigor was not rigor for its own sake. Rigor rarely is, as Philip Kitcher has emphasized [Kit83, pp. 213–217].

Cauchy needed a more rigorous development of analysis to continue his research. Euler, who had been interested in series of numbers, could discard the anomalous results divergent series sometimes caused: It was easy enough to test his results by summing a few terms. That procedure is not adequate in the problem context of concern to Cauchy—series of functions, not series of numbers. One would have to sum many terms of a series at many points to show anything, and even then there was the distinct possibility of not trying the right points. Cauchy was prepared to reject some of Euler's techniques to make progress on problems of his own. (See [Kit83, pp. 249–250].) Note that Cauchy's foundational worries primarily concerned series, not infinitesimals. His work was not motivated by the worries of Newton and Berkeley.

The rigor of Bernard Bolzano's *Rein analytischer Beweis des Lehrsatzes, dass zwischen je zwei Werthen, die ein entgegengesetztses Resultat gewähren, wenigstens eine reele Wurzel der Gleichung liege* of 1817 anticipated Cauchy by about four years and in some respects surpassed him. But Bolzano's work was neglected for fifty years. Perhaps that is in part because it came out of a simple desire for rigor, without any subsequent mathematical application [Kit83, p. 264].

Cauchy defined a limit thus:

> When the successive values attributed to a variable approach indefinitely a fixed value so as to end by differing from it by as little as one wishes, this last is called the limit of all the others.   (*Cours d'analyse algébrique*, 1821; [Kli72, p. 951].)

He gave essentially the modern definition of what it is for a function to be continuous,[10] not the one current in his day (which was, "defined by a single analytic expression"). He defined the derivative as a limit, and employed only differentials defined in terms of the derivative. He defined convergence and divergence of series and stated baldly that divergent series have no sum. He

---

10. The modern definition of continuity states, roughly, that a continuous function is one with a graph that is free of jumps and wild oscillations, and more precisely that the value of the function at each value of x is the limit of the values of the function at nearby values of x.

developed many useful tests for convergence, and, perhaps most important, he made no use of divergent series. He gave what is now called the Cauchy convergence criterion for a sequence: the sequence $S_0, S_1, \ldots$ has a limit if $|S_{n+r} - S_n|$ is less than any given quantity for every value of r and sufficiently large values of n. He proved the condition necessary but could only assert that it is sufficient on a geometric basis. He found that basis sufficient. A proof of sufficiency had to await the need for one—and a theory of real numbers [Kit83, p. 262].

In his *Résumé des leçons sur le calcul infinitésimal* (1823), Cauchy defined the integral as the limit of a sum of rectangles. That made it possible to make rigorous sense of Fourier's use of integrals of functions that did not have simple analytic expressions, functions for which the integral could not just be defined—in the way that had become usual—as the inverse differential [Haw80, p. 154]. Cauchy gave the first *proof* of the fundamental theorem of the calculus—that the derivative of the integral of a function is the function itself, when the function is continuous. He also discussed the situation in which the function is not continuous. He *defined* length, area, and volume as the values of certain integrals. The old geometric foundation of analysis was turned on its head, and infinitesimals were demoted to a secondary, dispensable role.

Cauchy's definition of the integral is, by modern standards, not very general, and he made false claims. His rigor was apparently adequate for his mathematical purposes, and so he did not pursue rigor further. Many of Cauchy's errors had a single source—Cauchy spoke of the limit of a variable, not the limit of a function, and he did not display the dependence of a variable on the independent variable. He thought of a variable that approached zero as an infinitesimal. His notation was therefore vague with respect to crucial further dependencies. (See [GG80a, p. 121] and [Kit83, pp. 254–255].) Some examples of the false claims: (1) He assumed that a continuous function had a derivative except possibly at a few points. (2) He said that if the sum of a sequence $u_0, u_1, \ldots$ of continuous functions converges everywhere on an interval to a function F then the function F is continuous on that interval. (3) Moreover, he said that under those circumstances one could integrate the function F by summing the integrals of the u_is. The last had been assumed by Fourier in deriving the coefficients b_ν of the Fourier series of a function from the assumption that a trigonometric series for the function exists. For claims (2) and (3) a stronger condition—not convergence but uniform convergence—is required. Nevertheless, Cauchy eliminated the use of divergent series, and he launched the effort to give general criteria for the range of applicability of various notions of analysis.

In a letter of 1826, Niels Henrik Abel complained about

> the tremendous obscurity which one unquestionably finds in analysis. It lacks so completely all plan and system that it is peculiar that so many . . . could have studied it. The worst of it is, it had never been treated stringently. There are very few theorems in advanced analysis which have been demonstrated in a logically tenable manner. Everywhere one finds this miserable way of concluding from the special to the general and it is extremely peculiar that such a procedure has led to so few of the so-called paradoxes.   [Kli72, p. 947]

In the same year, in a paper on the convergence of binomial series [Kli72, p. 947], Abel praised Cauchy and corrected his claim that the sum of a convergent series of continuous functions is continuous. He gave as a counterexample essentially the sawtoothed function of Figure 3: it is discontinuous, and yet it is the sum of continuous sine curves. He also made some progress toward the required concept of uniform convergence.

Gustav Lejeune Dirichlet, who would be the first to apply the techniques of analysis to obtain results about the natural numbers [Ste88, p. 242], met Fourier in Paris during the years 1822–1825. In 1829 he proved that the Fourier series of any function f meeting a certain condition would always converge to the function. The condition was sufficient but not necessary. It allowed f to have finitely many *exceptional points,* such as bounded discontinuities (for example, jumps). (See [Kli72, p. 966].) Moreover, his proof that the condition was sufficient did not really use the restriction to only finitely many discontinuities. That restriction was only used to ensure that the integrals defining the b_νs were well defined. (See [Haw80, p. 156].)

In the same article Dirichlet concocted the function $f(x)$ that has the value c for rational x and the value d for irrational x [GG80a, p. 126]. It was intended to be an example of a function that could not be integrated and which would therefore not have a well-defined Fourier series—since the coefficients b_ν would not be defined. By 1837, in an article on Fourier series, Dirichlet gave a precursor of the modern definition of a function. That definition is as follows: A function f associates a single value $f(x)$ with each member x of its domain. The association may be perfectly arbitrary—no rule, description, method of computation, or the like is required.[11] Dirichlet said that y

---

11. It is a matter of some dispute how much credit Dirichlet deserves for the modern

is a continuous function of x on an interval if there is a single value of y associated with every value of x on the interval in a continuous manner. Both in the definition of his attempted example of a nonintegrable function and in the definition of a continuous function, Dirichlet clearly viewed a function as being given by its graph alone—by its values—without the need for any associated law. He also extended his sufficient conditions for a function to have a Fourier series to allow more kinds of exceptional points, though still only finitely many. His work posed a problem that is important for our story: can one allow infinitely many exceptional points and still obtain sufficient conditions for the convergence of Fourier series? (See [GG80a, pp. 126–127] and [Bot86, p. 197].)

Dirichlet's standard of rigor exceeded Cauchy's. That is to be expected. As Abel's example showed, Cauchy's techniques were not reliable for deriving general results about the convergence of arbitrary Fourier series. Dirichlet was aware, for example, that the result of taking two limits is sensitive to the order in which they are taken.

Dirichlet's student Georg Friedrich Bernhard Riemann in his *Habilitationsschrift* of 1854 (published posthumously in 1866 by Richard Dedekind) took up Dirichlet's problem. Instead of following Cauchy in showing that certain nice (in today's terminology, piecewise uniformly continuous) functions have well-defined integrals, he gave a definition of an integral more or less like that of Cauchy, but then he sought general conditions, based on an analysis of that definition, under which a function would have an integral. His approach—the chief novelty of which lay in finding general conditions under which functions would be integrable instead of just showing that familiar functions were—was thought for many years to be the most general possible. In particular, he gave an example of a function with infinitely many discontinuities in every interval that could still be integrated in his extended sense. (See [Haw80, pp. 157, 159] and [Haw75].) That work was instrumental in showing that the sort of general definition of an arbitrary function that Dirich-

---

definition of a function, in part because he restricted his definition to continuous functions. See [You76, pp. 78–79], [Bot86, p. 197], [Vol86, pp. 200–201, 207–209], and [Med91, Chapter 2]. See [You76, pp. 77–80] on the issue of who first actually gave the modern definition of function suggested by Dirichlet's work. I believe Dirichlet defined the notion of a *continuous* function to make it clear that no rule or law is required even in the special case of continuous functions, not just in general. That would have deserved special emphasis because of Euler's *definition* of a continuous function as one given by single expression—or law. But I also doubt that there is sufficient evidence to settle the dispute.

let's ideas suggested has a genuine mathematical point: one could prove interesting things about classes of arbitrary functions that included "pathological" ones. Riemann raised many problems concerning Fourier series, including the one studied by Cantor that led to set theory. (See below.) Some of the problems are still unsolved and of great interest. (See [GG80a, pp. 132, 138].)

It was Karl Weierstrass who developed the methods needed to attack Riemann's problems. Most of analysis in the latter part of the nineteenth century consisted of applying Weierstrass's methods to Riemann's problems. (See [GG80a, p. 132].)

Weierstrass did important research on the representation of functions via power series. In the process, he put analysis into its modern rigorous form during the years 1841–1856, when he was a high-school teacher. That work did not become known until the late 1850s, when he finally obtained a professorship. Thus, for example, though Weierstrass understood the uniform convergence of series by 1842 (that was necessary for his work on power series), the concept actually became known through the work of George Gabriel Stokes and Ludwig Philipp Seidel, who in 1847 and 1848 independently arrived at formulations of closely related notions that are not quite as generally useful. Seidel was a student of Dirichlet. Dirichlet was aware that a convergent sum of continuous functions (sines) could be discontinuous, contrary to Cauchy's stated result. That is what got Seidel started. (See [GG80a, pp. 127–128] and [Med91, pp. 88–91].) Cauchy corrected himself in 1853. But he did not pursue the matter even to the extent of seeing where else he had illicitly assumed uniform convergence earlier. (See [Bot86, pp. 207–208].)

Weierstrass was one of the many mathematicians who corrected Cauchy's belief that continuous functions have derivatives except possibly at a few points by giving examples of functions that are continuous but fail to have derivatives at many points. Weierstrass's example was continuous everywhere but differentiable nowhere.

Weierstrass replaced Cauchy's definition of a limit that involved such notions as "approach indefinitely" and "differ by as little as one wishes" by the following definition, in which "as little as one wishes" has become ϵ: a function $f(x)$ has limit L at $x = x_0$ if for every $\epsilon > 0$ there is a δ such that if $|x - x_0| < \delta$ and $x \neq x_0$, then $|f(x) - L| < \epsilon$. The "variable," that is the function, in Cauchy's terminology "ends" in the interval defined by $|x - x_0| < \delta$ by differing from the "fixed value" L by "as little as one wishes," that is, by less than ϵ. The replacement enabled Weierstrass to distinguish convergence from uniform convergence and to make allied distinctions, cleanly and naturally. That was at least part of the motivation for the new rigor [Kit83, p. 257].

Weierstrass also gave a theory of irrational numbers around 1860 [Kli72, p. 979]. Dedekind had developed a theory in 1858 [Ded72, p. 2]. Their predecessors for the most part, when they defined irrational numbers at all, had defined them as certain limits of sequences of rational numbers. That procedure, as Weierstrass's precise definition of a limit makes clear, does not suffice: As Cantor pointed out in 1883, the number L must already exist for it to be the limit of a sequence. If we start with only rational numbers, a sequence that "converges to an irrational number" will not have a limit L.

In the early 1830s, William Rowan Hamilton gave a different treatment of the irrationals, taking time as a basis [Kli72, p. 983]. Weierstrass would not have approved. Weierstrass defined a variable simply as a letter that may be assigned various values. He banished the old idea of a variable quantity, which often in some metaphorical sense varied with time.

As Hilbert put it in 1925,

> As a result of his penetrating critique, Weierstrass has provided a solid foundation for mathematical analysis. By elucidating many notions, . . . he removed the defects which were still found in the infinitesimal calculus . . . If in analysis today there is complete agreement and certitude in employing the deductive methods which are based on the concepts of irrational number and limit, and if in even the most complex questions of the theory of differential and integral equations . . . there . . . is unanimity with respect to the results obtained, then this happy state of affairs is due primarily to Weierstrass's scientific work.   [Hil26, p. 183]

Cauchy and Weierstrass had eliminated time, infinitesimals, and infinite quantities from the foundations of analysis, and by so doing they made possible a standard of rigor surpassing that of the Greeks.

§6. Infinity Embraced

In 1817 Bolzano tried to prove that a continuous function that is both negative and positive in an interval takes on the value zero in that interval. He made use of the fact, which he also tried to prove, that every bounded set of values has a least upper bound. An adequate proof had to await an adequate theory of the real numbers. Weierstrass used his own theory of the irrationals and techniques suggested by those of Bolzano to prove in the 1860s that every bounded infinite set of points has a *limit point*—that is, a point such that every interval around it contains infinitely many members of the set. (For example, 1 is a limit point of the set $\{0, \frac{1}{2}, \frac{3}{4}, \frac{7}{8}, \ldots\}$. Intuitively one sees that the members of the set crowd against 1.) The result is now called the Bolzano–Weierstrass theorem. (See [Kli72, p. 953].)

In the years 1869–1872, Charles Méray, Cantor, Heinrich Eduard Heine, and Dedekind all published theories of the irrational numbers [Kli72, p. 983]. Cantor's was a modification of Weierstrass's earlier theory [Jou15, p. 26]: Weierstrass defined real numbers in terms of series of rational numbers, while Cantor used sequences. Dedekind published his theory in response to Cantor's publication [Dau79, p. 48]. In 1886 Stolz showed that one can identify the irrationals with the nonrepeating decimals [Kli72, p. 987]. Every single one of the theories of irrationals defines irrationals in terms of some actually infinite sets or sequences. A nonrepeating decimal involves an infinite sequence of digits. Dedekind's theory of cuts defines $\sqrt{2}$, for example, in terms of the infinite set of all positive rational numbers p such that $p^2 > 2$. (That set and the set of remaining rational numbers—that is, those rational numbers p that are negative or such that $p^2 < 2$—cut the rational numbers into two parts, an initial segment and a final segment, hence the name "cut.") Cantor's theory of Cauchy sequences defines a real number to be associated with an infinite set of infinite sequences of rational numbers. And so forth.

Dedekind's theory closely resembles that of Eudoxus for incommensurable ratios. Roughly speaking, the upper and lower parts of the cut correspond to the commensurable ratios greater than and less than a given incommensurable ratio. Indeed, Dedekind gave credit to Book V of Euclid's *Elements*. Cantor felt his own theory to be superior to that of Dedekind because it makes use of sequences of rational numbers—objects familiar to analysts—instead of the unfamiliar "cuts" [Kli72, p. 986].

The definitions of the irrational numbers provide one of the great ironies in the history of mathematics: Cauchy and Weierstrass had eliminated infinitely small and infinitely great numbers from analysis and replaced them by limits. But the theory of limits that thereby became so central required a clearer theory of the real line, that is to say, a theory of the irrational numbers. And that theory promptly reintroduced the infinite into analysis. The old infinity of infinitesimal and infinite numbers was simply replaced by the new infinity of infinitely large collections.[12]

---

12. See Russell's *Principles of Mathematics* [Rus03, p. 304] for a related sentiment.

In 1831 Carl Friedrich Gauss said [Kli72, p. 994], "I protest against the use of an infinite quantity as an actual entity; this is never allowed in mathematics. The infinite is only a manner of speaking, in which one properly speaks of limits to which certain ratios can come as near as desired, while others are permitted to increase without bound." But only 52 years later, we find this in Cantor's *Grundlagen* [Can76, p. 75]: "The idea of considering the infinitely large not only in the form of the unlimitedly increasing magnitude and in the closely related form of convergent infinite series . . . but to also fix it mathematically by numbers in the definite form of the completed infinite was logically forced upon me, almost against my will since it was contrary to traditions which I had come to cherish in the course of many years of scientific effort and investigations."

Fourier, among others, had given a proof that if a function is representable by a trigonometric series, then the series is unique, that is, that any two trigonometric series that converge to the function are the same. We exhibited the main parts of such a proof above: If there is a series, its coefficients must be the b_νs we gave, and the series is therefore unique. The proof does not work, because the formula for the b_νs was obtained by integrating the series term by term. As noted earlier, even Cauchy believed such a procedure legitimate, but it works only if the series is uniformly convergent.

Weierstrass emphasized the importance of uniform convergence, and Heine became interested. He may have learned about it from Cantor, who had studied with Weierstrass before becoming Heine's colleague at Halle. In a paper of 1870, Heine noted the gap in the proof of the uniqueness of a trigonometric expansion. What had actually been proved was that if a function has a uniformly convergent trigonometric series, then that series is the Fourier series and it is the only uniformly convergent trigonometric series that sums to the function. But it was known that even Fourier series need not be uniformly convergent. The Fourier series for the sawtoothed function discussed earlier provides an example. Heine gave some positive results concerning uniqueness.

Influenced by Heine, Cantor proved that the trigonometric-series representation of a function is unique. Uniform convergence is not required. That result applies to a trigonometric series that converges everywhere.

Cantor began to extend his result to allow exceptional points. In 1871 he showed, verifying a belief of Riemann, that if two trigonometric series converge to the same function everywhere except possibly at finitely many points, then that is enough to ensure that they are the same series.

In 1872 Cantor obtained results that allowed infinitely many exceptional points, answering a question of Riemann: He defined the derived set S' of a set S of real numbers to be the set of limit points of S. For example, if S is the set $\{0, \frac{1}{2}, \frac{3}{4}, \frac{7}{8}, \ldots\}$, then S' is the set $\{1\}$ whose only member is 1. One can form the derived set of a derived set, that is, a second derived set, and so forth. Cantor proved a generalization of his earlier result: Suppose a set S of real numbers is such that for some n the nth derived set is finite. If two trigonometric series converge to the same function except possibly at points in S, then they are the same. Cantor gave his definition of the real numbers in the paper in which the generalization appeared. He needed it to show that for every n there is a set S whose nth derived set is finite and nonempty while the $n - 1$st derived set is infinite. That is, he needed it to show that allowing the iteration of his derived-set operation actually led to new possible sets S of exceptional points.

Cantor's paper was one of the first in which infinite sets of points received careful, explicit consideration. Dirichlet, in 1829, had proposed a condition for a function to be integrable that was a condition on the set of points of discontinuity of the function, a condition that was clearly concerned with infinite sets of points of discontinuity. The condition is worth stating, since it will come up again later. It is that the set of points of discontinuity be *nowhere dense*—that is, that within every interval there is contained an interval that includes no points of discontinuity. The intuition is supposed to be that nowhere-dense sets are in some sense small. But Dirichlet published no results [Haw80, p. 156]. In his doctoral thesis of 1864, Rudolph Lipschitz developed a condition under which a function would have a convergent Fourier series even when it had infinitely many points of oscillation. But the proof made substantial use of the structure of trigonometric series [GG80a, p. 137]. In 1870 Hermann Hankel, who had studied under Riemann, developed a condition under which a function would be integrable (in Riemann's sense) that involved sets of points of discontinuity. But his primary focus was on integrability [Haw80, p. 166].

Cantor's theorem of 1872, unlike the other results just mentioned that had involved infinite sets of points, was proved using nothing new other than a careful study of the structure of the relevant sets of points. The only other ingredient was an absolutely straightforward application of his earlier results concerning trigonometric series. Moreover, though Cantor used only finite iterations of the derived-set operation in 1872, he was already aware of the possibility of infinite iterations: Given a set P, let P' be its derived set, and in general let $P^{(k+1)}$ be the derived set of $P^{(k)}$. We so far have the sequence

$$P^{(0)} = P, \, P^{(1)} = P', \, P^{(2)} = P^{(1)\prime} = P'', \, P^{(3)}, \ldots.$$

Now let $P^{(\infty)}$ be the set of points that are in $P^{(k)}$ for every finite k—the points that the operation of taking a derived set has not yet eliminated. Then we can continue

$$P^{(0)}, \, P^{(1)}, \ldots, P^{(\infty)}, \, P^{(\infty+1)} = P^{(\infty)\prime}, \, P^{(\infty+2)}, \ldots, P^{(\infty\cdot2)}, \ldots,$$

$$P^{(\infty\cdot3)}, \ldots, P^{(\infty^2)}, \ldots, P^{(\infty^3)}, \ldots, P^{(\infty^\infty)}, \ldots, P^{(\infty^{\infty^\infty})}, \ldots.$$

(See [Dau71, pp. 211–213].) Cantor discovered the sequence in attempting to analyze the structure of complicated sets of real numbers, not, as is sometimes said, by attempting to generalize the natural numbers. That is not to say that he did not come to view the sequence as a generalization of the natural numbers. He did. But the sequence arose from attempts to consider complicated functions on the real numbers (those with complicated sets of exceptional points), not from a study of the infinite or an attempt to generalize the natural numbers. The modern theory of the infinite evolved in a contiguous way out of the mathematics that preceded it. Though I disagree with Kitcher in detail about the origins of Cantorian set theory [Kit83, p. 207], I fully endorse his main thesis that new mathematics by and large—and in particular Cantorian set theory—evolves out of old mathematics [Kit83, Kit88].

Cantor had obtained mathematical results by focusing on the structure of infinite sets of points, and he knew that there were sets of points with more complicated structures. (His theorem, recall, concerned only sets P such that $P^{(k)}$ was finite for some finite k—no use was made of sets such that, say, $P^{(\infty^2+6)}$ was infinite, but $P^{(\infty^2+7)}$ was finite.) Understanding more complicated sets was connected with understanding more complicated and arbitrary functions—ones with more complicated sets of exceptional points. Cantor made the fateful decision to turn his attention to the study of sets of points in their own right.

Sets of Points

§1. Infinite Sizes

Cantor began by studying the two most obviously interesting sets of points: the set of rational numbers and the set of real numbers. He looked for differences between the two that are relevant to the fact that the real numbers are continuous while the rational numbers are not. In 1874 he published a paper in which he demonstrated the remarkable result that the algebraic numbers (and hence their subset, the rational numbers) can be placed into one-to-one correspondence with the natural numbers, while the real numbers cannot. The set of rational numbers is thus shown to have the same size as the set of natural numbers—they can be paired off—but the set of real numbers is shown to be bigger than the set of rational numbers. The proof that Cantor gave that the real numbers cannot be placed into one-to-one correspondence with the natural numbers is not the one that is most familiar today. In particular, it did not show that there were any other infinite sets that were of other sizes, and Cantor knew of nothing larger than the set of real numbers. It follows from Cantor's results that one cannot define the real numbers in terms of finite sets of rational numbers: there aren't enough finite sets of rational numbers. The use of actual infinity in the various definitions of the irrational numbers had been no accident.

Cantor began to investigate whether he could put the points on a plane into one-to-one correspondence with the points on a line, perhaps as part of a search for larger infinite sizes. In 1878 he published the unexpected result that one can indeed put the points on a plane, that is, a two-dimensional space, or indeed any n-dimensional space, into one-to-one correspondence with the points on a line. The techniques he used suffice to show that the points of an ∞-dimensional space can be put into one-to-one correspondence with the

42

points on a line. Cantor also stated that he could show that every infinite set of points on the line could be placed into one-to-one correspondence with either the natural numbers or the real numbers—that there are no intermediate possibilities [Jou15, p. 45]. His proof turned out to be incorrect, and a proof of his claim, known today as the continuum hypothesis,[1] continued to elude Cantor, though much of his work was apparently motivated by attempts to prove it. The problem is still open today. It has been shown that the truth or falsehood of the continuum hypothesis cannot be settled on the basis of the set-theoretic principles we accept today (assuming they are consistent): In 1938, Kurt Gödel showed that the continuum hypothesis cannot be disproved on that basis [Göd90, p. 26]. In 1963, Paul J. Cohen showed it cannot be proved on that basis (see [Jec78, p. 176]). It therefore could not have been settled on the basis of the similar principles that Cantor employed.

Since spaces of different dimensions can be placed into one-to-one correspondence, Cantor's work posed the problem of seeing how spaces of different dimensions differ. Dedekind observed that spaces of different dimensions cannot be placed into one-to-one correspondence by a *continuous* function—Cantor's correspondences were discontinuous. Luitzen Egbertus Jan Brouwer was the first to provide a satisfactory proof of Dedekind's observation, in 1911 [Dau80, p. 188].

Leopold Kronecker, Cantor's former teacher, was an editor of the journal to which Cantor submitted his paper on dimension. Kronecker believed that all of mathematics should be based on the natural numbers. That is an early version of a view that we shall refer to as *finitism*. He also believed that every definition of a property should give a method for determining whether or not an object has the property. That is an early version of a view that we shall refer to as *constructivism*. Note that, although Kronecker was both a finitist and a constructivist, there is no logically necessary association between the two: one can be a finitist without being a constructivist, or vice versa. That point requires emphasis since finitism and constructivism are so frequently associated that they are often not clearly distinguished. When Cantor's paper did not appear immediately, Cantor suspected Kronecker of delaying it ([Dau80, pp. 188–189], but see also [Edw88]).

In 1879 Cantor published the first of a series of papers about subsets of the real line. In that series and in related papers he defined many notions still

---

1. The source of the name is Felix Bernstein's Ph.D. dissertation, which discussed the "continuum problem." See [Moo82, p. 56].

in use today concerning subsets of the real line and other spaces. Though I have modernized the notation, the following definitions are Cantor's: A set P is *everywhere dense in the interval* (a, b)—that is, in the set of real numbers greater than a and less than b—if $(a, b) \subseteq P'$, where P' is the set of limit points of P. A set P is *perfect* if $P = P'$. A set P is *isolated* if $P \cap P' = \varnothing$. A set P is *closed* if $P \cap P' = P'$. I give the definitions to emphasize that the papers are very much concerned with the real line, since I am—somewhat misleadingly so far as historical summary is concerned—going to concentrate on the aspects of them that led to set theory in a more abstract form. Cantor's research centered around two ideas—that of the derived set and that of the transfinite symbols. I am going to concentrate on the work connected to the transfinite symbols. For more details of the other aspect of Cantor's work, and references, see [Dau79].

§2. Infinite Orders

In 1879 Cantor defined two sets to be of the *same power* if they can be placed into one-to-one correspondence. He noted that the concept generalizes that of whole number, and that power "can be regarded as an attribute of any *well-defined* collection, whatever may be the character of its elements." In 1880 Cantor published his transfinite symbols for iteration of derived sets— $\infty^{\infty^3} + 1$, and so forth—for the first time. He said, "we see here a dialectic generation of concepts which always continues further and thus is free of any arbitrariness."

By 1882 the "symbols" were an object of study in their own right in *Grundlagen einer allgemeinen Mannigfaltigkeitslehre* (Foundations of a general theory of manifolds, published in 1883 [Can83]). To separate his transfinite ordinal numbers from the notion of increasing without bound symbolized by ∞ in analysis, he began to use ω instead of ∞. The use of the symbol[2] ω has been standard ever since. Cantor introduced what was to become the distinction between cardinal and ordinal numbers: The sets $(a_1, a_2, \ldots)$ and $(b_2, b_3, \ldots, b_1)$ have the same power, or cardinality, but their numberings, their orders, are different.[3] The first one has the order ω, while the second

---

2. The symbol ω, a lower-case omega, is the last letter of the Greek alphabet.

3. The notation used to indicate orders does not meet modern standards of rigor. It is nonetheless clear enough what is meant. I use it here and below in this chapter since it is indicative of the way in which Cantor thought about orders.

has the order $\omega + 1$. Indeed, the very same set can be numbered or counted in more than one way: consider $(a_1, a_2, \ldots)$ and $(a_2, a_3, \ldots, a_1)$. If a set is finite, there is only one order to give it even though one can vary which element of the set occurs at which point in the order, and so finite ordinal and cardinal numbers coincide. Cantor defined operations of addition and multiplication on the ordinal numbers, which is part of what justified thinking of them as *numbers* [Kit83, p. 174].

In the *Grundlagen* Cantor declared for the first time that there are many infinite sizes: He showed how to produce a set of power greater than the natural numbers, namely, the set of all ordinal numbers of the power of the natural numbers. (As illustrated above, ω and $\omega + 1$ are such ordinal numbers.) The proof he offered is a straightforward generalization of the one he used to show that there are more real numbers than rational numbers. (The two proofs are given and then compared in detail in §IV.2.) He called the power of the natural numbers (I), the new power, (II). The power (III) is the power of the set of all ordinal numbers of power (II). And so forth. He said that for every ordinal number γ there is a new power (γ). He did not have full control of all of the details, but with hindsight we can see that the proof was essentially correct. The construction starts with the natural numbers. It is an iteration of much the same sort that led to the "symbols" for successive derived sets. Indeed, as Philip E. B. Jourdain said, it may have been the primary reason for Cantor to have considered the "symbols" on their own, separately from derived sets [Jou15, p. 51].

The proof that the powers are distinct provides no way to make contact with the power of the real numbers. For all Cantor knew, the powers he had constructed were all smaller than that of the real numbers, or even all incomparable with that of the real numbers. Cantor made an additional assumption in the *Grundlagen* that guaranteed that the new powers were comparable with that of the real numbers: he assumed that the real numbers form a set and that they can therefore be well ordered.[4] That ensures that the power of the real numbers is less than, equal to, or greater than each of the new powers, but it gives no information about which. Cantor felt less than certain about the new

---

4. As we shall discuss in detail below, in §IV.2, Cantor had good reason for thinking he would eventually be able to prove that. The formulation in the text here is in one respect a bit misleading: as we also discuss in §IV.2, for Cantor being well-orderable was in effect constitutive of being a set, and so the right way to put the point would be, "He assumed that the real numbers can be well ordered and that they therefore form a set."

assumption. Indeed, at one point Cantor worried briefly that the power of the continuum was not comparable with any of the infinite powers (γ) [Hal84, pp. 42, 73, 76–77].

During the period in which Cantor was working out the theory of ordinal numbers, many mathematicians, including Cantor himself, came up with examples of nowhere-dense (§II.6) subsets of an interval of real numbers that are not small in an important sense: They cannot be covered by finite unions of intervals of arbitrarily small total length. That is, for example, there is a nowhere-dense set P such that for any n if P is contained in a union $[a_1, b_1] \cup \cdots \cup [a_n, b_n]$ of n intervals, then the sum $|a_1 - b_1| + \cdots + |a_n - b_n|$ of the lengths of the intervals is greater than 1. It seems reasonable in some sense to say that P has length at least 1, and that it is therefore not small despite the fact that it is nowhere dense. The appropriate sense is called *outer content*. Outer content was introduced by Cantor and, independently, by Stolz in 1884 [Haw80, p. 168]. As we shall see in §3, outer content was to be important in devising a notion of integral sufficiently general for the purpose of studying Fourier series. Thus, set theory was not only the product of problems within analysis; it also gave rise early on to fruitful new ideas for solving problems within analysis.

In 1885 Cantor prepared a paper in which he studied general linear orders, defined independently of the rational numbers or the real numbers.[5] The paper was not published in Cantor's lifetime, but it is worth mentioning from our perspective, since it was probably the earliest study of abstract structure independent of some familiar intended mathematical model. Another likely candidate is Dedekind's work that led to [Ded88]. Since Cantor and Dedekind were frequent correspondents, it would be worthwhile to know more about their discussions related to the present topic.

In the 1885 paper Cantor said that pure mathematics is just set theory, in the sense that all of mathematics can be understood in purely set-theoretic terms [Dau80, p. 202]. The year before Gottlob Frege had published *Die Grundlagen der Arithmetik*, deriving arithmetic from logical principles. The later development of Fregean foundations of arithmetic, along with Dedekind's

---

5. A *linear order* is a set M with a binary relation $<$ on it such that no m in M is such that $m < m$ (irreflexive); for all l, m, and n in M, if $l < m$ and $m < n$, then $l < n$ (transitive); and for all m and n in M, $m < n$, or $m = n$, or $n < m$ (connected). Examples include the natural numbers, the rational numbers, the real numbers, the ordinal numbers, or any subset of any of those, in each case with the obvious order relation.

work, was part of what led to the eventual widespread acceptance of Cantor's insight.

I have been describing in detail the extent to which set theory is intertwined with analysis, particularly with the theory of trigonometric series, but there is another sense in which set theory has become important to mathematics: Mathematics is today thought of as the study of abstract structure, not the study of quantity. That point of view arose directly out of the development of the set-theoretic notion of abstract structure.

The motives of Dedekind and Frege were rather different from those of Cantor: Dedekind was seeking to give a foundation of arithmetic "entirely independent of the notions or intuitions of space and time" [Ded88, p. 31]. Frege was studying logic as part of a philosophically motivated program of giving an explicit foundation for arithmetic—the development of logic was needed to ensure that no assumptions were going unnoticed. Both wished to provide a foundation for analysis. It is arguable that each is an exception to the usual rule that rigorization is not undertaken for its own sake. Even if they are exceptions, they are not harmful ones: they had ample motivation for increased rigor in the recent great success of the rigor of Cauchy and Weierstrass, and in their perception that the program of arithmetization of analysis advocated by Dirichlet [Ded88, p. 35] and Weierstrass had not been completed.

It is important to realize that though rigor and the systematization of analysis were the motives of Dedekind and Frege, they were not Cantor's motives, despite the usual account. Cantor was studying sets of real numbers for mathematical reasons that grew out of the study of the Fourier series of increasingly arbitrary functions. He did not work axiomatically. He believed in the reality of his ordinal numbers and sets, and he saw himself as discovering their properties. Therefore, no axioms were necessary. The fact that Cantor did not work axiomatically shows that, in contrast to Dedekind and Frege, he did not see his project as that of working out the consequences of a system of assumptions or as that of systematizing a body of knowledge. When a fact seemed obvious or elementary, Cantor just stated it without proof.

In Cantor's *Grundlagen*, powers were not associated with cardinal numbers. In 1883 Cantor made such an association for the first time. In 1886 he introduced a notation for cardinal numbers, and in 1887 he gave definitions of operations of addition and multiplication for cardinal numbers. The ordinal numbers came first, and they were always more important than the cardinal numbers for Cantor. (See [Dau79, pp. 179–181].)

In 1888, Dedekind published his theory of the natural numbers [Ded88]. A year later, Peano, with reference to Dedekind's work, gave a semiformal version of what has become the standard axiomatization of the natural numbers.[6] Peano's paper gave the first statement of the need to distinguish between the relation of set membership (for which Peano introduced the symbol $\in$) and set inclusion [Pea89, p. 86]. Peano also introduced what has become known as a comprehension principle ([Pea89, p. 90], see also [Ken80, p. 26]): every "condition" (that is, every "proposition containing the indeterminate x") determines a class, namely the "class composed of the individuals that satisfy [the] condition." Peano also gave foundations for the rational and irrational numbers, and even discussed Cantor's set theory. In 1890, Peano defined a continuous curve that hits every point in the unit square at least once. He also introduced the distinction between an individual and the class composed of that individual alone, and denied that one can select members from infinitely many classes without a determinate rule. (See [Kli72, pp. 988, 1018], [Ken80, p. 33], and [Moo82, p. 76].)

In 1891, Cantor published his diagonal argument.[7] It yielded a new proof that there are more real numbers than natural numbers, and, much more important, it was the first completely worked out argument that showed that there are infinitely many infinite powers. Indeed, it shows more: it shows that given any set, there is another of greater power. By applying that fact to the set of real numbers, Cantor showed for the first time that there is an infinite power strictly greater than that of the set of real numbers.

In 1892 Frege published a review of Cantor's 1887 paper, the paper that had introduced cardinal arithmetic. He fully endorsed Cantor's acceptance of the actual infinite and saw that Cantor's work had important consequences for analysis. Nonetheless, he devoted the bulk of the review to castigating Cantor

---

6. Peano arithmetic, PA, will be a useful example in later chapters. I shall therefore describe a convenient version. It is not exactly the one given by Peano. Here it is: Every number has a successor. The number 0 is no number's successor. Numbers with the same successor are equal. The sum of any number x and 0 is x. For any number x, the sum of its successor and any number y is the successor of the sum of x and y.

The product of any number x and 1 (the successor of 0) is x. For any number x, the product of its successor and any number y is the sum of the product of x and y with y. Any property that holds of 0 and that is such that if it holds for any number x then it holds for the successor of x, holds for every number. The last axiom listed is known as the induction axiom.

7. The paper is translated as Appendix B to Chapter IV.

for his reliance on intuition and an ill-defined notion of "abstraction." He advocated logical rigor and explicit definition. He had, however, no doubt that Cantor's theory could be developed in a satisfactory manner. One year later, the first volume of Frege's *Grundgesetze der Arithmetik, begriffsschriftlich abgeleitet* was published. In it he began to carry out in detail the program of his *Grundlagen*: developing arithmetic within a formal system, giving fully formal proofs to show that nothing was being smuggled in on the basis of intuition. In the *Grundgesetze,* Frege introduced a theory that brought his program into contact with Cantor's theory of sets. Presumably that theory was what Frege thought was the appropriate background in which to develop Cantorian set theory in a rigorous manner. We shall be discussing that and related issues in some detail in the next chapter. The second volume of the *Grundgesetze* (published in 1903) contains a theory of cardinal numbers. (See [Dau79, pp. 220–225] and [Dum67].)

§3. Integration

In 1892 Camille Jordan gave a definitive formulation of the Cauchy-Riemann integral. The integral of a function in two dimensions—a surface integral— was at the time usually defined over the region bounded by a closed curve. The surface integral was defined in terms of limiting values of sums taken over arbitrary partitions of the plane into rectangles. There was an obvious problem about what to do with rectangles on the boundary of the region— should rectangles neither wholly within nor wholly outside of the region of integration be included or excluded from the sums? That problem was finessed by claiming that the sum of the areas of the rectangles on the boundary went to zero in the limit and that it therefore didn't matter whether the rectangles on the boundary were included or excluded. But Peano's curve—which had every point in a region on the boundary—suggested that the claim was in trouble.[8]

Jordan solved the problem by moving from rectangles to subsets of the plane, in a Cantorian sense. The Cantor–Stolz notion of outer content extends

---

8. See [Haw75] for a detailed chronicle of the modern theory of integration, including Cantor's contribution and the strong impact of set theory on the development of the theory. The history of integration presented here is abbreviated and simplified in important respects, since my only aim is to give an example of how Cantor's set theory influenced the subsequent development of analysis.

in an obvious way from the line to the plane: just use rectangles instead of intervals. Jordan introduced a notion of inner content suggested by that of outer content. The outer content of a set is defined using the areas of finite sets of rectangles that contain the set. It is the greatest lower bound of such areas. The inner content of a set is the least upper bound of the areas of finite sets of pairwise disjoint rectangles that are contained within the set. Jordan said that a subset of the plane is *measurable* if it has outer content equal to its inner content. Naturally the *measure* of a measurable set is its outer, or, indifferently, its inner content. One easily sees that the measures of familiar sets of points on the plane are just their areas.

Now any measurable set has a well-defined "area" or measure, and so one no longer needs to have rectangles play a special role. Jordan defined the integral in terms of limiting values of sums over arbitrary partitions of the plane into measurable sets instead of over partitions into rectangles. It was then natural to allow the region of integration to be an arbitrary measurable set, instead of just the interior of a curve. Jordan showed that a set is measurable if and only if the outer content of its boundary is zero. That is what he needed to show that the sum of the areas on the boundary went to zero in the limit, and that the integral was therefore well defined. In 1893 Jordan incorporated his approach into his text, the *Cours d'analyse*. The next generation of French mathematicians learned Jordan's set-theoretic formulation of analysis. (See [Haw80, pp. 169–171].)

In 1895 Cantor defined cardinal exponentiation and observed that the power of the set of real numbers is $2^{\aleph_0}$. He could then have given the algebraic formulation of the continuum hypothesis that is standard today: $2^{\aleph_0} = \aleph_1$. In fact he did not. (The symbol $\aleph_0$, aleph naught, denotes the cardinal number of the set of natural numbers—(I) in the old notation. The symbol $\aleph_1$ denotes the next larger cardinal number, (II). And so forth. In particular, $\aleph_\omega$ denotes the ωth cardinal number.)

In 1902 Henri Lebesgue, building on important intermediate work of Emile Borel and others, built set theory into the very foundations of analysis. He changed the definition of outer content of subsets of the unit interval to allow actually infinite *denumerable* sets of intervals instead of just finite ones. (I have switched from the plane to the unit interval for simplicity—his definitions on other domains are based on the one sketched here.) That is, he allowed $[a_1, b_1], \ldots, [a_n, b_n], \ldots$ in addition to $[a_1, b_1], \ldots, [a_n, b_n].$[9] He then

---

9. As exemplified in the text, a *denumerable* set is one that can be indexed by the natural numbers.

defined the inner content of a subset E of the unit interval to be 1 minus the outer content of the complement of E. The corresponding notion of measurability is known as Lebesgue measurability.

The Lebesgue integral may be defined in just the same way that Jordan defined the Cauchy-Riemann integral, except using Lebesgue measure instead of Jordan measure. The Lebesgue integral has many convenient properties that the Cauchy-Riemann integral does not. For example, if a sequence of Lebesgue integrable functions on a set of finite measure is uniformly bounded (which just means that there is a B such that all of the values of all of the functions are less than B) and converges to a function, then that function is Lebesgue integrable and the value of the integral is the limit of the sequence of values of the integrals of the functions:

$$\int_a^b \lim_{n \to \infty} f_n(x)\, dx = \lim_{n \to \infty} \int_a^b f_n(x)\, dx.$$

The Lebesgue integral has many applications in the theory of Fourier series. For example, Lebesgue showed in 1903 that if a bounded function is represented by a trigonometric series, then the series must be the Fourier series (where, of course, the Fourier coefficients are now defined using Lebesgue integration). (See [Kli72, pp. 1044–1048] and [Haw80, pp. 172–179].)

In 1915 Jourdain translated Cantor's *Beiträge zur Begründung der transfiniten Mengenlehre* into English (published as *Contributions to the Founding of the Theory of Transfinite Numbers*). He replaced *sets* [*Mengen*] in the title by *numbers*, "since these memoirs are chiefly occupied with the investigation of the various transfinite cardinal and ordinal numbers and not with investigations belonging to what is usually described as . . . 'the theory of sets' . . . —the elements of sets being real or complex numbers which are imaged as geometrical 'points' in space of one or more dimensions" [Can15, p. v, preface].

§4. Absolute vs. Transfinite

Cantor's study of sets began with his work on arbitrary functions and the discovery of the transfinite symbols, and it always remained tied to that beginning. Cantor believed he had discovered that between the finite and the "Absolute," which is "incomprehensible to the human understanding," there is a third category, which he called the *transfinite*. Cantor's initial reasons for postulating the Absolute were primarily theological, and theology continued to play an important part in his notion of the Absolute throughout his life. We shall discuss some of his mathematical ideas about the Absolute below, but our primary focus will be on Cantor's understanding of the transfinite. The Absolute enters into the discussion largely so that we can see how Cantor contrasted it with the transfinite. In the *Grundlagen* of 1883 he said that "the Absolute can only be acknowledged and admitted, never known, not even approximately," and that he was convinced "that the domain of definable quantities is not closed off with the finite quantities, and that the limits of our knowledge may be extended accordingly without this necessarily doing violence to our nature." In 1887 he characterized the transfinite as "in itself constant, and larger than any finite, but nevertheless unrestricted, increasable, and in this respect thus bounded." Cantor, from the beginning, devoted his efforts to understanding only the increasable infinite. (See [Hal84, pp. 13, 14].)[10]

In fact, Cantor's notion of the transfinite is even more specific, as he makes clear in the *Grundlagen:*

> The assumption that, besides the Absolute, which is unreachable by any determination, and the finite, there are no modifications which I call actually-infinite, that is to say which are determinable through numbers—this assumption I find to be quite unjustified . . . What I assert and believe to have demonstrated in this and earlier works is that following the finite there is a *transfinite* (which one could also call the *supra-finite*), that is an unbounded ascending ladder of definite modes, which by their nature are not finite but infinite, but which just like the finite can be determined by definite well-defined and distinguishable *numbers.*   [Hal84, p. 39]

Cantor defined ordinal but not cardinal numbers in the *Grundlagen* (cardinal numbers came later in the same year), and so it is ordinal numbers that Cantor takes to be basic in this passage.

Since ordinal numbers play such a central role for Cantorian set theory, it is worthwhile to see how Cantor conceived of them. The ordinal numbers are, according to Cantor, generated by two principles: each ordinal number has an immediate successor, and each unending sequence of increasing ordinal numbers has an ordinal number as its limit (that is, there is an ordinal that is

---

10. In this section, I have relied heavily on Michael Hallett's illuminating work [Hal84], from which most of the translations of Cantor's words are taken. My own analysis of Cantor's work, which in many respects takes Hallett's as a starting point, is presented in §IV.2.

the next after such a sequence). He made this precise using the notion of a well-ordered collection: a collection M is *well-ordered* by a relation $<$ if $<$ linearly orders M with a least element and every subset of M that has an upper bound not in it has an immediate successor.[11] Two well-ordered sets are of the same type if they "can be related to one another one to one and uniquely in such a way that the *sequence of elements is reciprocally preserved.*" Finally, an (ordinal) number is defined to be "the symbol or concept for a definite type of well-ordered set." (See [Hal84, pp. 49–52].)

It is reasonably clear that, on the one hand, the two principles represent an attempt to characterize the process that generated Cantor's transfinite symbols, and that, on the other, the notion of a well-ordering isolates key features of the sequence generated using the two principles. In the *Grundlagen,* Cantor regarded the process of bringing a set into the form of a well-ordered set, thereby specifying a definite succession of the elements of the set, as giving a way of *counting* the members of the set [Hal84, p. 146]. It is not hard to see why, though Cantor did not, so far as I know, explain in detail: When one counts a set, the order in which the members are counted linearly orders the set in such a way that there is a first member, each member has an immediate successor, and there is a next member after every sequence of members counted that has not exhausted the set [Hal84, p. 63]. That is, counting a set produces a well-ordering of it.[12] Conversely, if one has a well-ordering of a set, one can count it by following that well-ordering: one counts off the first element of the set (in the sense of the well-ordering) first, the successor of each element after that element, and if a sequence of members of the set has not exhausted the set and therefore has an upper bound not in the sequence, one counts the immediate successor of the sequence next.

Cantor clearly affirmed his commitment to the equivalence between counting and well-ordering when he said (in the *Grundlagen*)

> that definite countings can be effected both on finite and on infinite sets, assuming that one gives a definite law according to which they become *well-ordered* sets. That without such a lawlike succession of the

---

11. A member m of M is an upper bound of a subset N of M if m is not less than any member of N. A member m of M is a least upper bound of N if m is an upper bound of N such that if l is any other upper bound of N, then m is less than l. A member m of M is an immediate successor of N if m is an upper bound of N not in N such that if l is any other upper bound of N that is not in N, then m is less than l.

12. I am not here endorsing Cantor's view that one can make sense of a notion of counting for infinite sets, merely describing how it seems to me the view must go.

elements of a set no counting with it can be effected lies in the nature of the concept *counting.*   [Hal84, p. 146]

The sense in which the ordinal numbers are basic is now easily stated: The transfinite sets are those that can be counted, or, equivalently given Cantor's analysis of counting, those that can be numbered by an ordinal or well-ordered. The extent to which that idea is central to Cantor's thinking is indicated by the fact that in the *Grundlagen* he refers to all sets as "countable" [Hal84, p. 150].[13] The sets are the well-orderable, increasable manifolds or classes. Unincreasable manifolds are *not* sets.

> **Technical Remark.** Cantor was well aware that the same infinite set can be counted in various ways. Indeed, as Hallett said [Hal84, p. 151],"Cantor remarks [in the *Grundlagen*] that the *only* essential difference between finite and infinite sets is that the latter can be enumerated (counted) in various ways while the former can be enumerated in just *one* way." Cantor associated a *number class* with each infinite (ordinal) number γ: the class of all numbers of the same power as γ. He then used the number classes to represent the various powers, which did not become associated with a new sort of number, a cardinal number, until later. As discussed in §2, he knew that each number class has a power greater than that of any of its members. That formed part of an argument that each power "is coordinated" with a number [Hal84, p. 41]. Thus Cantor's theory of powers was based on his theory of ordinal numbers [Hal84, pp. 62, 65]. We shall have no need to discuss Cantor's theory of powers. The interested reader would do well to consult [Hal84].

Cantor's heuristic provides an immediate counterexample to the idea that his theory can be a theory of all classes: the class of ordinal numbers obviously cannot be counted, that is, assigned an ordinal number, since any ordinal number has many successors by Cantor's principles of generation. Briefly, the class of all numbers cannot be numbered. As Cantor himself put it in the *Grundlagen*,

> I . . . now consider the only problem to be to investigate the relations of these supra-finite numbers, not just mathematically but also quite generally in tracking down and demonstrating where they appear in nature. I

---

13. "Countable" has come to mean finite or denumerable. I avoid that usage because of the clash with the terminology of the *Grundlagen*.

have no doubt at all that in this way we extend ever further, never reaching an insuperable barrier, but also never reaching any even approximate comprehension of the Absolute. The Absolute can only be recognized, never known, not even approximately. For just as . . . given any finite number, no matter how large, the power of the finite numbers following it is *always* the same, so following each supra-finite number . . . there is a totality of numbers . . . which has lost nothing as regards power with respect to the whole of the absolutely infinite totality of numbers beginning with 1 . . . The absolutely infinite sequence of numbers therefore seems to me in a certain sense a suitable symbol of the Absolute.   [Hal84, p. 42]

Cantor's procedure was the reverse of what one might expect, and that is the source of much of his success, as Hallett convincingly argued [Hal84]. To obtain a theory of number that applies to both the finite and the infinite, Cantor restricted himself to those infinites that are like the finite in that they can be counted. Of course, we cannot actually count infinite sets. Frege pointed out that we can't even count sufficiently large finite sets, and so Cantor was using an idealization or extension of the usual notion of counting. Most attempts at giving a rationale for Cantorian set theory run head-on into the problem of justifying a suitable idealization. That was not a problem for Cantor, whose position was straightforwardly theological: for him "countable" meant countable by God [Hal84, pp. 15, 35–36, 44].

Frege, in contrast, took the expected approach: since infinite collections *cannot* be counted, he sought a theory of number that is independent of counting. He therefore took one-to-one correspondences to be basic, not well-orderings. That resulted in a theory in which the cardinal numbers are basic, not the ordinal numbers. (See [Hal84, pp. 151–153] for a detailed discussion.) No restriction seems necessary: a Fregean theory will not obviously need to exclude "the Absolute."

Cantor's argument that "the absolutely infinite sequence of numbers" is "a suitable symbol of the Absolute" can be turned into a mathematically precise proof that the class of all ordinal numbers is not a set. Cantor gave such a proof in a letter to Dedekind dated 3 August 1899 [Can32a].[14] According to a letter he wrote to Jourdain (dated 4 November 1903, see [GG71, p. 117]), Cantor knew that proof by 1895. Since he could now obtain a contradiction

---

14. See [GG74, pp. 127–128] for the reason the letter is dated incorrectly in [vH67].

from the assumption that the system of all numbers is a set, he began to call it an *inconsistent* absolutely infinite multiplicity.

Technical Remark. Here is the proof. (I have departed inessentially from Cantor's version.) As we have seen, to every well-ordered set F corresponds an ordinal number, $\overline{F}$. Thus, to use our earlier examples, $F = (a_1, a_2, \ldots)$ has ordinal number ω, that is, $\overline{F} = \omega$, while $G = (b_2, b_3, \ldots, b_1)$ has ordinal number $\omega + 1$. There is a natural sense in which ω is less than $\omega + 1$: there is an initial segment of G, namely the part of G that comes before b_1, that has order type ω, as the one-to-one correspondence that pairs off a_i with b_{i+1} shows.

As Cantor had proved, the ordinal numbers are well-ordered by the following natural order, which generalizes what was just illustrated for ω and $\omega + 1$: If α and β are both ordinal numbers, we say that α is less than β if there is a well-ordered set F such that $\overline{F} = \beta$ and a member a of F such that *the initial segment of F determined by a* has order type α, where the initial segment of F determined by a is just the subset of F that consists of members of F less than a, with the same order they had in F.

The natural order on the ordinal numbers has the following convenient property: For any ordinal α, the set of ordinal numbers less than α ordered by the natural order form a well-ordered set of order type α. Thus, for example, the set of ordinal numbers less than 3 ordered in the usual way, that is, $(0, 1, 2)$, is a well-ordered set of order type 3.

Let Ω be the class of all ordinal numbers, and suppose that Ω is a set. Then Ω is a set well ordered by the natural order, and so it has a corresponding ordinal number, say $\overline{\Omega} = \delta$. Thus, Ω is a well-ordered set of type δ. But, by the definition of Ω, the ordinal number δ must be a member of Ω. By the convenient property mentioned above, the initial segment of Ω determined by δ has order type δ, and so, by the definition of the natural order, δ is less than $\overline{\Omega}$. That is, δ is less than δ, which is impossible. The contradiction shows that our initial assumption that Ω is a set must be false. As Cantor put it [Can32a, p. 115], "The system Ω of all numbers is an inconsistent, absolutely infinite multiplicity."

Since Cantor had already argued that to every ordinal number corresponds a distinct cardinal number,[15] it followed from the fact that the system of all

---

15. The number classes inherit their ordering from the ordinal numbers, and they

ordinal numbers is an inconsistent, absolutely infinite multiplicity that the system of all the cardinal numbers that correspond to ordinal numbers is also an inconsistent, absolutely infinite multiplicity.[16] Cantor went on to employ those results to prove certain theorems he had been working on for a long time. The details belong in §IV.2.

§5. Paradoxes

In 1895, the year in which Cantor discovered the arguments just presented, Bertrand Russell wrote his fellowship dissertation, which was to become *An Essay on the Foundations of Geometry* (1897). It was a neo-Hegelian work: Russell believed that every science (except the universal science, metaphysics) necessarily contains contradictions that require a dialectical transition to another science for their resolution. For example, geometry is the science of pure spatial relations. But relations need something to relate. Thus geometry must postulate something beyond pure spatial relations: spatial points. The contradiction is transcended by moving toward physics. (See [Gri88, pp. 20, 24–26].)

In 1896, Russell learned of Cantor's work. Russell later said [Rus67a, p. 200], "At that time I falsely supposed all his arguments to be fallacious, but I nevertheless went through them all in the minutest detail. This stood me in good stead when later on I discovered that all the fallacies were mine." At the time, Russell believed that [Gri88, p. 32] "the continuum as an object of thought is self-contradictory." In 1897, Russell reaffirmed his doubts about the mathematical infinite, but in 1898 he tentatively accepted it, in an early draft of what was to become *Principles of Mathematics* [Rus03] (see [Moo88b, pp. 49, 50]).

Russell's acceptance of the infinite did not last long. In 1899, around the time Cantor was writing the letter to Dedekind discussed in the previous section, Russell was lecturing and writing about Leibniz. Leibniz accepted the actual infinite but argued against infinite number. In the second draft of what was to become the *Principles,* written in 1899, Russell accepted that infinite number is contradictory but worried that a class, the extension of a concept (that is, the collection of things to which the concept applies),

---

are therefore well-ordered. Since every sequence of cardinal numbers has an upper bound, they are absolutely limitless, and so they must form a well-ordered class similar to Ω.

16. The argument bears some relation to the later Axiom of Replacement: the range of a function on a set is a set. See §V.2.

is a totality, which should therefore have a number. The simplest version of the contradiction arises, he observed, when one considers the totality of numbers. He quoted Leibniz as saying, "the number of all numbers implies a contradiction," and he wrote: "There is, and is not, a number of numbers." (See [Moo88b, p. 50] and [MG81, p. 325].) The problem whether infinite collections have infinite number or no number continued to dog him in the subsequent draft, which he worked on into the next year.

In the summer of 1900, Russell met Peano at a conference and was favorably impressed. He began to study Peano's work. He later said,

> The time was one of intellectual intoxication. My sensations resembled those one has after climbing a mountain in a mist, when, on reaching the summit, the mist suddenly clears, and the country becomes visible for forty miles in every direction. For years I had been endeavouring to analyse the fundamental notions of mathematics, such as order and cardinal numbers. Suddenly, in the space of a few weeks, I discovered what appeared to be definitive answers to the problems which had baffled me for years . . . Intellectually, the month of September 1900 was the highest point of my life.   [Rus67a, pp. 232–233]

As a result of studying Peano's work, apparently in September, Russell came to accept that every collection has a cardinal number. By November, he had found an "error" [Cof 79, p. 33] in Cantor: Cantor's diagonal argument proved that there is no largest cardinal number. But the number of individuals is the largest number, since every class is included in the class of individuals. (Russell counted classes and numbers as individuals.) At around the same time he also noted that if the ordinal numbers are, as Cantor claimed, well ordered, then there is a maximum ordinal number, namely, the order type of the class of all ordinal numbers. He also described the error as involving the class of classes instead of the class of individuals.[17] (See [Moo88b, pp. 52–53].)

---

17. I do not know why Russell shifted from the class of individuals to the class of classes. It is not hard to argue that the two classes have the same cardinal number, and conclude that if the cardinal number of one of them is the largest cardinal number, then so is the cardinal number of the other. One argument, which Russell definitely gave later [Rus03, p. 367], goes like this: The class of classes is contained in the class of individuals, and so it is no larger. Conversely, the class of classes with exactly one member is the same size as the class of individuals (since each individual corresponds with the class that has it and nothing else as a member), and the class of classes with exactly one member is

Russell did not yet suspect any paradox, though he had found a contradiction. He believed that Cantor's conclusions were not quite so general as they seemed. In detail, he doubted Cantor's assertion that the ordinal numbers are well ordered, and he supposed that Cantor's diagonal argument, which he took to show that the class of all subclasses of a class is of strictly greater power (cardinal number) than the class is not quite so general as it seemed—it does not apply to the class of all individuals. The last supposition is one Cantor later endorsed:[18] Cantor viewed his results as applying only to "countable" *sets*—a notion we shall discuss in detail in §IV.2—not to arbitrary collections. Russell's work concerned classes, the notion introduced here and discussed in more detail in §IV.1. But during the period under discussion—and usually even later—Russell just interpreted Cantor's work as if it concerned Russellian classes. When Cantor saw Russell's later work, he concluded that the class of all individuals was not a set at all but an inconsistent, absolutely infinite multiplicity.

Russell's suspicion that Cantor's argument did not apply to the class of classes was based on the following, quite reasonable grounds. The class of classes has as members all classes, including those that have individuals other than classes as members. The class of all sub*classes* of the class of classes, in contrast, is composed entirely of classes that have only classes as members. It must therefore be a proper part of the class of classes, and so it cannot be of larger power than the class of classes. (See [Cof 79, p. 34].)

Russell took Cantor's argument to show that the class of all subclasses of a class has greater power than the class, and he recast the argument in essentially the following way:[19]

(1) He first showed for any function k from a class u to the class of all subclasses of the class u, that the class of all members x of u such that x is not in k_x is a subclass of u not in the range of the function k.

(2) He then observed that the class of subclasses of a given class has power at least as great as that of the class, since the function that takes each member x of the given class to the class whose only member is x is a one-to-one correspondence between the given class and some of its

---

contained in the class of classes, and so it is no larger. Neither class is larger than the other, and so they have the same cardinal number, as required.

18. In a letter translated in part as Appendix A to Chapter IV.

19. Cantor's point of view was different. His paper is translated as Appendix B to Chapter IV.

subclasses. To show that the class of subclasses of a class has greater power than the class, it therefore suffices to show that the two do not have equal power.

(3) Finally, he supposed for the sake of contradiction that a class and the class of its subclasses have equal power. Then there is a function from the class to the class of its subclasses that establishes a one-to-one correspondence between them. But that contradicts item 1: a one-to-one correspondence cannot omit a member of the range.

Russell thought the argument was in error when the "class" was the class of classes. Following Russell, let *class* be the class of classes. We can, it seems, define a function k from *class* to the class of subclasses of *class* that includes every subclass of *class* in its range as follows: when x is in *class* and x is a class of classes, let k_x be x, and when x is in *class* and x is not a class of classes, let k_x be the class whose only member is x. But that violates item 1 of the Cantorian proof: According to the proof of item 1, the class u' of classes x such that x is not a member of k_x should not be in the range of k. But, Russell observed, u' is $k_{u'}$ (presumably since k is the identity function on classes of classes), and so, contrary to the Cantorian argument, u' *is* in the range of k. Thus, Russell concluded, item 1 of Cantor's argument is incorrect when the class involved is *class* and the function is k, and so Cantor had not shown that there is no largest cardinal number. Russell gave this analysis by November of 1900. He added, seemingly as an afterthought, that "in fact, the procedure is, in this case, impossible; for if we apply it to u' itself, we find that u' is a $k_{u'}$, and therefore not a u'; but from the definition, u' should be a u'." (See [Cof 79, pp. 35–36].)

Russell seems to have maintained the view that Cantor's argument is defective and that there is a largest cardinal number at least through the middle of January 1901 [Cof 79, p. 33]. But the u' of the above argument is readily seen to be the class of classes that do not belong to themselves, and the above afterthought just shows that u' both is and is not a member of itself. That is, the definition of u' leads to a contradiction. There is no class of all classes that do not belong to themselves.[20] Russell discovered that[21] by May [Moo88b,

---

20. Zermelo discovered the paradox independently, but little is known about the details. See [RT81].

21. Actually, so far as I know, he did not at the time discuss the class of all classes that are not members of themselves, but only the class of all predicates that cannot be

p. 53]. There is no class u', and so Russell had not produced a counterexample to Cantor's argument. In October 1901, Russell wrote to Louis Couturat that Cantor is irrefutable [Cof79, p. 37].

Russell did not know what to make of his contradiction:

> It seemed unworthy of a grown man to spend his time on such trivialities, but what was I to do? There was something wrong, since such contradictions were unavoidable on ordinary premisses. Trivial or not, the matter was a challenge. Throughout the latter half of 1901 I supposed the solution would be easy, but by the end of that time I had concluded that it was a big job.   [Rus67a, p. 236]

Russell finally wrote to Peano about "the matter" and to Frege in June 1902. The letter to Frege [Rus67b] introduced the argument with some diffidence: "There is just one point where I have encountered a difficulty." But Frege's attitude was clear. He replied [Fre67], "not only the foundations of my arithmetic, but also the sole possible foundations of arithmetic, seem to vanish." That is how Russell's "conundrum" became Russell's paradox. By September, the paradox was a central problem for Russell. (See [MG81, p. 328].)

Russell had uncovered two more paradoxes by the time the *Principles* appeared:[22] the paradox of the largest ordinal and the paradox of the largest cardinal. The paradox of the largest ordinal is this: The class of all ordinal numbers is apparently well ordered, and so it has an ordinal number as order type, which must be the largest ordinal. But there cannot be a largest ordinal number since every ordinal number can be increased by 1. (See [Rus03, p. 323].) The similarity between that argument and the one Cantor used to show that the ordinal numbers form an inconsistent multiplicity should be clear. The paradox of the largest ordinal has come to be known as Burali-Forti's paradox, since Russell attributed it to Cesare Burali-Forti. In fact, the paradox is due to Russell, though it was apparently suggested to him by his reading of a paper by Burali-Forti [MG81].

The paradox of the largest cardinal is this: The class of classes can be no

---

predicated of themselves. The class version appears in his letter to Frege [Rus67b], which he wrote a year later.

22. My method of counting paradoxes is somewhat arbitrary. For example, I am counting the paradoxes of the class of classes that are not members of themselves and of the class of predicates that are not predicable of themselves as one, because of their evident similarity.

larger than the class of individuals, since it is contained in the class of individuals. But the class of classes is the class of all subclasses of the class of individuals, and so Cantor's diagonal argument shows it to be larger than the class of individuals. Russell introduced that paradox as follows [Rus03, pp. 366–367]: "[Cantor's] argument, it must be confessed, appears to contain no dubitable assumption. Yet there are certain cases in which the conclusion seems plainly false." The paradox is often called Cantor's paradox, presumably because it is based on Cantor's argument. It may be rephrased as follows: Cantor's argument shows that there is no largest cardinal number. But the cardinality of the class of all individuals must be the largest cardinal number, since every other class is included in that one.

Russell's paradox seems the most important, because it is so much more direct than the others. The paradox of the largest cardinal in some sense already involves Russell's paradox, as we have seen from the manner in which Russell discovered his paradox. The paradox of the largest ordinal involves the machinery of well-ordered sets and ordinal numbers, and so Russell thought it might be dissolved in some technical way.

What Are Sets?

§1. Russell

Why did Russell find paradoxes where Cantor found none? Because Russell accepted a principle that Cantor did not, one that conflicted with principles on which they agreed. The extra principle, which seems to have originated with Peano, was the Comprehension Principle. In Russell's words [Rus03, p. 20], "a class may be defined as all the terms satisfying some propositional function." In that and most other respects concerning the notion of class Russell followed Peano, as he said quite clearly. In particular, for Russell, a class is "composed of terms."

The Peano–Russell notion of class is essentially what Penelope Maddy has called the logical notion of collection. The characteristic mark of the notion is that according to it each collection is associated with some kind of a definition or rule that characterizes the members of the collection.[1]

Frege had a notion in the *Grundgesetze der Arithmetik* that is formally equivalent to that of a class, and a principle analogous to the Comprehension Principle—a principle that subjects his system to the paradoxes. Nonetheless, as Russell noted [Rus03, p. 513], Frege did not countenance classes in the now familiar form that comes down to us from Peano via Russell.[2]

---

1. The term is Maddy's, but she used it slightly differently [Mad90, pp. 103, 121]: "The logical notion . . . takes a number of different forms depending on exactly what sort of entity provides the principle of selection, but all these have in common the idea of dividing absolutely everything into two groups according to some sort of rule." Compare [Göd47, p. 475].

2. Frege took concepts to be basic. He was interested in a particularly important equivalence relation between concepts, that of extensional equivalence: two concepts are extensionally equivalent if they hold of the same objects. He postulated that to each con-

Russell was at least dimly aware that Cantor's conception of a set was different from his own:

> when mathematicians deal with what they call a manifold, aggregate, *Menge* [Cantor's term], *ensemble,* or some equivalent name, it is common, especially where the number of terms involved is finite, to regard the object in question (which is in fact a class) as defined by the enumeration of its terms . . . Here it is not predicates and denoting that are relevant, but terms connected by the word *and,* in the sense in which this word stands for a *numerical* conjunction.   [Rus03, p. 67]

Cantor's conception, which is discussed in detail in the next section, forms the basis for the one almost universally used by mathematicians. The main evidence for that claim is presented in §V.1. After that point, we give a detailed history of the Cantorian conception with practically no need to refer to

---

cept corresponds a logical object, the *extension* of the concept, in such a way that extensionally equivalent concepts correspond to the same object, while concepts that are not equivalent do not.

Frege did not have much more to say about the nature of his logical objects—his "extensions." They are often confused with Peano's classes because propositional functions that are satisfied by the same objects determine the same class—a property formally analogous to the one postulated by Frege. They are not the same: classes are composed of terms, and so the membership relation was basic for Peano, but Frege's logical objects were defined without reference to membership. To be sure, Frege later defined a notion formally equivalent to membership as follows: x is a "member" of the logical object y if there is some concept F such that y is the logical object that corresponds to F and x falls under the concept F. But that was clearly not the basis for his logical objects. Indeed, Frege said [Fre95, p. 228], "the concept is logically prior to its extension; and I regard as futile the attempt to take the extension of a concept as a class, and make it rest, not on the concept, but on single things." He summed up: "The extension of a concept does not consist of objects falling under the concept, in the way, e.g., that a wood consists of trees; it attaches to the concept and to this alone. The concept thus takes logical precedence of its extension."

Moreover, Frege wanted everything in his system to be one of his logical objects, so he just arbitrarily stipulated that the object that is the extension of the concept "x is the true" is the true, *not* the class of truths, and, similarly, that the object that is the extension of another concept is the false. That would not have been possible if he had intended the Peano–Russell notion of a class composed of members. (See [Rus03, pp. 510–512] and [Res80, pp. 204–220].)

I am under the impression that as late as 1903 Frege did not fully understand the Peano–Russell notion: Frege attempted to "avoid the contradiction" (Russell's paradox) by permitting two concepts to correspond to the same object even though that object "falls

the Peano–Russell conception. That in itself will help to show that the Cantorian conception is dominant. But one can also find an explicit statement of that dominance in an article by Skolem to be discussed in detail in §V.3:

> Until now, so far as I know, only *one* such system of axioms [for set theory] has found rather general acceptance, namely, that constructed by Zermelo [Zer08b]. Russell and Whitehead, too, constructed a system of logic that provides a foundation for set theory; if I am not mistaken, however, mathematicians have taken but little interest in it.   [Sko23b, p. 291]

As we shall see in §V.1, Zermelo's system is an outgrowth of Cantor's. I do not wish to leave the impression that Frege and Russell are or were unimportant. It is only that their mathematical work was for the most part concerned with logic, formalization, axiomatization, and related issues, not with the theory of sets. In particular, the paradoxes are important for what they tell us about our conceptions of properties and of truth, but they are not important for the theory of sets, as Gödel had already observed in 1947 [Par86, p. 105].

It is necessary to discuss Russell's conception of classes as it developed in response to the paradoxes before turning to Cantor's conception of sets, if only to make sure that the two are clearly distinguished. Russell was a logicist. He wished to show that mathematics and logic are one by showing how to develop all of mathematics within a framework free of any special

---

under the one and not under the other." (See [Fre80, p. 150].) (Frege called that object the extension and Russell used the term *range of values*, which was his translation of a term of Frege's that includes but is more general than Frege's *extension*. Since what is at issue is whether their use of the terms is similar to ours, I have been avoiding those terms.) Russell queried [Fre80, p. 155], "Do you believe that the range of values remains unchanged if some subclass of the class is assigned to it as a new member?" Frege replied [Fre80, p. 157], "I do not believe that a class remains in general unchanged when a particular subclass is added to it. All I mean is that two concepts have the same extension (the same class) when the only difference between them is that this class falls under the first concept but not under the second." Whatever Frege's extensions are, their members are not constitutive of them. The fact that he identified them with classes in the passage shows that he had not understood the notion of class. According to Charles Parsons, Frege always took classes to be either (Fregean) extensions or aggregates made up of parts, that is, mereological sums [Par76, p. 268].

conditions or empirical and psychological assumptions. That is a program substantially similar to Frege's for arithmetic and analysis.[3]

Frege and Russell faced a common problem: mathematics is apparently about objects (numbers and so forth), and yet the assumption that objects of one sort or another exist apparently goes beyond pure logic. They arrived at formally analogous solutions, which I explain with an example. Definable relations *are* within the province of logic, including definable one-to-one correspondences, and so, without going beyond logic, one can use Cantor's method to define *equinumerosity:* two systems[4] are equinumerous if there is a one-to-one correspondence between them. Suppose one postulates as a *logical* principle that every equivalence relation (equinumerosity in the example) determines logical objects and a logical relation such that entities in the field of the equivalence bear the relation to the same logical object if and only if they are equivalent. Then the logical objects will be suitable to play the role of mathematical objects. Russell used classes as the logical objects and membership as the logical relation (see [Rus03, pp. 166–167]), as was suggested by the work of Peano, Burali-Forti, and Mario Pieri (see [Con87]). Thus, for Russell a number was a class of all systems equinumerous to any member of the class. For example, on Russell's account, the number 2 is the class of all pairs. Thus, to be a system of two objects is just to be a member of the number 2, that is, a member of the class of all pairs. In general, the number of a system was simply the number of which the system was a member. For Frege, the number of a system was the extension of the concept "being equinumerous with that system."

The Comprehension Principle was what provided the mathematical objects on Russell's early logicist account of mathematics. It played a central role. But the Comprehension Principle was the source of the paradoxes. Russell therefore had two options: restrict the propositional functions to which the principle applied or restrict the propositional functions themselves, so that the principle still held outright. Whatever restrictions he adopted had to be purely logical in character. He tried both options, and gave many variants on the second one. I shall only discuss one of his theories, which is of the second kind: his 1908 theory of types, as presented in [Rus08]. I have focused on that

---

3. Russell's motivation was rather different from that of Frege. See [Hyl90].

4. I am using *system* here as a neutral word for whatever has an associated number. Frege and Russell disagreed.

theory since I understand it the best, and since it raises some points that will be useful in subsequent chapters.[5]

In the 1908 theory of types, individuals and propositions are taken to be basic,[6] and sentences that apparently mention propositional functions and classes are analyzed as involving only basic entities. In that sense, the theory is a no-class theory—propositional functions and classes are not taken to be "part of the ultimate furniture of the world."

I shall discuss propositional functions in some detail before getting to classes, since most of the work is done by the theory of propositional functions. The propositional function "x is mortal," for example, can be represented by the pair of the proposition "Socrates is mortal" and the individual Socrates. Russell took as primitive the quaternary relation exemplified by: the result of substituting *Plato* for *Socrates* in *"Socrates is mortal"* is *"Plato is mortal."* That has the following advantage: If we take propositional functions as basic, we run straight into paradoxes, as is seen by substituting the propositional function "x is not self-predicable" into itself. However, if we adopt Russell's stratagem, then "x is not self-predicable" must seemingly be represented by some pair like "Socrates is not self-predicable" and Socrates or " 'Socrates is mortal' is not self-predicable" and "Socrates is mortal," depending on whether we take the variable in "x is self-predicable" to range over individuals or propositions. Clearly, neither captures our intent— the variable was supposed to range over propositional functions. The one variable must therefore become two: "(p, a) is self-predicable," where now 'self-predicable' must mean something concerning a pair. But "(p, a) is self-predicable" is a propositional function that is to be represented (since there are two free variables) by a triple: "('Socrates is mortal,' Socrates) is self-predicable," "Socrates is mortal," Socrates. We needn't worry about how to make sense of 'self-predicable' because there is a different kind of trouble— 'self-predicable' was to be defined for propositional functions represented by pairs, but the propositional function we intended to substitute to obtain a para-

5. The article [Urq88] provides a useful brief history of Russell's attempts to solve the paradoxes. The articles [Lan87] and [Lan89] are very helpful in understanding the development of Russell's theory of types, and I have relied on them heavily.

6. According to Peter Hylton [Hyl90, pp. 151, 155], Russell had just postulated classes, but he had an argument for the existence of individuals and propositions, namely that they are required for logic.

dox is represented by a triple, and so it cannot be substituted. The paradox is blocked!

The analysis of propositional functions into propositions and individuals creates in effect a hierarchy of types, with propositions and individuals at the bottom, with single variables ranging over them; propositional functions of propositions or individuals next, with pairs of variables ranging over them; propositional functions of propositional functions of propositions or individuals next; and so forth. (The types are not in strict linear order: there can be propositional functions of both individuals and propositional functions of individuals, and so forth. See [Rus08] or [Lan87] for details.) According to Henri Poincaré, a vicious circle of definitions is the source of the paradoxes. Here, no propositional function can have itself in its own range, and an analogous circle is blocked.

Note the great ingenuity of Russell's device. It falls out of Russell's system of representation that the old notion of propositional function was incoherently wide. Eliminating the incoherence, of course, has the effect of restricting which propositional functions are allowed. But the restriction, while it suffices to block the paradoxes, allows one to retain the air of perfect generality: eliminating the use of incoherent propositions, while it is the required restriction on what came before, is not a defect in logical purity, and when one begins with the new system of representations, it need not be presented as a restriction. All the usual devices of logic can be allowed, and the types arise without any special pleading. The theory can still lay claim to being a part of pure logic, and so mathematics might still be one with logic, if no additional modifications were required.

Unfortunately, the typed system just described is still subject to paradox, as Russell had realized by 1906 [Lan89, p. 37].[7] The present theory allows quantification over all propositions and individuals, and hence, derivatively, over all propositional functions of a single type, including propositional functions of a type that are specified using quantification over propositional functions of that very same type.

Technical Remark. The paradox Russell discovered involved propositional functions of propositions. He worked directly in his basic sys-

---

7. I am simplifying the story somewhat. In 1906, Russell was working with a type theory of complexity intermediate between that of the one described in the text and the 1908 version toward which I am working.

tem, in which quantification over pairs of propositions stands in stead of quantification over such propositional functions. I shall use quantification over propositional functions instead, since the argument becomes easier to follow in that notation, but such quantification is just an abbreviation for something more complicated in the base notation. The simplification is one Russell had adopted by 1908. All of my Latin variables will range over propositions and Latin constant symbols stand for propositions, while my Greek variables range over propositional functions and Greek constant symbols stand for propositional functions. Since we shall explicitly consider only propositional functions of one variable, we can use, for example, ϕ to indicate a propositional function and $\phi(x)$ to indicate the value of that function at x. Let ψ be the propositional function (of y)

$$(\exists \phi)(y = [\phi(b) = q] \wedge \neg\phi(y)).$$

I have respected modern scruples about use and mention to the extent of forming a name for a proposition by enclosing the proposition in square brackets. Russell had no such scruples.[8] Now consider the proposition $\psi([\psi(b) = q])$, which reads as follows:

$$(\exists \phi)([\psi(b) = q] = [\phi(b) = q] \wedge \neg\phi([\psi(b) = q])).$$

The equality $[\psi(b) = q] = [\phi(b) = q]$ is between propositions, which are intensional. Thus, when the equality holds, it follows that $(\forall y)([\psi(y)] = [\phi(y)])$ and hence that $(\forall y)(\psi(y) \leftrightarrow \phi(y))$. We can therefore derive the contradiction $\psi([\psi(b) = q]) \leftrightarrow \neg\psi([\psi(b) = q])$.

Once more we have a circle of substitutions—ψ has been substituted into itself.

To block the new paradox, Russell introduced "orders" of propositions. At the bottom, there are the first-order propositions: "elementary" propositions and those that involve quantification only over individuals. Next come

---

8. Since Russell was happy to allow truth predicates in his base notation, it is possible to reformulate the paradox in a way that meets modern standards, though I shall not stop to do so here.

second-order propositions, which may also involve quantification over first-order propositions. And so forth. Quantification is not permitted over all propositions, but only over propositions of a given order.

We have now described Russell's "ramified" hierarchy. Propositional functions of any given type may be defined by propositional functions of various orders. The system, while less natural than before, is still motivated exclusively by logical concerns. After all, paradoxes cannot be permitted, and so the argument above might be taken to show that orders of propositions (or of whatever may serve as a surrogate for propositions) are logically necessary. The ramified hierarchy is indeed Russell's proposed logical system, though he modified it in various ways subsequent to 1908. Unfortunately, additional assumptions are required to do mathematics. I shall explain after showing how Russell handled classes.

Propositional functions may be intensional. For example, someone (call her Caila) might believe that all humans are mortal without believing that all featherless bipeds are, despite the fact that "x is human" and "x is a featherless biped" are coextensive.[9] Thus, the propositional function "Caila believes that for all x if $\phi(x)$ then x is mortal" depends for its truth value on the particular function ϕ, not just on what satisfies ϕ. Say that a propositional function Θ is extensional if

$$(\forall \phi \psi)((\forall x)(\phi(x) \leftrightarrow \psi(x)) \rightarrow (\Theta(\phi) \leftrightarrow \Theta(\psi))),$$

that is, if its truth value is the same on coextensive propositional functions ϕ and ψ.[10] For extensional propositional functions used only within extensional propositional functions, which are all we need for mathematical purposes, we can simply identify each function with the class of things that satisfy it and then define, for example, $\phi \in \Theta$ to be $\Theta(\phi)$, $\phi \cap \psi$ to be $\phi \wedge \psi$, and so forth. We make analogous definitions for propositional functions of individuals, and also for relations and functions as well as classes.

But now we are in trouble, because relations, classes, and functions, since after all they are just certain propositional functions, have orders. Thus, for example, a finite class is one such that there is no function whatsoever that maps it one-to-one to a proper subclass, but we cannot use that fact to define

---

9. I shall follow tradition in ignoring the existence of plucked chickens and other counterexamples to the supposed coextensiveness.

10. All of the lower-case Greek variables here range over propositional functions of some fixed order, though it makes no difference which order. Similar remarks apply below.

the finite classes, because we shall only be able to quantify over functions of some order or other. We cannot express "no function whatsoever," only "no function of such and such an order." We are faced with the possibility of a class being finite with respect to functions of some order, but infinite with respect to functions of some higher order, and the possibility of doing ordinary mathematics has apparently been lost. Russell's logicist program failed after all as a result of the paradoxes—his ingenious repair was unsuccessful.

Russell did not give up so easily as I have just made it seem. What is missing from his theory so far is a notion of class that does not have an associated order. The paradox that led us to introduce orders was an intensional one, while classes are extensional. That suggests a way out: We cannot assume that all propositional functions (of a given type) are of the same order, because of the paradox, but we can fix an order for any given type and then assume that every propositional function (of whatever order) of the given type is coextensive with one of the fixed order. That is to say, we can assume that some order includes so many propositional functions that they can represent all the classes. Since all the classes will then be of the same order, we shall be able to quantify over all of them. The assumption that allows that, and which therefore makes it possible to do mathematics within the theory of types, is Russell's Axiom of Reducibility. Details follow.

Technical Remark. Recall that a propositional function is analyzed as a proposition (the *prototype*) plus one or more propositions or individuals that in effect indicate the argument places. Let the order of a propositional function be (the least number that is) at least the order of the prototype and greater than the orders of the arguments. (Individuals have order zero; elementary propositions, order one. The definition differs from that of [Rus08, p. 164]. It makes, for example, $(\Theta(\phi), \phi)$, where ϕ is elementary, at least second order, which brings the orders of propositional functions into line with those of [WR57, pp. 163–164].) Say that a propositional function is *predicative* if its order is one greater than the orders of the propositions and individuals indicating the argument places. Following Russell, we shall indicate that a propositional function is predicative with an exclamation point, thus: $\phi!$.[11] The Axiom of Reducibility reads

---

11. Alfred North Whitehead and Russell [WR57, pp. 164–165] defined a predicative propositional function to be one that is quantifier-free, which is fine for the statement of the Axiom of Reducibility, perhaps even more natural than what I have done here.

$$(\exists \phi)(\forall x)(\phi!(x) \leftrightarrow \psi(x)),$$

where ψ is a schematic letter that may be replaced by any propositional function of any type and order with one free variable, and x is a variable of the appropriate type and order. The axioms for two or more free variables are analogous. Russell also refers to the axiom for one free variable as the Axiom of Classes and to the axiom for two free variables as the Axiom of (Binary) Relations.

We are not quite out of the woods yet: we cannot take the class associated with a propositional function to be the corresponding coextensive predicative propositional function, since there may be many coextensive predicative propositional functions. We therefore give a contextual definition that shows how to reinterpret formulas involving classes as formulas without them. They are to be eliminated from our official vocabulary much as propositional functions have already been eliminated.

With any propositional function Θ of a predicative argument $\phi!$, we associate a schematic formula $\Theta(\{z : \psi(z)\})$ that is defined as follows:

$$(\exists \phi)((\forall x)(\phi!(x) \leftrightarrow \psi(x)) \wedge \Theta(\phi!)),$$

where ψ is a schematic letter that may be replaced by any propositional function of a variable x appropriate to $\phi!$. For example, we define $x \in \{z : \psi(z)\}$ to be the formula that arises from our scheme using the propositional function Θ defined by $\Theta(\phi!) = \phi(x)$. Then $x \in \{z : \psi(z)\}$ is an abbreviation for

$$(\exists \phi)((\forall x)(\phi!(x) \leftrightarrow \psi(x)) \wedge \phi!(x)).$$

But they also claimed that it is possible to restrict quantification to predicative—that is, quantifier-free—propositional functions. The reason they offered is not correct, and so I have offered a definition of predicative here that makes their claim correct. In a special case, their reason amounts to this: The formula $(\forall x)\phi(x)$, where $\phi(x)$ is $(\forall y)\psi!(x, y)$, is just $(\forall x)(\forall y)\psi!(x, y)$. (In this note the exclamation point indicates that ψ is quantifier-free.) We can therefore replace $(\forall \phi)(\forall x)\phi(x)$ by $(\forall \psi)(\forall x)(\forall y)\psi!(x, y)$. But that doesn't work, since $(\forall \phi)$ includes in its range $(\forall y_1)\psi!(x, y_1)$, $(\forall y_1)(\forall y_2)\psi!(x, y_1, y_2)$, $(\forall y_1)$ $(\forall y_2)(\forall y_3)\psi!(x, y_1, y_2, y_3)$, and so on, while the quantifier $(\forall \psi)$ that is supposed to replace it accommodates only a fixed number of y_is. Since the orders play no role in mathematical considerations after the introduction of the Axiom of Reducibility, the rest of [WR57] is unaffected.

As a side effect of the contextual elimination of classes, every propositional function gets associated with one that is extensional, and so we can drop the previous restriction to extensional propositional functions and contexts.

With the Axiom of Reducibility, it becomes possible to develop mathematics within the theory of types (with two notable exceptions, to be discussed just below). However, as Russell himself put it,

> Viewed from this strictly logical point of view, I do not see any reason to believe that the axiom of reducibility is logically necessary, which is what would be meant by saying that it is true in all possible worlds. The admission of this axiom into a system of logic is therefore a defect, even if the axiom is empirically true.[12] . . . There is need of further work on the theory of types, in the hope of arriving at a doctrine of classes which does not require such a dubious assumption.   [Rus19, p. 193]

The earlier verdict is correct after all: Russell's logicist program failed as a result of the paradoxes.

Even if we allow the Axiom of Reducibility, there are still two gaps in the development of mathematics within the ramified theory of types: it is impossible to prove the Axiom of Infinity, which says that there is a class with infinitely many members,[13] and it is impossible to prove the Axiom of Choice (an assumption we shall be discussing in detail in §V.1). Moreover, as Russell recognized [Moo82, p. 131], the Axiom of Choice seems dubious when it is construed as an axiom concerning classes with membership specified via a rule. Those problems are serious since, for example, the Axiom of Infinity is invoked even to show that the sum of two real numbers is a real number, but they are less serious than the fundamental problem posed by the Axiom of Reducibility, since the Axioms of Infinity and Choice can simply be taken as hypotheses of every theorem in whose proof they are employed.[14] The Axiom of Reducibility must be used so universally that even the theory of the

---

12. For Russell, it is an empirical matter what classes there are. Only the ones that are definable in a suitable sense exist (to the extent that classes exist at all) necessarily.

13. Russell's axiom of infinity is that there are infinitely many individuals. It follows immediately from that in Russell's system that there is an infinite class, for example the class of individuals, and so his axiom of infinity has what we are calling the Axiom of Infinity as a consequence.

14. Since truths about the real numbers, for example, were supposed to turn out to

natural numbers—let alone the rest of mathematics—would depend on it as
a hypothesis, since it is required to give an adequate definition of finiteness.
Indeed, since all mathematical objects turn out to be classes, and Reducibility
is needed for the definition of class, not even a single mathematical object can
be defined as it was defined by Whitehead and Russell without an appeal to
the Axiom. To put it as starkly as possible, even the definition of the number
1 depends on the Axiom of Reducibility. That, of course, poses no problem
from a technical point of view, but it hardly suffices to establish the Russellian
thesis that logic and mathematics are one.

So much for the logical notion of collection. As far as the primary purpose
of this section is concerned, our discussion should end forthwith. But Russell
introduced a distinction that will play an important role in some of our later
considerations, a distinction intimately related to the theory of types, and so
we go on to present it here.

Every variable permitted in the ramified theory of types is restricted by type
and order. Nonetheless, definitions, axioms, and theorems must be available
at all types and orders, and so it is necessary to have devices that permit
generalization of some sort across types and orders.

The first device has been much discussed. It is that of "systematic ambiguity." In actual practice we never care about the absolute types and orders
of variables, but only the relative types and orders. Anything we assert about
orders 1 and n holds equally about orders $1 + m$ and $n + m$, and analogously
for types. Thus, we can use ambiguous symbols whose relative orders and
types are fixed by the context of use, and which can be applied at any absolute type and order. We normally read the variables of lowest type in a context
as ranging over individuals, but, when convenient, we can reinterpret them
as ranging over some higher type. Systematic ambiguity has been exploited
above when we allowed Greek variables to "range over propositional functions of some fixed order, though it makes no difference which order," and
when in stating the Axiom of Reducibility we took ψ to be "a schematic letter that may be replaced by any propositional function of any type and order
with one free variable," and x to be "a variable of the appropriate type and

---

be logical truths, not conditional truths requiring an assumption—Infinity, Choice—that
might be false, the procedure of taking such assumptions as hypotheses was not adequate
to Russell's task. We take up the question whether such a procedure can be part of an
adequate philosophy of mathematics of a different sort in §VI.3.

order."[15] In effect the device of systematic ambiguity lets us specify schemas in which the base types and orders may take on any permissible values. For example, the theorem $(\forall\phi)\phi = \phi$ is actually infinitely many distinct theorems (one for each type and order of the variable ϕ) involving infinitely many distinct equality relations.

Systematic ambiguity makes it possible to express certain general facts about the formalism of ramified type theory, facts that cut across types and orders, but it does not permit us to express those facts within the formalism. After all, every quantified variable must be of a particular type and order, and so, despite appearances, $(\forall\phi)\phi = \phi$ is not a sentence expressible within ramified type theory—it is a host of separate, unconnected sentences. The trick of taking the lowest order within a formula to be that of individuals and allowing the possibility of adding m to all the orders is no help, since it is not expressible within the formalism.

To make it possible to express general facts within ramified type theory, Russell introduced the distinction between "all" and "any," which he discussed in detail in Section 2 of [Rus08]. "Given a statement containing a variable x, say '$x = x$', we may affirm that this holds in all instances, or we may affirm any one of the instances without deciding as to which instance we are affirming" (p. 156). The affirmation of $x = x$ for *all* values of x is represented $\vdash (\forall x)x = x$, and the fact that x is quantified forces us to use a variable x of fixed type and order. But the affirmation of $x = x$ for *any* value of x is represented by $\vdash x = x$. No quantifier is involved, and so we may allow a new type of variable that is free of type and order restrictions, a new type of variable that cannot be quantified over. It then becomes possible to express generality across types and orders. As Russell remarked,

> we may admit "any value" of a variable in cases where "all values" would lead to reflexive fallacies . . . the fundamental laws of logic can be stated concerning *any* proposition, though we cannot significantly say that they hold of *all* propositions.   [Rus08, p. 158]

---

15. Whitehead and Russell use the term *systematic ambiguity* only for cases in which all of the types and orders are determined by the choice of the base—that is, by the reinterpretation of the individual variables. It seems they may have regarded the use of a variable like ψ in the Axiom of Reducibility, which does not have its order determined by the surrounding context, as a new device. (See [WR57, p. 165].) But my extension of their terminology seems to me to be a natural and harmless one.

Russell's idea that we affirm an instance without deciding which, which he calls an ambiguous assertion, is not very clear, but it suggests taking the new variables as schematic variables that admit of substitution by variables or constants of any type and order. Some of what Russell said suggests that reading: "We can only truly assert a propositional function if, whatever value we choose, that value is true; similarly we can only truly deny it if, whatever value we choose, that value is false" (p. 157).[16]

In the first edition of the *Principia Mathematica* more or less the same distinction between "any" and "all" appeared, but at some points what I have distinguished as schematic variables are allowed and exploited [WR57, pp. 128–129], while at others, even "any" is allowed only with variables of fixed type and order [WR57, pp. 17–18]. There is no point to using "any" with variables of fixed type and order. By the second edition, the use of "any" had been repudiated [WR57, p. xiii] on the grounds that any free variable used to express generality can simply be universally quantified, replacing the "any" by "all." As we have seen, that is not exactly correct, since it leaves no way of expressing certain facts that cut across types and orders within the formalism.[17]

§2. Cantor

The paradoxes posed no problem for Cantor's theory of sets—transfinite objects that can be counted. Indeed, in 1904 Cantor queried Jourdain about the availability of the second volume of Russell's *Principles*. Jourdain replied that it would not be available for some time, since Russell wished to present a "solution" of his "contradiction" in it, and he had not yet found one. Cantor replied with a discussion of the "difficulty" that Russell had described: Russell slightly extended Cantor's proof that $2^{\aleph_0} > \aleph_0$ to show that $2^a > a$ when a is the cardinality of any *set* $\mathfrak{M}$. The extended proof shows, given a *set* $\mathfrak{M}$, how to form a *totality* $\mathfrak{G}$ of greater power. But Russell tried to apply the proof

---

16. Hylton has argued [Hyl90, pp. 152–154] that Russell could not have employed schematic letters, because his conception of logic as universal blocked anything like an ascent to a metalanguage. Perhaps that is why Russell's pronouncements on "any" were confusing.

17. Hylton has argued that the inability of the formalism of the *Principia* to express its own formulation is yet another fatal blow to Russell's logicism, given Russell's conception of logic as universal [Hyl90, pp. 159–161]. Even if the notion of "any" turns out to be compatible with Russell's conception of logic, I do not know whether it is strong enough to enable the formalism to be used to formulate itself.

with an inconsistent multiplicity or totality in place of $\mathfrak{M}$. But since this $\mathfrak{M}$ is not a set, a totality corresponding to $\mathfrak{G}$ cannot be formed, and no contradiction arises.[18] Thus, in Cantor's eyes, the difficulty is avoided. Russell had said [Rus03, p. 368] that "the application of Cantor's argument to the doubtful cases yields contradictions." Cantor had never accepted those cases.

Cantor's notion of a set is that of a collection "defined by the enumeration of its terms," as Russell said (see §1). I shall refer to that as the *combinatorial* notion of a collection.[19]

Cantor started investigating combinatorial collections of exceptional points in order to extend the results of Fourier analysis to as many functions as possible, building on the general definition of a function usually attributed to Dirichlet. The work was part of a program of freeing analysis of the restriction to functions given by analytic expressions—that is, to functions given by rules.

The values of a function are determined by the collection of points that form the graph of the function. The logical notion of a collection, that is, the notion of a collection determined by a rule, therefore goes hand in hand with that of a function determined by a rule or analytic expression: a function is determined by a rule, an analytic expression, if and only if its graph is given by a corresponding rule, and the graph is therefore a logical collection.

We see that Cantor's work that led to his set theory and to the notion of a combinatorial collection grew in a natural way out of what amounted to the attempt to free mathematics of the restriction to logical, rule-based collections. The whole point of the combinatorial notion is that combinatorial collections may exist whose members *cannot* be characterized by any rule. The Cantorian notion of a combinatorial collection is not merely different from the Peano–Russell notion of a logical collection—it arose in opposition to it.

I have just presented combinatorial collections as more general than logical collections. There are two main marks of the additional generality: First, on any fixed infinite domain there are more combinatorial collections than logical collections. That bald assertion presupposes some suitable clarification of what a permissible rule is for forming a logical collection. After all, given any

---

18. The full correspondence I have just described is published in [GG71, pp. 118–119]. I have translated the relevant passage from Cantor's letter as Appendix A.

19. The term is suggested by [Ber35b, pp. 259–260], compare [Mad90, pp. 102–103].

combinatorial collection C, one could consider the instruction to collect the members of C to be a permissible rule. In that case, every combinatorial collection is trivially a logical one. But for any specification of allowable rules of which I am aware that does *not* presuppose combinatorial collections, the assertion holds. Second, combinatorial collections obviously obey the Axiom of Choice, while it is at best dubious whether logical collections do. I shall, however, postpone a discussion to §V.1.

There is a different sense in which things go the other way: combinatorial collections are restricted, while logical collections are not. Since combinatorial collections are enumerated, some multiplicities may be too large to be gathered into a combinatorial collection. We have already seen Cantor's example—the multiplicity of all ordinal numbers. In contrast, the size of a multiplicity seems absolutely irrelevant to whether it forms a logical collection. Since there is a property characterizing the ordinal numbers—just that of being ordinal numbers—it seems that they do form a logical collection. That is part of why Russell's theory of logical collections led to paradox while Cantor's theory of combinatorial collections did not. Any restriction on logical collections motivated by the notion of a logical collection would have to be a restriction on allowable rules, a restriction like that imposed by the ramified theory of types, not a simple restriction on size.

Though Cantor's theory was free of contradictions, it had other problems. In order to make it clear what those problems were, I shall give an axiomatic reconstruction of what I take to have been Cantor's mathematical theory in the period from somewhere around the time he arrived at the ideas in the *Grundlagen* of 1883 to the time he realized that the ideas of [Can91], the paper in which he first published his "diagonal argument" (translated as Appendix B to this chapter), could be used to show that the continuum of real numbers had cardinality $2^{\aleph_0}$. That period encompassed the main part of the development of his theory. It began with his acceptance of the ordinal numbers as objects of study in their own right, and it ended at a time when he was to publish only two more works on set theory, albeit important ones. Those last two works show some awareness of the problems with the theory as developed earlier. Note that that period of development came before the diagonal argument that led to Russell's paradox.

As we have seen, Cantor did not work axiomatically. He was working out the facts on the basis of a picture or conception, not on the basis of stipulated assumptions. Nonetheless, he did take each of the principles I take as Cantorian axioms to be, in one or another sense, basic.

I shall engage in some simplification. For example, Cantor did not take 0 to be an ordinal number; he started with 1. I shall start with 0 anyhow. Cantor identified things other than sets (and sometimes, it seems, perhaps sets as well) with their singletons.[20] It is not clear whether Cantor took either ordinal numbers or cardinal numbers to themselves be sets. I shall remain neutral about whether ordinal numbers are sets in the way that I reconstruct Cantor's view. I shall leave cardinal numbers out of account altogether. They do not play a central role in Cantor's view, and so adding them on is not especially illuminating. Besides, later developments have made it clear that initial ordinal numbers will serve perfectly well as cardinal numbers.[21] I shall give a definition of the least infinite ordinal number, ω, that does not rely on a prior knowledge of the natural numbers. Cantor tended to rely on such knowledge. I have also modified Cantor's notation for number classes.

Technical Remark. Cantor worked in German, not in a formal language, and he worked years before anyone had distinguished between first- and second-order logic. (See §V.3.) He made free use of the notions of function and relation, taking them to be part of an antecedently given background.

Russell (and Frege before him) could not have adopted Cantor's procedure: Russell's notion of a class is so intimately connected to that of a relation that to introduce it without carefully specifying what is meant by a relation would have begged the question, and an analogous comment applies to Frege. But Cantor's notion of a set is a quite different one, as I have emphasized beginning in §III.4, and his notion is sufficiently far from those of relation and function that the present procedure is a reasonable one. It is true that one of the principal reasons that Cantor introduced set theory was to understand real-valued functions. But he did not regard the general notion of a function as problematic—his problems concerned special aspects of real-valued functions. Cantor's background theory of functions could not have led him into paradoxes in the way

---

20. A singleton is a set with one member. Given an object a we can form the singleton $\{a\}$, a set that has a as its only member. Since a need not be a set, it is clear that a and its singleton need not be the same, and we shall, in general, take them to be distinct.

21. An ordinal number α is an *initial ordinal number* if α is no greater than any ordinal number β such that the set of predecessors of β has the same power as the set of predecessors of α.

that Russell's and Frege's theories led them into paradoxes, since Cantor only considered functions on specified domains, and he never considered domains that consisted of functions of any very general sort. The problems of circular reference and self-application were therefore far from anything he considered.

When I claim that something or other follows from Cantor's axioms, it becomes necessary to specify what the assumptions of the background second-order logic are. All that is assumed is the usual quantifier rules plus the axiom scheme of comprehension for first-order formulas with parameters—which says that such formulas define legitimate relations and functions, not sets! (See, for example, [Sha91, p. 66].)

Here are the Cantorian axioms:[22]

Axiom 2.1. *The ordinal numbers are linearly ordered by* $<$.

Axiom 2.2. *There is a least ordinal number,* 0.[23]

Axiom 2.3. *Every ordinal number* α *has an immediate successor* $\alpha + 1$.[24]

Axiom 2.4. *There is an ordinal number* ω *such that* $0 < \omega$; *for every ordinal number* α, *if* $\alpha < \omega$, *then* $\alpha + 1 < \omega$; *and for every nonzero ordinal number* $\alpha < \omega$ *there is an ordinal number* β *such that* $\alpha = \beta + 1$.

---

22. The less mathematically sophisticated reader may wish to skip directly to the discussion of the axioms. Axioms, definitions, theorems, and lemmas will be numbered in a single numbering system within each section of the book. Thus, the number 2.1 in Axiom 2.1 is the first numbered item of § 2 of the present chapter. A reference to, for example, Axiom 1, is always a reference to the first numbered item in the present section; a reference to Axiom 2.1 will be a reference (made outside of § 2) to the first item of § 2 of the present chapter; and a reference to Axiom IV.2.1 (made outside of Chapter IV) is a reference to the first numbered item of § 2 of Chapter IV. The system has the virtue that it is easy to locate items to which reference has been made. It has the defect that the numbering does not always reflect the logical grouping of the numbered items.

23. More precisely, every ordinal number α other than 0 is such that $0 < \alpha$.

24. More precisely, the axiom says that for every ordinal number α, there is an ordinal number β such that $\alpha < \beta$ and for any ordinal number γ, if $\alpha < \gamma$, then $\beta \leq \gamma$. It is easily seen that the ordinal number β is unique. Call it $\alpha + 1$.

That is, ω is a nonzero ordinal number whose predecessors are closed under successor and whose nonzero predecessors all have predecessors.

DEFINITION 2.5. A *set* is the range of a one-to-one function with domain a proper initial segment of the ordinal numbers.

The definition says that a set is whatever can be counted. Note that it follows from the definition that the predecessors of any ordinal number form a set.

AXIOM 2.6 (EXTENSIONALITY). *Sets with the same members are equal.*

AXIOM 2.7. *Every set of ordinals has a least upper bound.*

Note that it follows from the axiom (and Definition 5) that every proper initial segment of the ordinal numbers is the set of predecessors of an ordinal. Using Definition 5 and Axiom 7 we see that ω is the least nonzero ordinal number with predecessors that are closed under successor.

AXIOM 2.8. *For every ordinal number α there is an associated set (α), the number class of α, such that β is in (α) if and only if β is an ordinal number and the set of predecessors of α is the range of a one-to-one function with domain the predecessors of β.*

A set can be counted by α if and only if it can be counted by any member of (α) and only by members of (α).

The above axioms serve to emphasize the primacy of ordinals in Cantor's conception of set theory: the only set-existence principles are Axiom 8, which postulates a set of ordinals, and the definition of set, which ties each set to an ordinal that counts it. The two set-existence principles cohere well: Since (α) is a set of ordinals, it has least upper bound, say β, by Axiom 7. But when α is infinite, it is not hard to prove that (α) is the range of a function with domain the predecessors of β, and hence that (α) is a set in the sense of the definition. In fact, one can clean things up a bit by replacing Axiom 8 by

AXIOM 2.9. *For every ordinal number α, there is an ordinal number $\beta > \alpha$ such that the set of predecessors of β is not the range of a one-to-one function with domain α.*

The resulting set of axioms is equivalent to the one above, but it has only one set-existence principle, Definition 5.

Either set of axioms requires supplementation by another principle that Cantor often in effect made use of, though he nowhere made anything like it explicit. Say that a one-to-one function F with domain the set of predecessors of an ordinal and with range S *witnesses* that S is a set. According to Definition 5, every set has a function witnessing that it is a set.

> AXIOM 2.10. *Let S be a set of sets, and let F be a function with domain the predecessors of some ordinal number α witnessing that S is a set, that is, such that every member of S is $F(\gamma)$ for some $\gamma < \alpha$. Then there is a binary function H such that for every $\gamma < \alpha$, the unary function $H(\gamma, \bullet)$ that remains after γ is plugged into H has domain the set of predecessors of an initial ordinal and witnesses that $F(\gamma)$ is a set.*

Axiom 10 is based on the rather natural principle that since every member of the set S has a function witnessing that it is a set, there is a set of such witnesses. Perhaps Cantor did not make anything like it explicit because it is a principle concerning functions, not sets, and he took himself to be working with an antecedently given notion of function.

Technical Remark. The mathematically sophisticated reader may be a bit bemused at this point, since Definition 5 and Axiom 10 are both closely related to the Axiom of Choice. But each seems to be independent of the other in the present setting, since the Axioms of Union and Power Set are absent. On the basis of what I have taken to be Cantor's axioms, Axiom 10 entails the Axiom of Union, as is not hard to check.

If we supplement my Cantorian axioms with the assumption that the ordinals are sets, for example by using the von Neumann ordinals, and assume (for simplicity) that there are no urelements, then the resulting second-order axioms are equivalent to second-order ZF minus the Foundation and Power Set Axioms plus the axiom that for every cardinal (initial ordinal) κ, a next cardinal κ^+ exists and the following axiom: for every set S there is a set T such that for every member s of S there is exactly one member of T that is a well-ordering of s with order type an initial ordinal. The final axiom entails the Axiom of Choice but does not follow from it on the basis of the first-order version of the theory just described, that is, in the absence of the Power Set Axiom.[25] Andrzej Zarach has investigated theories closely related to the one just outlined [Zar82].

The axiomatization given here relies heavily on the notion of a witnessing function, which does not appear in Cantor's work. It is my way of expressing that a set can be "counted," in Cantor's own terms. Of course, it is immediate from Definition 5 that every set can be well-ordered, and Cantor relied on the notion of well-ordering instead of on the witnessing functions used here. But that involved him in the following detour: He had both well-order types [*Anzahlen*] and ordinal numbers [*Zahlen*]. Each well-ordered set is associated with the ordinal number such that the predecessors of that ordinal number in the natural order ($<$) have the same well-order type as the well-ordered set. (See [Can83, p. 168] or [Can76, p. 72]. The English translation does not always distinguish between *Anzahl* and *Zahl*.) There was some point to that added complexity, since the existence of well-ordered sets of one or another order type was part of Cantor's argument for the reality of the corresponding ordinal numbers. For example, the ordinal number ω is introduced in terms of the order type of the sequence of natural numbers. I have simplified Cantor's theory by making the association between ordinal numbers and well-ordered sets directly via witnessing functions, avoiding the mathematically superfluous detour through well-order types.

Let me briefly document that each of the proposed axioms is in fact a principle that Cantor accepted. Each pair of page numbers here is a reference to the *Grundlagen,* the first to [Can83] and the second to the English translation [Can76].[26]

Axiom 1. The numbers are in a "natural succession" [168, 72], and they are "comparable to each other" [177, 77].

Axiom 2. Actually, as mentioned above, Cantor started with 1. He spoke of the ordinal numbers as "an extension or rather a continuation of the sequence of real[27] whole numbers [$1, 2, 3, \ldots, \nu, \ldots$] beyond the infinite" [165, 70].

---

25. Zbigniew Szczepaniak has shown that the Well-Ordering Principle does not follow from ZFC minus Power Set (ZFC$^-$) if ZFC$^-$ is consistent [Zar82, p. 339]. His proof establishes the claim in the text.

26. All but the briefest of the translations are, except as noted, my own.

27. Cantor distinguished between "*reellen*" numbers—the continuum—and "*realen*" numbers—genuine ones—which include not only the real and complex numbers but also

Axiom 3. Cantor's *"first principle of generation"* is "the principle of the addition of a unit to an existing, already formed number" [195, 87, translation from [Can76]].

Axiom 4. Cantor held that there is

> nothing objectionable in conceiving of a *new* number—we shall call it ω—which is to be the expression for this: that the whole domain [*Inbegriff*] (I) [of the positive real whole numbers 1, 2, 3, . . . , ν, . . .] be given in its natural succession according to law. (Similar to the way in which ν is an expression for this: that a certain finite type [*Anzahl*] of units is unified into a whole.)   [195, 87]

Definition 5. First, I shall discuss the reason for only allowing *proper* initial segments of the ordinal numbers. In investigating the suprafinite numbers, Cantor said that "we will get ever farther ahead, never reaching an unsurmountable limit, but also attaining not even an approximate grasp of the absolute. The absolute can only be acknowledged, but never known," and that "the absolutely infinite sequence of numbers . . . seems to me in a certain sense a suitable symbol of the absolute" [205, 94, the first translation is essentially that of [Can76]].

Now on to the central point of the definition.

> The concept of a *well-ordered set* shows itself to be fundamental for the whole theory of sets. That it is always possible to bring any *well-defined* set into the *form* of a *well-ordered* set seems to me to be basic and rich in consequences and through its general validity an especially remarkable law of thought . . .   [169, 72]

Though that quote provides convincing evidence, I must discuss a certain passage in the *Grundlagen* that is often taken to be evidence that Cantor's set theory is so-called *naive* set theory,[28] a contradictory theory that is chiefly distinguished by the fact that it has as a postulate a Comprehension Principle much like that of Russell. The passage is in an endnote for Section 1 of

---

the transfinite ordinals. I have not preserved Cantor's distinction in the translations. The reader will have no trouble sorting things out.

28. As I learned from [Moo82, p. 260], the term is due to John von Neumann, who, following Ernst Zermelo, attributed the theory to Cantor. See, for example, [Zer08b, p. 200] and [vN25, p. 394].

the *Grundlagen*, not in the text, so it would be curious if it had the central importance often claimed.[29] The passage opens as follows:

> By "set" [*"Mannigfaltigkeit" oder "Menge"*] I understand in general every many that can be thought of as one, i.e., any domain of definite elements which by means of a law can be bound up into a whole . . . [204, 93]

If one takes a "law" to be something like Peano's "condition," Frege's "concept," or Russell's "propositional function," then the passage is a classic statement of a Comprehension Principle, and that is how it is usually taken. That reading is, perhaps, encouraged by the fact that the passage continues[30] "and I believe that in this I am defining something which is related to the Platonic $\epsilon\tilde{\iota}\delta o\varsigma$ or $\iota\delta\epsilon\alpha$ . . . " But Cantor's typical use of the word *law* in the *Grundlagen* is "natural succession according to law," which suggests quite a different picture: a "law" is, for Cantor, a well-ordering or "counting," and so the passage suggests Definition 5, not a Comprehension Principle. That reading is strongly supported by the fact that the passage continues

> as well as to that which Plato in his dialogue "Philebus or the Highest Good" calls $\mu\iota\kappa\tau\acute{o}\nu$. He counterposes this to the $\H{\alpha}\pi\epsilon\iota\rho o\nu$, i.e., the unlimited, indeterminate, which I call the non-genuine-infinite, as well as to the $\pi\acute{\epsilon}\rho\alpha\varsigma$, i.e., the limit, and explains it as an ordered "mixture" of the two latter.

Let me just string together some relevant quotations from the *Philebus*.[31] The "$\mu\iota\kappa\tau\acute{o}\nu$" is the subject of the third quote. What is at issue here is Cantor's understanding, not Plato's. It is not clear what Plato meant or what Cantor made of it, but there is no question that numbering, not conditions, concepts, or propositional functions, is the central idea.

---

29. A similar passage *is* the opening of Cantor's *Beiträge* of 1895 [Can95, p. 481], but, as I shall argue below, Cantor had modified his theory by that time. The passage there is ambiguous in much the same way the one described here is. For a discussion of that passage and an explicit statement that Russell, not Cantor, is the originator of naive set theory, see [Hal84, p. 38].

30. I am here and below using the translation in [Can76].

31. It is interesting to note that the *Philebus* was likely written in response to Eudoxus, both his ethics and, what is relevant to the passages here cited, his theory of incommensurable ratios [Gos75, pp. 166–181, 196–206]. The translations are taken from [Gos75].

But one should not attribute the character of indeterminate to the plurality until one can see the complete number between the indeterminate and the one. Then one can consign every one of them to the indeterminate with a clear conscience.   [16D]

If a person grasps any one, then, as I say, he must not turn immediately to its indeterminate character but rather look for some number. Similarly the other way round, when one is forced to start with what is indeterminate, one should not immediately look to the unitary aspect, but again note some number embracing every plurality, and from all these end up at the one.   [18A]

That of equal and double, and whatever puts an end to opposites being at odds with each other, and by the introduction of number makes them commensurate and harmonious.   [25E]

Axiom 6. The Axiom of Extensionality, which is often taken to be constitutive of the notion of set, is curiously difficult to locate in Cantor's writings. Extensionality is present in the beginning of the passage discussed immediately above, in the idea that a set is a "domain of definite elements." The clearest statement is perhaps, as Hallett suggests, to be found in an article published in 1887 [Can87, p. 387] in which Cantor writes of a set "consisting of clearly differentiated, conceptually separated elements $m, m', \ldots$ and which is thereby determined and delimited" [Hal84, p. 34].

Axiom 7. Cantor defined [196, 87] the "*second principle of generation* of whole real numbers" to be that "if any definite succession of defined whole real numbers is given, of which no greatest exists, then on the basis of this second principle of generation a new number is created, which is thought of as a *limit* of those numbers, i.e., which is defined as the number next greater than all of them." As Cantor's application of the second principle of generation in the *Grundlagen* makes clear, a "succession" is intended to be an initial segment of the ordinal numbers. It is not clear what "definite" means, but the applications make it fairly clear that the principle is intended to apply to initial segments that are sets.

Suppose that given any set of ordinals one can form the set of all ordinals less than or equal to any member of the set. The set formed will be an initial segment, and so the second principle of generation guarantees that it has a least upper bound, which is therefore a least upper bound of the original set. The axiom then follows from the second principle of generation. The axiom is frequently more convenient to apply than the second principle of generation since it does not require forming an initial segment as an intermediate step.

The argument I just gave for Axiom 7 rested on assuming that given a set of ordinals one can form the set of all ordinals less than or equal to any member of the set, or, what is the same thing, that one can form the union of the members of the set of sets of predecessors of members of the first set. Cantor expressed no such idea in the period under discussion, but it—or something like it—seems to be required in order to make sense of Cantor's assertions about number classes and powers. Cantor rarely mentioned any number class beyond the first three explicitly. Though he explicitly declared that for every ordinal number γ there is a γth power [205, 94], he did not even make explicit mention of the ωth power in print until 1895 [Can95, p. 495], when he promised to prove its existence. He never did so in print. In 1899, in the letter to Dedekind in which he proved that the ordinal numbers form an inconsistent multiplicity [Can32a], he introduced $\aleph_\omega$, the ωth power, as the cardinality of the set of predecessors of the least ordinal number that does not have $\aleph_\nu$ predecessors for any finite ν and, what is essentially the same, also as the cardinality of the union of the first, second, third, and so forth, number classes. Thus, he allowed forming the union of a set of sets of ordinals when the result is an initial segment of the ordinals.[32] Such a union is just what is needed to justify my assumption. A more cautious reconstruction might replace Axiom 7 with "Every initial segment of the ordinals that is a set has a least upper bound." But, by a lemma of Azriel Levy [Lev68, p. 763], the resulting system has the full Axiom 7 as a consequence. I have used the apparently stronger version largely for perspicuity, but also because, as I just argued, Cantor seemed to use something like it, and he almost certainly did not know Levy's derivation of it.

Axiom 8. Along with the two principles of generation, Cantor stated a third principle, which he called a "stopping or confining principle" [*Hemmungs- oder Beschränkungsprinzip*]. He said that the principle or condition satisfied by all the ordinal numbers defined to that point in the *Grundlagen* is that their predecessors can be placed in one-to-one correspondence with the natural numbers, that is, that they are of the first infinite power. He went on to

---

32. He used a denumerable disjoint union, while my proposed analysis allows an arbitrary union. He characteristically preferred disjoint unions, but he did consider nondisjoint unions from time to time [Can82, p. 152; Can84, p. 226], and so I think there can be no objection to allowing that in my reconstruction. Similarly, though he usually employed only finite or denumerable unions, he did make use of unions that cannot always be finite or denumerable to define multiplication of ordinal numbers [170, 73] and of cardinal numbers [Can87, p. 414].

define the second number class, which I have called (ω), as *"the domain of all numbers α that can be formed with the help of the two principles of generation"* such that the set of predecessors of α is of the first infinite power. (See [197, 88].) Now all that only gives the second number class, but he said earlier [167, 71], "In the same fashion the third number-class yields the definition of the third power, or the power of the third class, and so on."[33] Moreover, he referred to the third number class at a couple of other points in the paper.

Axiom 10. Axiom 10 is the hardest to defend, since Cantor nowhere stated anything like it. Various alternative formulations may be just as appropriate as the one I have given. But something like the proposed axiom is needed to prove the theorems Cantor does. In modern set theory, the counterparts of Definition 5 and Axiom 10 are equivalent—they are variants of the Axiom of Choice. But even when Cantor came to have doubts about Definition 5 (see below), he continued to use something like Axiom 10 unhesitatingly, which provides evidence that there was a principle at work that Cantor thought of as independent of Definition 5.

The first published theorem that cannot be rigorously proved without some form of the Axiom of Choice seems to have been a theorem of analysis proved by Cantor, published by his colleague Heine in 1872 [Moo82, p. 14].[34] There was no apparent recognition that a new principle was involved in the proof. Many theorems of set theory that Cantor subsequently published require some form or other of the Axiom of Choice. Virtually all are stated without proof, usually as elementary lemmas involved in the proof of other theorems. (See [Moo82] for a thorough survey.) Many of those theorems are immediate consequences of Definition 5. The theorem that every infinite set S has a denumerable subset provides an example: Let F witness that S is a set, and say the domain of F is the predecessors of α. Then $\alpha \geq \omega$, since S is infinite. Let F' be F with domain restricted to the predecessors of ω. Then the range of F' is a denumerable subset of S. Cantor did not publish that theorem until 1895. But he did state earlier that the powers are well ordered, that a set is finite if it has no proper subset equal in power to itself, and other immediate consequences of Definition 5.

In addition to the results just outlined that flow from Definition 5, Cantor also made use of the following results in the years indicated, which he re-

---

33. The translation is taken from [Can76].

34. Here is the theorem: A function f from the real numbers to the real numbers is continuous at a point p if and only if it is sequentially continuous at p.

garded as too elementary to require explicit proof.[35] I have written them in modern notation, using $\sim$ for equivalence in power, $\cap$ for intersection, and $\bigcup$ for indexed union. (Thus, for example, $\bigcup_{i<\omega} A_i$ is the set of a such that for some $i < \omega$ a is in A_i.)

THEOREM 2.11 (1878, 1880). *Suppose that $A_i \sim B_i$ and $A_i \cap A_j = B_i \cap B_j = \varnothing$ for all $i < j < \omega$. Then $\bigcup_{i<\omega} A_i \sim \bigcup_{i<\omega} B_i$.*

THEOREM 2.12 (1878, 1882, 1883, 1884, 1885). *A finite or denumerable union of finite or denumerable sets is finite or denumerable.*

THEOREM 2.13 (1885). *A finite or denumerable union of sets of cardinality $\aleph_\alpha$ has cardinality $\aleph_\alpha$.*

THEOREM 2.14 (1887). *Suppose that for all $i \neq j \in S$, $A_i \sim B_i$ and $A_i \cap A_j = B_i \cap B_j = \varnothing$. Then $\bigcup_{i \in S} A_i \sim \bigcup_{i \in S} B_i$.*

My argument for attributing to Cantor a commitment to Axiom 10 is simply that those results do not, so far as I can see, follow without it and that they have natural direct proofs with it.

Technical Remark. I shall prove Theorem 12 as an example: Let S be a finite or denumerable set of finite or denumerable sets. If any of the sets involved is finite, without loss of generality add more members to make it denumerable. By Axiom 10 there is a function H of two variables, each of which ranges over the natural numbers, such that the range of H is the union of the members of S. Let f and g be functions such that every pair of natural numbers (m, n) is of the form $(f(x), g(x))$ for exactly one natural number x. (In 1874 Cantor in effect showed that such functions exist, in the paper in which he proved that the algebraic numbers are denumerable.) The function $H(f(x), g(x))$ witnesses that the union of the members of S is a finite or denumerable set, as required, if the members of S were pairwise disjoint. Otherwise, one must delete duplications from the range of the function to obtain the required witness. The proof simultaneously shows that the union is a set and that it is finite or denumerable.

---

35. See [Moo82, pp. 30–37] for references and a detailed discussion.

I have now introduced my reconstruction of Cantor's theory and argued for it. It is therefore time to admit that it does not suffice for formalizing Cantor's work during the period indicated. Such a formalization requires the additional assumption that the real numbers form a set and also, perhaps, the additional assumption that the functions from the real numbers to the real numbers form a set. (I say perhaps because the latter assumption is used only for offhand remarks that occur in endnotes and the like.) Cantor can be argued to have derived the additional assumptions from a guiding principle that is not a part of my reconstruction, perhaps one that says something like "every domain of a mathematical variable is a set." I shall call that the Domain Principle.[36] Indeed, some such assumption forms part of Cantor's argument for the existence of ω, as is indicated by the quote used to justify Axiom 4 above. The Domain Principle is not suitable for axiomatic formulation, but that is not why I have omitted it from account.

During the period under discussion, Cantor believed that he would be able to prove on the basis of his theory, essentially as I have reconstructed it above, that every mathematical domain is a set. Thus, the Domain Principle, or anything like it, was superfluous as an additional mathematical assumption. In particular, Cantor believed that he would be able to prove that the real numbers and the functions from the real numbers to the real numbers form sets within a framework like the one I have outlined, even though he could not yet even prove that the real numbers were not an absolutely infinite totality. Let us see, following [Hal84, pp. 74–81], why he thought so.

Cantor first showed that there are more real numbers than natural numbers or algebraic numbers in 1874. In 1891 he proved that $\aleph_0 < 2^{\aleph_0}$. (See Appendix B.) Thus, when he noted in 1895 that the real numbers have cardinality $2^{\aleph_0}$ [Can95, p. 488], the result of 1891 yielded a new proof that there are more real numbers than natural numbers. The later proof is the one in common use today. In fact, the earlier proof is no longer well known to mathematicians.[37] It was the earlier proof that shaped Cantor's thinking in the period under consideration.

---

36. Hallett formulated what he calls the "domain principle" or "principle (a)" in much the same spirit, though his principle is not identical to mine. He argued along more or less the indicated lines in some detail.

37. The basic technique of the earlier proof (1884) is still the one used to prove Cantor's theorem that no denumerable set is perfect [Can84, p. 215]. Since the real numbers form a perfect set, it follows that the real numbers are nondenumerable.

Here is the earlier proof: It suffices to show that given any sequence $u_1, u_2, \ldots$ of real numbers there is a number that is not in the sequence in any interval of real numbers (a, b)—that amounts to showing that any sequence $u_1, u_2, \ldots$ that supposedly lists all of the real numbers, thereby showing them denumerable, must omit at least one in every interval. Let a' and b' be the first two members of the sequence $u_1, u_2, \ldots$ that are in the interval (a, b), with $a' < b'$; let a'' and b'' be the first two members of the sequence that are in (a', b'), with $a'' < b''$; and so forth. If the sequence of nested intervals only goes on for finitely many steps, then there is at most one member of the last interval that appears in the sequence $u_1, u_2, \ldots$. Any other member of that interval is omitted from the sequence, as required. If the sequence of nested intervals is infinite, then the left endpoints converge to a point a^∞ and the right endpoints to b^∞. If $a^\infty = b^\infty$, then that point is as required. Moreover, as Cantor noted, if $u_1, u_2, \ldots$ is an enumeration of all the algebraic numbers, then that will be the case, and so a^∞ will not be algebraic. Finally, if $a^\infty < b^\infty$ then any member of $[a^\infty, b^\infty]$ (including the endpoints) is as required. That completes the proof.

Next, I shall present Cantor's proof in the *Grundlagen* that there are more numbers in the second number class, (ω), than there are natural numbers, or, as Cantor actually put it, that the second number class is of greater power than the first number class. Note how similar this proof is to the proof that there are more real numbers than natural numbers. (I have changed Cantor's notation to emphasize the parallelism between the two proofs.)

It suffices to show that given any sequence $u_1, u_2, \ldots$ of members of (ω) there is a member of (ω) that is not in the sequence. Let a be the first member of the sequence, let a' be the first member of the sequence that is greater than a, let a'' be the first member of the sequence that is greater than a', and so forth. If the increasing sequence $a, a', a'', \ldots$ only goes on for finitely many steps, then its last member is the greatest number that occurs in the sequence $u_1, u_2, \ldots$. Its successor is as required. If the increasing sequence $a, a', a'', \ldots$ is infinite, there is a least ordinal number greater then all its members, call it a^∞, which is in (ω) (since its predecessors have in effect been specified as a denumerable union of finite or denumerable sets). Thus, the ordinal number a^∞ is as required. That completes the proof. (Actually, Cantor went on to show that the ordinal number I have called a^∞ is the least upper bound of $u_1, u_2, \ldots$.)

Each of the two proofs shows that a set does not have the power of the natural numbers by showing how, given a sequence of members of the set, to find a member of the set that is not in the sequence. Moreover, that is done in each proof by considering appropriate subsequences, and in the main cases the number that is outside the sequence is obtained as a "limit" of the subsequence. Given the strong parallels between Cantor's analysis of the real numbers and his analysis of the second number class, it is small wonder that Cantor believed that the real numbers and the second number class (ω) were intimately connected and that he would eventually be able to prove a strong form of the Continuum Hypothesis: the real numbers have the power of the second number class. Note that a proof that the real numbers have the power of the second number class would show that they form a set, according to Definition 5, and so obviate the need to apply a domain principle. That was, I believe, the chief importance of the Continuum Hypothesis for Cantor—it would show that the real numbers form a set, and hence that they were encompassed by his theory.

Cantor repeatedly thought that he had proved that the real numbers have the power of the second number class, and he announced countless times that he hoped to publish such a proof.[38] He also announced that the functions from the real numbers to the real numbers have the power of the third number class [207, 95]. That would once more make an appeal to a Domain Principle unnecessary by showing that the functions from the real numbers to the real numbers form a set.

Cantor took the analogy between the real numbers and the second number class very seriously. In §9 of the *Grundlagen* he outlined three methods of introducing the real numbers: Weierstrass's method relying on series, Dedekind's method relying on cuts, and his own method relying on sequences. He gave various reasons to prefer his own over the others, including this one: his, unlike the others, generalizes to the case of transfinite numbers [190, 84]. What he had in mind is that just as irrationals are introduced to play the role of limits of Cauchy sequences of rational numbers, new ordinal numbers are introduced to play the role of limits of sequences of ordinals. That is strikingly clear in this passage:

> Indeed ω can in a certain way be viewed as the limit that the variable finite whole number ν aims at, though only in the sense that ω is the *smallest* transfinite ordinal number, i.e., the smallest *fixed* number that

---

38. See, for example, [192, 86]. Detailed histories may be found in [Hal84], [Dau79], and [Moo82], among others.

is greater than *all* finite numbers ν; in the same way $\sqrt{2}$ is the limit of certain variable, increasing, rational numbers, though here in addition the difference between $\sqrt{2}$ and these approaching fractions becomes arbitrarily small, whereas $\omega - \nu$ always equals ω; this difference does *not* alter that ω is to be recognized as just as definite and complete as $\sqrt{2}$, nor does it alter that ω has in it some traces of the numbers ν that aim at it, just as $\sqrt{2}$ does of the approaching rational fractions.

The transfinite numbers are in a certain sense themselves *new irrationalities* and in fact in my opinion the best method of defining the *finite* irrational numbers is wholly similar to, and I might even say in principle the same as, my method described above of introducing transfinite numbers. One can say unconditionally: the transfinite numbers *stand or fall* with the finite irrational numbers; they are like each other in their innermost being; for the former like the latter are definite delimited forms or modifications ($\dot{\alpha}\phi\omega\rho\iota\sigma\mu\acute{\epsilon}\nu\alpha$) of the actual infinite.   ([Can87, pp. 395–396], my translation, but see also [Hal84, p. 80])

Cantor had produced a simple but powerful theory in which he could formulate a lot of new and interesting mathematics, but a proof that the real numbers have the power of the second number class—a proof that they form a set—continued to elude him. In 1891 he published a new proof that there are infinite powers other than that of the natural numbers, a proof that did not "depend on considering the irrational numbers."[39] What he had in fact shown is that for any set L and some fixed pair of distinct elements, the set of functions from L to that pair has power strictly greater than that of L. Thus, the infinite powers have no maximum, a result he had shown quite differently in the *Grundlagen*. First, he stated and proved the theorem in the case in which L is the set of natural numbers. Next, he stated the fully general form of the theorem that I just did, but he did not give a proof. Instead, he illustrated the method by showing that the theorem holds in the case in which L is the set of real numbers between 0 and 1. To avoid misunderstandings, let me state explicitly that there is no mention in the article of any set of subsets of a set, and that there is no proof in the article that there are more real numbers than natural numbers (though Cantor does mention that that was proved in an earlier article).

Cantor had reason for wishing to avoid any use of the irrational numbers in

---

39. See Appendix B for a full translation of his article.

proving the theorem that there is an infinite power other than that of the natural numbers, but one should notice that he already had a proof that avoided the irrational numbers: the proof in the *Grundlagen*. I can only think that, since he thought of the transfinite numbers as "new irrationalities," he thought of that proof as involving irrational numbers in an extended sense. Thus, what is important is that the new proof avoids both the irrational and the transfinite numbers.

As Joseph Warren Dauben noted, the irrational numbers were controversial [Dau79, p. 165], and so, of course, were the transfinite numbers. Kronecker in particular was an influential figure who denied the existence of irrational numbers. (See, for example, [Dau79, p. 69].) It was therefore worthwhile, for polemical purposes, to avoid their use in proving that there is more than one infinite power. The two earlier proofs of that showed that sets defined in a similar way were large—sets defined by introducing limits to sequences or successions. But Cantor had no mathematical proof with which to confront his opponents that one could legitimately introduce such limits either in the case of the real numbers or in the case of the second number class. In the case of the irrational numbers he seems only to have had the argument that in analysis numbers present themselves in the form of limits of sequences.[40] In the case of the transfinite numbers he used the second principle of generation discussed above. Whatever the justification for those assumptions that suitable sequences have limits—and Cantor clearly believed them to be justified— a proof that there are infinitely many infinite sizes that is independent of them would clearly be more convincing than one that did depend on them. Moreover, an independent proof of that result could serve to bolster those assumptions.

The new proof is independent of any notion of limit or of transfinite number. It does, however, require a new set-existence principle: if L is a set, then so is the domain of all functions from L into an arbitrarily fixed pair. That principle can clearly be justified using the Domain Principle, but unlike the Domain Principle it is mathematically precise and provides a basis suitable for proving theorems.

As Russell had realized by 1900 (see [Cof79, p. 33] and [Rus03, p. 366]),

---

40. Cantor didn't actually say that. What he did do is take it as an argument against Dedekind's definition of irrational real numbers in terms of "cuts" and as an argument in favor of his own definition in terms of sequences that numbers do *not* present themselves in the form of cuts [185, 81].

any function from a set L to a pair is fully determined by the subset of L that is taken by the function to a fixed member of the pair: in effect, we think of the two members of the pair as meaning "yes, this element is in the subset" and "no, this element is not in the subset." The affinity with Russellian propositional functions is obvious. Thus, the domain of all functions from L into some fixed pair is canonically identifiable with the domain of subsets of L. It follows that the new principle is equivalent to the Power Set Axiom in common use today: the subsets of a set form a set. I shall therefore, at the cost of a slight anachronism, refer to the Cantorian principle above as the Power Set Axiom.

By 1895 [Can95, p. 488], Cantor had realized that the Power Set Axiom had another important consequence: he could show that the power of the set of real numbers (the continuum) was that of the set of functions from the natural numbers to a pair or, as he now wrote, that $c = 2^{\aleph_0}$. The proof of that fact is sufficiently easy, given what Cantor knew, that it would surprise me if he had not already discovered it in 1891, but the exact date doesn't matter.

The point is *not* that the new consequence that $c = 2^{\aleph_0}$ provided a new proof of the nondenumerability of the set of real numbers when combined with the 1891 result that, in the 1895 notation, reads $\aleph_0 < 2^{\aleph_0}$. Though that is the Cantorian proof familiar today, Cantor never actually gave it. Rather, the point is that the new Power Set Axiom enabled Cantor to prove for the first time that the real numbers form a set, instead of just taking that as an additional assumption. Cantor may well have seen that as a victory of consequence for more than just the extension and development of his theory: it provided at least the beginning of a new independent argument for the existence of the real and transfinite numbers. The Power Set Axiom had become vital for Cantor.

The Power Set Axiom, however, was not easily integrated with the conception of a set as anything that can be counted. For the first time, Cantor needed to allow for the existence of a set that he did not know how to introduce explicitly via a counting, or, more precisely, that he did not know how to well-order in a definable way. (I am excepting Cantor's earlier use of the real numbers—though he did not know how to count them, he had good reason to think that he would eventually be able to do so.)[41]

---

41. As mentioned in §III.2, as a result of considering the real numbers Cantor did very briefly doubt that every set can be well ordered. That supports the point that Cantor's

Technical Remark. Since I am talking here as if Cantor were only interested in *definable* well-orderings, let me point out an ambiguity in Cantor's theory. The axioms are second order, and I have so far just been acting as if "function" meant arbitrary function in the sense of today's standard second-order logic, and so forth. But given the lack of clarity of Cantor and his contemporaries about definability, it would not be terribly implausible to reinterpret what Cantor had in mind as involving not arbitrary functions in the modern sense but instead α-recursive functions for some α. (See, for example, [Hin78, p. 377] for a definition and discussion.) In that case, Cantor's theory becomes the theory of α-finite sets for an infinite, recursively regular (that is, admissible), recursively inaccessible ordinal α.[42] This idea has its attractions for Cantor exegesis. For example, his "unconscious" use of a principle like Axiom 10—a choice principle—no longer involves an additional assumption, and his great interest in notations for ordinals ("normal forms" for denumerable ordinals and the like) becomes better motivated. Moreover, in the *Grundlagen* and elsewhere, Cantor gave a number of recursive definitions of larger and larger initial segments of the recursive ordinals and seemed to define the second number class as consisting of the set of those. That is, of course, easily explained away, since he could hardly have given nonrecursive examples, but it does cohere well with the present proposal.

Cantor always intended his theory of sets to be, in some none-too-clear sense, as comprehensive as possible. That strongly militates against the kind of interpretation of his work given in this Remark. His extra-mathematical intentions went beyond this interpretation and in that sense rule it out. The point is rather that up until 1891 nothing in his mathematical work even suggested that there might be any possibility of a set that did not have a definable well-ordering on it.

At some point after 1891 Cantor's mathematical thinking began to include the set of all functions from a set to two elements as an important example of a set, and so it was no longer part of his conception of sets that every set

---

doubts about well-ordering and related topics were induced by the need to allow power sets—which are not canonically well ordered.

42. The only remaining vestige of full second-order logic in this interpretation of Cantor's theory is that one must interpret the notion of a proper initial segment of the ordinals in a standard way, to guarantee that $<$ is a well-ordering.

is born well-ordered. In Cantor's subsequent and final two publications on set theory, his *Beiträge* [Can95, Can97], the Power Set Axiom enters in in §4 [Can95, p. 487], when the exponentiation of cardinal numbers is defined, but many of the principles he had earlier taken to be obvious had become conjectures in need of proof [Moo82, pp. 44–46]. In particular, he no longer assumed that the powers are linearly ordered [Can95, p. 484]—that is "by no means self-evident and can hardly be proved at this stage." He announced in 1895 that he would show that the powers are even well-ordered [Can95, p. 495], presumably in the second (1897) article, but he did not. Since those results would easily follow from the principle that every set can be well-ordered, presumably he had come to doubt that too. Indeed, ordinal numbers and well-orderings had lost their primacy: in the *Grundlagen,* ordinal numbers were mentioned in the very first sentence, and well-ordered sets were defined in the second section, immediately after the introduction, but in the *Beiträge* well-ordered sets are not defined until §12, which is the beginning of the second article, and ordinal numbers not until §14. Cantor's theory was in trouble, but it was not trouble caused by the paradoxes. It was trouble caused by trying to fit the Power Set Axiom into a theory that took well-orderings to be primary.

In the letter to Dedekind that was discussed in §III.4, Cantor used his result that the ordinal numbers form an inconsistent multiplicity to show that the cardinal numbers are well-ordered, indeed to show that every cardinal number is an $\aleph$, that is, the power of the set of predecessors of an ordinal number: Suppose a multiplicity does not have any $\aleph$ as its cardinality. Then, Cantor wrote, the whole system of all ordinal numbers is projectible into that multiplicity. But then the multiplicity is inconsistent, and hence not a set.

Cantor must have had mixed feelings about the above proof. Jourdain discovered a similar one about which he wrote to Cantor. Cantor put his own in a letter to Jourdain, but subsequently refused Jourdain permission to publish the letter [GG71, pp. 115–118]. Of course, the proof would not have been necessary in Cantor's earlier theory. There one could just argue as follows: every set can be counted and hence has the same power as the set of predecessors of some ordinal number—indeed any ordinal number that can be used to count it. But the power of the set of predecessors of an ordinal is an $\aleph$.

A variant of the proof *does* show something about Cantor's earlier conception of sets that was to be important later. Namely, it shows that if a multiplicity is not a set, then it is larger than every set: Begin counting the multiplicity off. If you succeed at some ordinal stage, then the multiplicity is a set. But if you do not succeed, then all the ordinals will have been used without exhausting the set, that is, the multiplicity has a part that is the size of all the ordinals. Thus, the multiplicity is larger than every set. That is the origin of the later "limitation-of-size hypothesis": a collection forms a set just if it is not too large. Note that the limitation-of-size hypothesis arose within a theory that is free of all known paradoxes. It did not arise as a solution to paradoxes.[43] Indeed, on the logical conception of collection, which is the source of the paradoxes, it is not at all clear how size could be relevant to the question whether a multiplicity forms a set—the elements are, after all, not gathered, they simply obey a rule. That suggests that on the logical conception, one would have to limit not the size of collections but the structure of the rules, just as Russell did.

Once Cantor had accepted the Power Set Axiom as a second set-existence principle, he no longer had a unitary conception of set. He could therefore no longer say which multiplicities were sets; the only way he had to show that some were not sets was by arriving at contradictions. In the absence of a positive account, that policy seemed *ad hoc*. It is a corollary of the result that the cardinality of every set is an $\aleph$ that every set can be well-ordered or counted. But that no longer serves as a criterion for what things are sets, because one can now show, for example, that the set of real numbers can be well-ordered only after the fact—by showing that it is a set. One cannot show that it is a set by showing that it can be well-ordered. The elegant theory of the *Grundlagen* was lost, and it was not clear what could replace it.

§3. Appendix A: Letter from Cantor to Jourdain, 9 July 1904[44]

Today I want to reply to only one point in your kind letter, that is the difficulty, which Mr. Russell describes in his work *The Principles of Mathematics* pp. 365–368.

He starts out from my proof of the theorem

---

43. The term *limitation of size* was, it must be admitted, introduced by Russell in 1906 to name a theory considered by him as one possible way of solving the paradoxes concerning logical collections. As he said [Rus05, p. 152], however, "This theory naturally becomes particularised into the theory that a proper class [that is, an allowable collection] must always be capable of being arranged in a well-ordered series ordinally similar to a segment of the series of ordinals in order of magnitude." That is, as we have seen, Cantor's theory developed before the paradoxes were known, and before Cantor had proved the existence of absolutely infinite collections.

44. I have based my translation on the text as published in [GG71, p. 119].

$$2^{\aleph_0} > \aleph_0$$

which can easily be extended to the case where $\aleph_0$ is replaced by some transfinite cardinal $\mathfrak{a}$.

One assumes here some *set* (that is, a *consistent* multiplicity) $\mathfrak{M}$ with the cardinal number $\mathfrak{a}$ and imagines the totality $\mathfrak{G}$ of all coverings of $\mathfrak{M}$ with two mutually exclusive symbols, perhaps with 0 and 1.

The *elements* of $\mathfrak{G}$ are therefore definite coverings of $\mathfrak{M}$, each therefore an individual set of the same cardinal number $\mathfrak{a}$.

Were we now, as Mr. Russell proposes, to replace $\mathfrak{M}$ by an *inconsistent* multiplicity (perhaps by the totality of *all* transfinite ordinal numbers, which you call $\mathfrak{W}$), then a totality corresponding to $\mathfrak{G}$ could *by no means be formed.* The impossibility rests upon this: an inconsistent multiplicity because it cannot be understood as a *whole,* thus as *a thing,* can*not* be used as an *element* of a multiplicity.

Only *complete things* can be taken as *elements* of a multiplicity, only *sets,* but not *inconsistent multiplicities,* in whose nature it lies, that they can never be conceived as *complete* and *actually existing.*

§4. Appendix B: On an Elementary Question of Set Theory[45]

In the article "Über eine Eigenschaft des Inbegriffs aller reellen algebraischen Zahlen" (*Journ. Math.* **77**, 258) [(1874), see [Can32b, pp. 115–118]] one finds, probably for the first time, a proof of the theorem that there are infinite sets that cannot be placed into one-to-one correspondence with the totality of all finite whole numbers $1, 2, 3, \ldots, \nu, \ldots$, or, as I like to put it, that do not have the power of the number sequence $1, 2, 3, \ldots, \nu, \ldots$. From what has been proved in §2 it follows without further argument that, for example, the totality of all real numbers within any interval $(\alpha \ldots \beta)$ are *not* representable in the form of a sequence

$$\omega_1, \omega_2, \ldots, \omega_\nu, \ldots.$$

It is possible, however, to produce a much simpler proof of that theorem that does not depend on considering the irrational numbers.

Take any two symbols m and w that are distinct from one another. Now we consider a domain [*Inbegriff*] M of elements

$$E = (x_1, x_2, \ldots, x_\nu, \ldots),$$

---

45. My translation of [Can91] as based on the reprinting [Can32b, pp. 278–280].

which depend on infinitely many coordinates $x_1, x_2, \ldots, x_\nu, \ldots$, such that each of these coordinates is either m or w. M is the totality of all elements E.

Among the elements of M belong, for example, the following three:

$$E^{\mathrm{I}} = (m, m, m, m, \ldots),$$
$$E^{\mathrm{II}} = (w, w, w, w, \ldots),$$
$$E^{\mathrm{III}} = (m, w, m, w, \ldots).$$

I claim that such a set M is not of the power of the sequence $1, 2, 3, \ldots, \nu, \ldots$.

That is shown by the following theorem: "If $E_1, E_2, \ldots, E_\nu, \ldots$ is any simply infinite sequence of elements of the set M, then there is always an element E_0 of M that corresponds to no E_ν."

To prove this, let

$$E_1 = (a_{1,1}, a_{1,2}, \ldots, a_{1,\nu}, \ldots),$$
$$E_2 = (a_{2,1}, a_{2,2}, \ldots, a_{2,\nu}, \ldots).$$
$$\cdot\ \cdot\ \cdot\ \cdot\ \cdot\ \cdot\ \cdot\ \cdot\ \cdot\ \cdot\ \cdot\ \cdot$$
$$E_\mu = (a_{\mu,1}, a_{\mu,2}, \ldots, a_{\mu,\nu}, \ldots).$$
$$\cdot\ \cdot\ \cdot\ \cdot\ \cdot\ \cdot\ \cdot\ \cdot\ \cdot\ \cdot\ \cdot\ \cdot$$

Here the $a_{\mu,\nu}$ are m or w in a definite manner. Produce now a sequence

$$b_1, b_2, \ldots, b_\nu, \ldots,$$

so defined that b_ν is *different* from $a_{\nu,\nu}$ but is also either m or w.

Thus if $a_{\nu,\nu} = m$, then $b_\nu = w$, and if $a_{\nu,\nu} = w$, then $b_\nu = m$.

If we now consider the element

$$E_0 = (b_1, b_2, b_3, \ldots)$$

of M, we see at once that the equality

$$E_0 = E_\mu$$

can be satisfied for no whole-number value for μ. Otherwise, for the μ in question and for all whole-number values of ν

$$b_\nu = a_{\mu,\nu},$$

therefore in particular

$$b_\mu = a_{\mu,\mu}$$

would hold, which is ruled out by the definition of b_ν. It follows immediately from this theorem that the totality of all elements of M cannot be brought into the form of a sequence $E_1, E_2, \ldots, E_\nu, \ldots$; we would otherwise be faced with the contradiction that a thing E_0 both is and is not an element of M.

This proof seems remarkable not only because of its great simplicity, but especially also because the principle that is employed in it can easily be extended to the general theorem, that the powers of well-defined sets have no maximum or, what is the same, that for any given set L another M can be placed beside it that is of greater power than L.

For example let L be a linear continuum, perhaps the domain of all real numerical quantities z that are ≥ 0 and ≤ 1.

Let M be understood as the domain of all single-valued functions $f(x)$ that take on only the two values 0 or 1, while x runs through all real values that are ≥ 0 and ≤ 1.

That M does *not* have *smaller* power than L follows from this: subsets of M can be specified that have the same power as L, e.g., the subset that consists of all functions of x that have the value 1 for a single value x_0 of x and the value 0 for all other values of x.

But then M does *not* have the *same* power as L either. For otherwise M could be put into one-to-one correspondence to the variable z [of L], and thus M could be thought of in the form of a single-valued function

$$\phi(x, z)$$

of the two variables x and z, in such a way that through every specification of z one would obtain an element $f(x) = \phi(x, z)$ of M and also conversely each element $f(x)$ of M could be generated from $\phi(x, z)$ through a single definite specification of z. This however leads to a contradiction. For if we understand by $g(x)$ that single-valued function of x which takes only values 0 or 1 and which for every value of x is different from $\phi(x, x)$, then on the one hand $g(x)$ is an element of M, and on the other it cannot be generated from $\phi(x, z)$ by any specification $z = z_0$, because $\phi(z_0, z_0)$ is different from $g(z_0)$.

Since the power of M is neither less than nor equal to that of L, it follows that it is larger than the power of *L*. (Cf. *Crelle's Journal* **84,** 242) [(1878) [Can32b, pp. 119–133]].

I have already shown by entirely other means in my *Grundlagen einer allgemeinen Mannigfaltigkeitslehre* (Leipzig, 1883; *Math. Ann.* **21**) [Can83] that the powers have no maximum. There it was even proved that the domain of all powers, when we imagine them ordered according to size, forms a "well-ordered set" so that in nature for each power there is a next greater, and that for each endlessly increasing set of powers there is a next greater one that follows.

The "powers" represent the only and necessary generalization of the finite "cardinal numbers"; they are nothing other than the actually infinitely large "cardinal numbers," and they possess the same reality and definiteness as the finite cardinal numbers; the only difference being that the law-like relations among them, their respective "number theory," is partly different in kind from that in the region of the finite.

The further exploration of this field is a job for the future.

The Axiomatization of Set Theory

§1. The Axiom of Choice

It was crucial for, indeed constitutive of, Cantor's early theory that the Well-Ordering Principle hold, that is, that every set can be well ordered. (The term "the Well-Ordering Principle" is taken from [Moo82].) That principle had, with the advent of the Power Set Axiom, become questionable, a conjecture that it was critical to prove in order to rescue Cantor's theory of powers. At the very least, it was necessary to prove that the power set of a well-orderable set is well-orderable. The following test case, which is also the central case, suggests itself: Prove that the set of subsets of the natural numbers can be well-ordered, or, equivalently, prove that the real numbers can be well-ordered. That problem, together with the apparently closely related continuum problem, was the first on Hilbert's tremendously influential list of problems presented to the Second International Congress of Mathematicians in 1900. Since the Cantorian theory had proved inadequate, there was the closely related problem of arriving at a workable notion of set with which to formulate set theory. The ordinal numbers could not, it seemed, play the defining role Cantor had assigned them.

In 1904, Julius König gave a purported proof that the continuum cannot be well ordered, a result that would have put Cantor's theory in doubt. Zermelo found an error in the proof: König had relied on a theorem of Felix Bernstein, and Bernstein's proof is incomplete in the relevant case. (See, for example, [Moo82, pp. 86–88] for an excellent account of that often-told tale.) Within a month, in conversations with Erhard Schmidt, Zermelo had solved the problem of well-ordering the continuum to his own satisfaction and that of present-day set theorists [Zer04]. Zermelo's contemporaries were not so sure.

Zermelo's work presumed familiarity with Cantorian set theory. He used cardinal numbers and their products, functions from sets to their members, well-ordered sets, and the Power Set Axiom without discussion. He introduced the "assumption," which he later called the Axiom of Choice,[1] that for every set of nonempty sets there is a function that takes each of the nonempty sets to one of its elements and used that assumption to prove the Well-Ordering Principle.

It is characteristic of combinatorial collections that they obey the Well-Ordering Principle and the Axiom of Choice. They obey Well-Ordering because combinatorial collections are gathered by enumerating their elements, and they obey Choice because combinatorial collections are gathered by picking their elements in an arbitrary way, not necessarily in virtue of a rule. That makes it possible to pick one member out of each set in a set of nonempty sets. Such picking does not give rise to a single rule that selects exactly one member from each set in the set of nonempty sets: it may be done in an arbitrary way. Thus, Choice is dubious for logical collections, which require a rule.

Given a set M Zermelo applied his assumption to the set of its nonempty subsets to obtain a function γ. When m is any nonempty subset of M, he called $\gamma(m)$ the *distinguished* member of m. He defined a γ-*set* to be a well-ordered subset S of M such that each member a of S is the distinguished member of the set of elements of S that do not come before a. He then defined a γ-*element* to be a member of any γ-set, and showed that (i) the set L_γ of γ-elements is a γ-set, and thus well-ordered, and (ii) that $L_\gamma = M$, and hence that M is well-ordered. In support of his assumption, he said only that it is a "logical principle" that is "applied without hesitation everywhere in mathematical deduction." As an example, he noted that it is used to prove that if a set is decomposed into parts then there are not more parts than members of the set.

Zermelo's article provoked a storm of criticisms. Bernstein and Arthur Schoenflies objected to the proof that $L_\gamma = M$. Poincaré objected to the definition of L_γ. Jourdain claimed to have proved the result earlier, in a simpler way. Peano, Borel, Lebesgue, and René Baire objected to the assumption. I have just listed the published criticisms that had come to Zermelo's attention by 1908 [Zer08a]. There were other published criticisms [Moo78, p. 320].

---

1. See [Moo82] for a thorough and careful discussion of precursors of the Axiom of Choice, equivalents of the axiom, early theorems whose proofs require the axiom, reactions to the axiom, and practically everything else about it.

Zermelo responded to his critics in a pair of papers [Zer08a, Zer08b], written within sixteen days of each other, that were virtually a single paper [Moo78, p. 319]. One of the papers [Zer08a] contains a new proof that every set can be well ordered, a detailed argument for the Axiom of Choice, and replies to critics. The other contains the first axiomatization of a Cantorian set theory.

Bernstein and Schoenflies were concerned, in somewhat different ways, with the "set" W of all ordinals and with Burali-Forti's paradox. Each wished to accept W as a set, and so they restricted set-theoretic principles in various ways in an attempt to block the paradox. In particular, they denied that, given any well-ordered set and an object not in the set, there is always a well-ordered set consisting of the members of the original one plus the new object ordered so that the new object comes after all the others. They pretty much had to do that, since the paradox arises immediately once one continues W with another object. They therefore criticized Zermelo for assuming that one could always extend a well-ordered set by a new element.

Zermelo's main reply was three pronged, though he also criticized the details of the theories of Bernstein and Schoenflies. First, he argued that any attempt to save W is pointless, since Russell's paradox shows that the problems that give rise to the Burali-Forti paradox go deeper than the theory of well-ordered sets and require "a suitable restriction of the notion of set" [Zer08a, p. 192], not merely a modification of the theory of well-ordered sets. Second, he claimed that set theories that admit W as a set will inherit its "inconsistent character" and that they are therefore doomed to failure. As evidence he cited Gerhard Hessenberg, who had noted that while Bernstein had used W to show that there are sets that cannot be well-ordered, Jourdain had used it to show the opposite. And third, and most important, he proved his result in a system in which W does not appear. The main point of giving his new proof is that that is even clearer than it was for the old proof:

> Already in my 1904 proof, having such reservations in mind, I avoided not only all notions that were in any way dubious but also the use of ordinals in general; I clearly restricted myself to principles and devices that have not yet by themselves given rise to any antinomy . . . Now I succeeded in completing my new proof without even the device of rank-ordering, and I hope thereby to have definitively cut off every possibility of introducing W.   [Zer08a, p. 192]

We have just seen Zermelo's main motive both for his new proof and for his axiomatization of set theory. It was *not* to secure set theory from paradoxes.

(See [Moo78] and [Moo82, pp. 155–160].) Paradoxes were a side issue. It was to secure his theorem from the criticism that the methods he had employed in its proof led to paradoxes.

The old proof used, in addition to sets, the separate notion of orders that are imposed on sets. Today we just take such an order, or indeed any binary relation on a set S, to be the set of ordered pairs of members of S between which the relation holds. (Indeed, we take an n-ary relation to be a set of n-tuples for any n, and in particular we take a unary relation to be a set of members of S.) Since we know today how to reduce ordered pairs to sets, that device reduces talk of orders to talk of sets and so avoids postulating two sorts of objects. That avoidance is a serious convenience when trying to analyze what principles are being used. The device of ordered pairs was not available to Zermelo. What he therefore did in the second proof was reduce well-orderings on sets to particular sets by using an *ad hoc* device in order to avoid the need to postulate two sorts of objects, sets and orders. That is what made it possible for Zermelo to derive the Well-Ordering Principle from principles that concern only sets.

Zermelo was not attempting to present a theory of what sets are. He emphasized that he had sought out "the principles required for establishing the foundations of" set theory as it was historically given and that he would not discuss their origin in his article [Zer08b, p. 200].[2] It is not in fact quite correct that his principles are adequate for set theory as it was historically given, as we shall see below. What *is* true is that they are the principles required to prove Zermelo's theorem. Every single one of the axioms, with the exception of the last one, the Axiom of Infinity, is used in Zermelo's new proof, and of course the theorem depends on the Axiom of Infinity for its interest.[3] Zermelo's hodgepodge of axioms extracted opportunistically from a proof forms the basis of the axioms of present-day set theory. The method of obtaining the axioms was appropriate to Zermelo's limited purpose, but it should hardly be

---

2. Zermelo took "Cantor's original definition of a set" to lead to paradoxes, apparently mistaking it for a statement of Comprehension in the manner discussed in §IV.2. He thus took Cantor's conception to be naive. But Zermelo ignored what he took to be Cantor's definition and took his principles from Cantorian proofs. Zermelo's theory is thus extremely Cantorian despite his lack of recognition of that fact.

3. The Union Axiom is used only to derive the Axiom of Choice—every set of nonempty sets has a choice function—from the Multiplicative Axiom—see the text below.

expected to yield an axiomatization of a coherently motivated theory based on some distinctive conception of what a set is.

Poincaré rejected the actual infinite. He viewed the mathematics that is apparently concerned with the actual infinite as actually concerning the finite linguistic definitions that putatively describe actually infinite objects. He therefore thought of definitions of such objects as giving them existence, rather than pointing them out or otherwise distinguishing them from other objects. The set L_γ, which, recall, is a γ-set, is defined as the set of γ-elements, that is, all members of γ-sets. But then, since L_γ is a γ-set, an object could be a γ-element, and hence a member of L_γ, in virtue of its membership in L_γ. Thus, Poincaré viewed the definition of L_γ as viciously circular because it is, in his terminology, impredicative, and therefore incoherent.

Zermelo's reply was simple, but in the end decisive. He noted that impredicative definitions are common in analysis. They are indispensable to the practice of ordinary mathematics. Moreover, they are unobjectionable on Zermelo's view, since they do not create the objects they define, but merely distinguish them from other objects. Thus, L_γ is indeed a γ-set, and so something can be shown to be a member of L_γ in virtue of its membership in L_γ without circularity. Showing that something is in L_γ from the fact that it is in L_γ is not very informative, unless we have somehow identified L_γ using a different description. But there is nothing contradictory or viciously circular about Zermelo's definition if it picks out L_γ from a collection of previously existing objects, instead of creating L_γ. To summarize, impredicative definitions are necessary for ordinary mathematics, and they are unproblematic if one adopts a realist attitude about the objects defined, realist in just the sense that the objects exist in advance of the definitions, that they are picked out by the definitions, not created by them. That imposes a substantial constraint on any acceptable philosophy of mathematics.

Jourdain claimed that his proof that every set has an ℵ as its power, outlined in §IV.2, accomplished all that Zermelo's did, but more simply. Zermelo pointed out that Jourdain's principles, which allow W, did not permit him to show that the continuum is a set. Thus, Zermelo's theorem entails that the continuum is well-orderable, while Jourdain's did not.

Zermelo went on to discuss Jourdain's proof. That proof made use of an arbitrary succession of choices—as Zermelo put it [Zer08a, p. 193], "after an arbitrary finite or infinite number of elements take an arbitrary element of the remainder as the next one; and continue in this way until the entire set is exhausted." Though Zermelo had emphasized earlier in his article (p. 186) that the Axiom of Choice involved simultaneous choices, here he accepted Jourdain's use of a temporal succession of choices. Perhaps he found simultaneous "choices" just as dubious as Jourdain's successive ones, since he replaced the axiom that "a simultaneous choice of distinguished elements is in principle always possible for an arbitrary set of sets" with an equivalent that is less "tainted with subjectivity and liable to misinterpretation" [Zer08a, p. 186], namely what is now known as the Multiplicative Axiom: for every set T of pairwise disjoint nonempty sets, there is a subset S of $\bigcup T$ that has exactly one member in common with each member of T. The Multiplicative Axiom is, as Zermelo showed, equivalent to the Axiom of Choice. It was also stated, apparently independently, by Russell, who gave it its name. (I shall not discuss Russell's independent discovery of the Axiom of Choice in any detail. He was working within the theory of types—based on the logical notion of collection—when he discovered that he needed a new assumption—the Multiplicative Axiom—to fill in gaps in certain proofs. He used it reluctantly since he thought it to be complicated and dubious—as indeed it is as a principle concerning logical collections. See [Moo82, pp. 121–132].)

Zermelo objected not to successive choices but to Jourdain's assumption that the entire set will be exhausted, that is, the assumption that the set of well-orderings of subsets of the set has a maximal element. He said that that requires proof. The assumption is essentially a special case of Hausdorff's Maximal Principle,[4] which is equivalent to the Axiom of Choice, though Zermelo does not seem to have recognized that fact. Zermelo was certainly right that Jourdain had used an unidentified additional assumption. With that assumption, Jourdain's proof is still vitiated by its use of W. There is, however, a proof of the Well-Ordering Principle from Hausdorff's Maximal Principle that is simpler than Zermelo's proof of the principle from the Multiplicative Axiom. In 1932, in his editorial comments to Cantor's letter to Dedekind discussed in §§III.4 and IV.2, in which Cantor gave the same proof as Jourdain, Zermelo criticized the use of successive choices [Can32a, p. 117]: "the intuition of time is applied here to a process that goes beyond all intuition, and a fictitious entity is posited of which it is assumed that it could make *succes-*

---

4. Hausdorff's Maximal Principle says that every partially ordered set has a linearly ordered subset that is $\subseteq$-maximal among such subsets. (It might seem more natural to invoke Zorn's Lemma here, but Jourdain would not have viewed well-orderings simply as sets.)

sive arbitrary choices." Zermelo went on to say that one would have to use simultaneous choices.

Peano, Borel, Lebesgue, and Baire all doubted the assumption, that is, the Axiom of Choice. Peano essentially rested content with noting that the axiom did not follow from the laws of logic. Borel admitted that a version of the axiom that was restricted in its application to denumerable sets of nonempty sets might be acceptable. Lebesgue rejected any form of the axiom. Baire rejected any form of the axiom, and he rejected the Power Set Axiom (for infinite sets) as well. (See [Moo82, pp. 92–96].) Borel explicitly accepted Zermelo's proof, but he viewed it as demonstrating (the difficult direction of) the equivalence between the Axiom of Choice and the Well-Ordering Principle. Zermelo had, in Borel's eyes, shown two problems equivalent without solving either one [Moo78, p. 312].

Zermelo began his reply by carefully noting that he could not *prove* the Axiom of Choice, and that in accusing him of failing to provide a proof for it, his critics endorsed his own view of the matter. But [Zer08a, p. 187] "every proof in turn presupposes unproved principles." He said that to "reject such a fundamental principle" one would have to show it false or contradictory, and none of his critics had attempted to do so.

Zermelo mentioned two sorts of support for the axiom, or indeed any mathematical principle. The first is intrinsic support—that the axiom is "intuitively evident"—the second is extrinsic support—that it is "necessary for science."[5] That is in apparent contrast to his immediately preceding claim that his opponents could not reject a principle unless it were shown to be false or contradictory. I take it the point is this: A principle cannot be definitively rejected, and it may therefore be used, unless it is shown to be false or contradictory. But a principle must be accepted, and it *must* therefore be used, if it is "intuitively evident and necessary for science." Zermelo argued [Zer08a, p. 187] that Peano had selected his fundamental principles by analyzing what ones mathematicians have used "and by pointing out that the principles are intuitively evident and necessary for science." He noted that he could marshal the same kind of arguments for his axiom. The fact that it did not happen to appear in Peano's list is not an argument against it. Zermelo also pointed out that while his proof can be carried out in a system that is free of all known paradoxes, Peano's system is subject to paradoxes and hence inconsistent. It

---

5. The useful terms *intrinsic* and *extrinsic* are taken from Maddy. See [Mad90, p. 118].

is simultaneously too narrow, in omitting the axiom, and too wide, in permitting paradoxes.

To show that the axiom is intrinsically motivated, indeed "self-evident," Zermelo noted [Zer08a, p. 187] that it had been, in effect, used by many mathematicians with a great deal of success "even though it was never formulated in textbook style." If a principle, possibly in variant forms, is independently and unquestioningly applied by many mathematicians, surely that shows that the principle is self-evident. Self-evidence is a psychological or perhaps sociological phenomenon, but, as Zermelo said,

> No matter if this self-evidence is to a certain degree subjective—it is surely a necessary source of mathematical principles, even if it is not a tool of mathematical proofs, and Peano's assertion that it has nothing to do with mathematics fails to do justice to manifest facts.   [Zer08a, p. 187]

The reader will surely object that the axiom was not self-evident to, for example, Peano, Borel, Lebesgue, and Baire, but of that, more below.

Another objection that may occur to the reader is that while the Axiom of Choice is indeed self-evident for finite sets of nonempty sets, the extension to infinite sets is not self-evident. The appearance of self-evidence could therefore turn out to be simply a case of unwarranted generalization.[6]

The Axiom of Choice certainly is evident in the finite case. But it is not new in that case: the Axiom of Choice restricted to finite sets of nonempty sets is a theorem of, for example, Peano's system, as Zermelo noted [Zer08a, p. 187]. Zermelo noted that fact immediately before arguing for the self-evidence of the axiom. It is precisely the version that applies to infinite sets of nonempty sets that was at issue, and it was the earlier uses of that version that provided the basis for Zermelo's argument. Thus, the axiom provides an example of a positive principle, distinctively about the infinite, that is self-evident.

Because the importance of claims of self-evidence is often dismissed on the basis of an overly simple account of their function, it is worthwhile to emphasize the sophisticated nature of Zermelo's claim that Choice is self-evident. He did not at any time claim that the truth of Choice was revealed to him through some mysterious faculty or that it could be revealed to others

---

6. See [Lav92, pp. 325–326] for a discussion of why the generalization would be unwarranted on the basis of a theory of what warrants generalizations that was advanced by Maddy [Mad90], a theory that is based on the notion of a natural kind.

in that way. As Kitcher has emphasized [Kit88, p. 297], that is just not how mathematics grows. Zermelo did not ask us to accept the axiom on the basis of some armchair introspection. Nor did he treat self-evidence as a sufficient condition for acceptance. Zermelo's evidence for the self-evidence of Choice was that many mathematicians had used principles or techniques equivalent to applications of Choice (1) with a great deal of success and (2) without any awareness that a new principle was being applied.

There is a tendency to view the lack of awareness that a new principle is being applied as an oversight on the part of the mathematicians applying it. Zermelo did not have that tendency, and I believe that it is one that should be resisted. Those who employed Choice-like principles were not working in a late-twentieth-century axiomatic setting. They saw themselves as discovering truths, mostly about the real numbers, on the basis of what they already knew about the real numbers. A proof technique or principle, whether new or old, was fully appropriate to their task just if it is a *correct principle.* A correct principle is, more or less, a principle that was true to their conception of the real numbers.

There would indeed have been an oversight if the Choice-like principles used had not been true to the conception of the real numbers. If the principles had been inconsistent with that conception, we would appropriately identify their use as a mistake. If the principles had been compatible with the conception though not derivative of it, we would appropriately identify the use of the principles without explicit mention as an oversight. But as a matter of fact we appropriately regard the theorems that we now prove using Choice, reconstructing the old proofs, as true of the real numbers—the very same real numbers that were under investigation all along. Choice is licensed by the idea of arbitrarily picking the members of a set. Its eventual use was therefore perhaps inevitable given the notion that a function is an arbitrary succession of values not subject to a common law. The subject matter was not changed by the use of Choice principles—it had already been changed by Fourier, who gave that characterization of a function.

The leading role, I take it, of the claim of self-evidence—though admittedly I cannot show that this is what Zermelo had in mind—was to establish that to use Choice is to go on with the exploration of the same old real numbers and the same old mathematical subject—analysis—as before. It is not to take up the investigation of something new or more special.

The role of self-evidence is not, or at least not always, to ensure *a priori* truth, and it is not independent of the goal of systematization. Historical explanations involving self-evidence are compatible with, and indeed an important part of, an understanding of the growth of mathematical knowledge, which takes place along the lines characterized by Kitcher's mathematical naturalism [Kit88, p. 295]. Taking the phenomenon of self-evidence seriously is compatible with the idea that mathematical knowledge evolves in a manner much like that in which scientific knowledge evolves—through the refinement and extension of theories, in a broad sense of the term.

To show that the axiom is extrinsically required, "necessary for science," Zermelo provided a list of important results that are intimately bound up with it, a list that included the basic results of cardinal arithmetic and also results important to analysis and algebra. Since Zermelo's time, that list has become far longer.[7]

Zermelo did not reply in any further detail to Borel, Lebesgue, and Baire, who had argued against the axiom and therefore had stated why they did not find it self-evident. One defense available to him was that Borel, Lebesgue, and Baire *did* despite their protestations find the axiom self-evident: each of them had used it—or rather various equivalent principles or consequences of it—in their work without question [Moo82, §1.7]. That work formed part of Zermelo's positive argument for the self-evidence of the axiom.

Perhaps Zermelo thought it unnecessary to reply to Borel, Lebesgue, and Baire: their objections to his work had given rise to a correspondence with Jacques Hadamard [BBHL05], who had ably defended Zermelo's axiom. Why did Borel, Lebesgue, and Baire doubt the axiom? Borel [BBHL05, p. 273] required that a theorem, to be "completely irreproachable," be "a precise result expressible in a finite number of words," and he said that the paradoxes arise "because sets that are not really defined are introduced." Lebesgue said "that to define a set M is to name a property P which is possessed by certain elements of a previously defined set N and which characterizes, by definition, the elements of M" and that

> The question comes down to this, which is hardly new: *Can one prove the existence of a mathematical object without defining it?*
>
> This is obviously a matter of convention. Nevertheless, I believe that we can only build solidly *by granting that it is impossible to demonstrate the existence of an object without defining it.*   [BBHL05, p. 265].

---

7. For a detailed history and discussion of the many theorems proved before and after Zermelo's work that entail some form of the axiom, see [Moo82].

And Baire expressed the opinion [BBHL05, pp. 263–264] that the infinite is "in the realm of *potentiality*" and so infinite objects are merely given or defined by convention. To go further "the meaning of these words [set, well-ordered set] must be extended in an extraordinary way and, I would add, a fallacious one." In each case, and indeed in the case of every mathematician of whom I am aware who has expressed reasons for doubting the axiom in print, the objection presupposes that every set is somehow associated with a definition of some kind. The objection is then simply that the axiom provides no means of *defining* a choice function. The great difficulty in defining such functions in standard cases, like the set of all nonempty subsets of the real numbers, suggested that supplying a definition would not be possible in general.

Hadamard's reply was straightforward:

> What is certain is that Zermelo provides no method to carry out *effectively* the operation which he mentions, and it remains doubtful that anyone will be able to supply such a method in the future. Undoubtedly, it would have been more interesting to resolve the problem in this manner. But the question posed in this way (the effective determination of the desired correspondence) is nonetheless completely distinct from the one that we are examining (does such a correspondence exist?). Between them lies all the difference, and it is fundamental, separating what Tannery calls a *correspondence* that can be *defined* from a correspondence that can be *described.*   [BBHL05, p. 262]

Whatever notion of set Zermelo and Hadamard were employing, it was expressly not one that required that every set be associated with a definition.

We may now briefly characterize the situation, following Maddy [Mad90, pp. 121–123] in important respects, in these terms: The opponents of the Axiom of Choice were employing the logical notion of collection. The Axiom of Choice is, at best, dubious for logical collections, and it is certainly not self-evident or otherwise suitable for adoption as a basic principle concerning such collections. The supporters of the Axiom of Choice were employing a quite different notion of collection, a combinatorial notion of the sort that originated with Cantor, that is to say a notion according to which collections consist of members enumerated in a perfectly arbitrary way. Such collections exist independently of our ability to give a defining principle, and the Axiom of Choice is indeed self-evident for them, even though it is dubious for logical collections. Cantor had even taken a close cousin of Choice, the

Well-Ordering Principle, to be the basic principle for combinatorial collections. The real disagreement was about whether mathematicians should employ logical collections or combinatorial collections, not about Choice. The verdict of history has been that mathematicians should employ combinatorial collections—and hence Choice—for reasons we discuss later in this section.

Cantor's early theory ran into difficulties because it is not clear that the collection of subcollections of a well-ordered collection can be well-ordered. As a result Cantor was not sure whether the power set of a combinatorial collection is itself a combinatorial collection.

Zermelo's substantial contribution was that he succeeded in making it plausible that the Power Set Axiom is compatible with the combinatorial notion of collection, and that it is in fact an illuminating supplement to that notion. First of all, as Zermelo's theorem shows, in the presence of Power Set (and other simple Cantorian principles), Well-Ordering and Choice—apparently two distinct characteristics of combinatorial collections—become provably equivalent. That is, they merge to become a single characteristic. Second, Zermelo's theorem provides some reason to believe the Well-Ordering Principle for the power sets of enumerative collections: We already believe Choice for at least one such collection, and Well-Ordering follows. That argument is based on the fact that the Axiom of Choice is self-evident for the set of real numbers, which is essentially the set of all subsets of the natural numbers and is thus a power set of an enumerative set. For the set of real numbers, Choice is part of the program of freeing the notion of a function from that of a rule.

There is a third notion of collection: the collections actually employed in mathematics, the *mathematical collections*.[8] I am not sure that deserves to be called a notion, since it is defined historically, but no matter. As Cantor basically discovered late in his career, the infinite mathematical collections can all be generated starting from the set of natural numbers using the power set operation and a few simple methods of combination. By showing that the Power Set Axiom is a plausible supplement to the Cantorian notion of a combinatorial collection, Zermelo made it possible to identify the mathematical collections with the combinatorial ones.[9]

---

8. Maddy used the term *mathematical collection* for pretty much what I have called combinatorial collections, and she associated such collections with the modern iterative conception of sets, often attributed to Zermelo, which is discussed in §5 [Mad90, pp. 102–103, 121].

9. Zermelo's collections are of a different sort from Cantor's. I see the one sort as a

In Cantor's early theory the Well-Ordering Principle had, in effect, served as the only criterion of sethood. That gave a clear characterization of what sets are. The Axiom of Choice is not suitable to serve as a criterion for sethood, and so Zermelo had to supplement it with an *ad hoc* list of supplementary principles. Each of his axioms is true to the combinatorial notion of collection—they all derive from the Cantorian theory. But Zermelo's arbitrarily selected list of axioms is not suitable for characterizing the notion of a combinatorial set.

The two sides of the debate about the Axiom of Choice were really disagreeing about which type of collection, logical or combinatorial, mathematical collections are. There was little disagreement that the Axiom of Choice is not self-evident, and indeed likely false, of logical collections or that the axiom is indeed self-evident of combinatorial collections. As Zermelo argued, and as subsequent mathematical developments have shown, the Axiom of Choice plays an important role in mathematics, and therefore the right notion is the combinatorial one.[10]

Zermelo had shown that the combinatorial notion of collection could be identified with the mathematical one, but that does not show that the logical notion cannot be identified with the mathematical one in some different way. One would lose the Axiom of Choice in so doing, but in the first decade of the century that was perhaps not yet decisive. Indeed, that seems to have been essentially the attitude of Russell in developing the theory of types (letter to Jourdain, 1905 [GG77, p. 55]): "I don't think the continuum of real numbers is upset by the multiplicative class difficulty. Also I have hopes that much will be discovered to circumscribe the difficulty; for all we have at present is a mere absence of proofs of propositions which are very likely to be true." But there are other substantial difficulties with identifying the logical collections with the mathematical collections, as the earlier description of the theory of types made clear. Logical collections are intimately connected with the incon-

---

natural outgrowth of the other, and so I have used the same term. The issue is, so far as I can see, merely a terminological one.

10. The opponents of the Axiom of Choice, all supporters of the logical conception of collection, did not just argue that the axiom is false of logical collections. Many also argued that the very idea of a combinatorial collection is incoherent—since infinite collections can only be introduced by means of definitions. I have omitted those arguments here. As we have mentioned, Poincaré had ideas along those lines. The most fully worked out position is that of Brouwer. Both Poincaré and Brouwer are discussed in Chapter VI.

sistent Comprehension Principle, and so one must confront the paradoxes in any successful attempt to identify logical with mathematical collections. But natural ways to handle the paradoxes also rule out impredicative definitions, as we saw in the case of the theory of types, where the need to allow such definitions led to the *ad hoc* Axiom of Reducibility. Since ordinary mathematics is impredicative, the use of logical collections is apparently blocked. Much progress had been made in this century in reconstructing parts of mathematics within one or another more or less predicative framework, but the fact remains that the ordinary practice of mathematicians cannot be captured in that way and many theorems cannot be reconstructed.[11] The mathematics of today, and of Zermelo's day, allows impredicative definitions and requires the Axiom of Choice. For those extrinsic reasons, we have come to recognize that ordinary mathematical collections are combinatorial collections. No mathematical theory of logical collections adequate to encompass the mathematical collections is as yet available.[12]

For various extrinsic reasons, the theory of mathematical collections is seen to be the theory of combinatorial collections, and the Axiom of Choice is self-evident for such collections. The controversy that surrounded the Axiom of Choice is usually assumed to cast doubt on the self-evidence of the axiom. But there was never a controversy about the axiom itself, which is in fact an uncontroversially self-evident principle about combinatorial collections. The controversy was over two notions of collection, logical and combinatorial, and the combinatorial notion has apparently won. As a modern set theorist, Donald A. Martin, put it,

> much of the traditional concern about the axiom of choice is probably based on a confusion between sets and definable properties . . . Once this kind of confusion is avoided, the axiom of choice appears as one of the least problematic of the set theoretic axioms.   [Mad90, p. 124]

The Axiom of Choice is *not,* as it is often taken to be, an example that shows that mathematicians disagree about "self-evidence," an example that casts

---

11. For more information, see, for example, [Fef 77].

12. Since the extensions of properties are logical collections, to the extent that they are taken to be collections constituted of members at all and not something more Fregean, the considerations in the text cast doubt on their mathematical utility and even their cogency. At the very least, the support that the use of such extensions is often thought to gain from the successful mathematical theory of sets is in fact strikingly absent.

doubt on that notion. On the contrary, the axiom is just one more example of the surprisingly broad agreement among mathematicians about what is, and what is not, self-evident, and of the real role such considerations play in mathematics.

As we have seen, the notion of combinatorial collection is in need of clarification. Logical collections need only be considered in an incidental way henceforth, since logical collections have never played more than an incidental role in mathematics in general and in set theory in particular. The paradoxes, which have often been thought to pose the central problem in making sense of the theory of sets, are, along with the logical collections for which they arise, a side issue. The real issue is how infinite collections can have combinatorial properties. Cantor's private answer, that God can manipulate them much as we manipulate finite collections, may not seem very helpful, but it is our historical starting point.

To see how set theory and our understanding of it have developed since 1908, it is necessary to look at Zermelo's axioms. Much of what has happened was in reaction to them. Here they are:

The first is Axiom IV.2.6, the Axiom of Extensionality. The second axiom is closely related to the Cantorian axioms IV.2.2 and IV.2.3.

AXIOM 1.1 (ELEMENTARY SETS). *There is an empty set. For any object a, there is a set {a} with a as its only member. For any two objects a and b, there is a set {a, b} that contains them and nothing else.*

AXIOM 1.2 (SEPARATION). *"Whenever the propositional function $\mathfrak{E}(x)$*[13] *is definite for all elements of a set M, M possesses a subset $M_\mathfrak{E}$ containing as elements precisely those elements x of M for which $\mathfrak{E}(x)$ is true." [Zer08b, p. 202]*

The Separation Axiom relies on the following definition [Zer08b, p. 201]:

A question or assertion $\mathfrak{E}$ is said to be *definite* if the fundamental relations of the domain, by means of the axioms and the universally valid laws of logic, determine without arbitrariness whether it holds or not. Likewise, a 'propositional function' $\mathfrak{E}(x)$, in which the variable term x

---

13. The symbol is a German capital E.

ranges over all individuals of a class $\mathfrak{K}$,[14] is said to be *definite* if it is definite for *each single* individual x of the class $\mathfrak{K}$.

The Separation Axiom is immediate from Cantorian Definition IV.2.5. At least, that is true for a suitable clarification of "definiteness." The notion of definiteness was the main source of controversy concerning Zermelo's axioms, and we shall discuss it in detail in §3.

AXIOM 1.3 (POWER SET). *Every set T has a power set, that is, a set that contains exactly the subsets of T.*

As we have discussed above, the Power Set Axiom does not follow from the Cantorian Axioms. It was, in fact, their downfall.

AXIOM 1.4 (UNION). *Every set T has a union set, that is a set that contains exactly the members of members of T.*

The Union Axiom follows from Cantorian Definition IV.2.5 and Axioms IV.2.7 and IV.2.10.

AXIOM 1.5 (CHOICE). *If T is a set of pairwise disjoint nonempty sets, then there is a subset of the union of T that has exactly one member in common with each member of T.*

The Axiom of Choice follows from Cantorian Axiom IV.2.10.

AXIOM 1.6 (INFINITY). *There is a set that contains the empty set and the set $\{a\}$ for each of its members a.*

The Axiom of Infinity follows from Cantorian Axiom IV.2.4 and Definition IV.2.5.

I have mentioned how most of Zermelo's axioms can be obtained from the Cantorian axioms. The other direction is more subtle, since Cantorian Definition IV.2.5 does not follow from Zermelo's axioms. That is partially why I said that Zermelo's axioms were not, as he claimed, "the principles required for establishing the foundations of" set theory. That is the story of the next section.

---

14. The symbol is a German capital K.

§2. The Axiom of Replacement

Fraenkel [Fra21, p. 97] and Skolem [Sko23b, p. 296] discovered that, in Skolem's words (Fraenkel's are similar), "Zermelo's axiom system is not sufficient to provide a complete foundation for the usual theory of sets." Each had noted that, if Z_0 is Zermelo's official counterpart for the natural numbers within his theory of sets and $\mathcal{P}$ is the power set operation, which takes each set to the set of all of its subsets, then one cannot prove in Zermelo's system that the set $\{Z_0, \mathcal{P}(Z_0), \mathcal{P}(\mathcal{P}(Z_0)), \ldots\}$ exists. Skolem gave a proof.[15] Without that set, as Fraenkel observed [Fra21, p. 97], one cannot prove the existence of $\aleph_\omega$.

Fraenkel [Fra22b, p. 231] and Skolem [Sko23b, p. 297] independently proposed the same remedy for the inadequacy in Zermelo's system, namely, introducing a new axiom, which Fraenkel named the Axiom of Replacement:[16]

AXIOM 2.1 (REPLACEMENT). *The range of a function on a set is itself a set.*[17]

Fraenkel observed [Fra21, p. 97] that it would have sufficed for the problem at hand to introduce an extended axiom of infinity, one that asserted that

$$\{Z_0, \mathcal{P}(Z_0), \mathcal{P}(\mathcal{P}(Z_0)), \ldots\}$$

exists. But if Zermelo's system were to be extended in that way, the resulting system could be shown inadequate in much the same way as before, and so a more general principle was required [Fra22b, p. 231].

The axiom solves the problem of ensuring the existence of

$$\{Z_0, \mathcal{P}(Z_0), \mathcal{P}(\mathcal{P}(Z_0)), \ldots\}$$

since there is a function that takes any natural number n (actually, its counterpart in Z_0) to the result $\mathcal{P}^n(Z_0)$ of applying the power set operation to Z_0 n times. The range of that function on Z_0 is the required set. That

---

15. He observed that $V_{\omega+\omega}$ is a model of Zermelo's axioms that does not contain the set. See §4 for the definition of $V_{\omega+\omega}$.

16. Both [Moo82] and especially [Hal84] provide useful histories of the Axiom of Replacement.

17. Skolem allowed partial functions. The two versions are equivalent.

proof mimics our intuitive reason for believing that if Z_0 exists then so does $\{Z_0, \mathcal{P}(Z_0), \mathcal{P}(\mathcal{P}(Z_0)), \ldots\}$: we just replace each member of Z_0, that is, each natural number n, with the corresponding member of the set, $\mathcal{P}^n(Z_0)$. Neither Fraenkel nor Skolem doubted the existence of the set, and they independently arrived at the same method of showing that it exists.

The axiom as stated is ambiguous since it is not clear what functions are to be allowed. Fraenkel did not publish an answer to that question until later, and so his version of Replacement was, initially, ambiguous. Skolem gave a precise answer. For both Fraenkel and Skolem the functions to be allowed are the "definite" functions. An answer to the question is thus a solution to the problem of making sense of Zermelo's notion of "definite." That is the topic of the next section.

Replacement is self-evident for combinatorial collections and an immediate consequence of Cantor's theory: If we form one collection by replacing the members of another, a well-ordering of the new collection is determined by a well-ordering of the original one. More formally, if we enumerate the members of S using F, and if we pick the members of a new collection by using f on S, then $f \circ F$ (with any duplications deleted) enumerates the new collection, which is the range of f on S, showing that range to be a combinatorial collection.

Replacement has various precursors. Cantor did not give any argument for Replacement, but he did state something like it [Can32a, p. 114], though as a truth not a postulate: "Two equivalent multiplicities are both 'sets' or are both inconsistent." That has as an immediate consequence that the range of a one-to-one function on a set is a set, which is a version of Replacement. (Though it is a special case, full Replacement follows from it using Zermelo's axioms.) Dimitry Mirimanoff [Mir17a, p. 49] stated the Axiom as a "Postulate," apparently because it was required to develop his theory of set-theoretic representatives of ordinal numbers.[18] That motivation does not have the clear intuitive character of the motivation of Fraenkel and Skolem, and the postulate was not part of a complete axiomatization of set theory.

Neither Fraenkel nor Skolem advocated adding the Axiom of Replacement to Zermelo's system. Neither investigated its consequences. Skolem just said

---

18. His representatives are essentially what are today generally known as "von Neumann ordinals," which will be described later in this section. The only difference is that Mirimanoff allowed urelements but apparently not an empty set and so he identified the number 0 with an arbitrary fixed urelement e instead of with the empty set.

that "we could introduce" the axiom [Sko23b, p. 297]. But Skolem had general doubts about the utility of axiomatic set theory. (See §3.)

Fraenkel in some ways opposed the Axiom of Replacement. (See [Hal84, pp. 296–297] for a summary.) The sets produced using Replacement are, according to Fraenkel, very large, while those that lead to paradoxes are too large. That suggested to Hallett that Fraenkel was "suspicious" of the axiom [Hal84, p. 296], presumably suspicious that it might lead to contradictions. But I do not think so, and to the best of my knowledge Fraenkel never said so.

Fraenkel's doubts seem to me to lie in another direction, since he said forthrightly that Zermelo's axioms "are not sufficient for the foundation of legitimate set theory" ([Fra21, p. 97], translation [Hal84, p. 280]), and he never seemed to doubt the existence of $\{Z_0, \mathcal{P}(Z_0), \mathcal{P}(\mathcal{P}(Z_0)), \ldots\}$. Moreover, he acknowledged, after the work of John von Neumann to be described next, that the axiom is necessary for the theory of ordinal numbers [Fra25, p. 251]. As Hallett put it, "What he seems to challenge is that any of this extra content which the axiom furnishes is set-theoretically important" [Hal84, p. 297]. (See, for example, [Fra25, pp. 251–252] or [Fra28, p. 310].)

Fraenkel believed that "general set theory" does not need the Axiom of Replacement, although some special results, including the theory of ordinal numbers, require it. That belief was entirely reasonable in the 1920s: Ordinary mathematics, including Fraenkel's general set theory, is as a matter of fact concerned exclusively with objects that have counterparts in the set $\bigcup\{Z_0, \mathcal{P}(Z_0), \mathcal{P}(\mathcal{P}(Z_0)), \ldots\}$. That set contains counterparts of the real and complex numbers, functions from real numbers to real numbers, function spaces, and so forth. All of those sets can therefore be proved to exist in Zermelo's theory. No consequence of the Axiom of Replacement that is a part of ordinary mathematics—that is, no consequence that could even be stated without making use of sets going beyond those—was discovered until 1975.[19] To be sure, other theorems may have been proved using Replacement, but they could perfectly well have been proved without it.[20]

---

19. The first such result is that every Borel game is determined. See [Mar75] for the relevant definitions and the proof of the result, and see [Fri71] for the proof that Replacement is required and the claim, made prior to 1975, that Replacement plays no role in ordinary mathematics. Further theorems of ordinary mathematics that cannot be proved without Replacement may be found in [HMŠS85].

20. Though modern mathematics is permeated with set theory, and set theory orig-

Fraenkel's hesitation to add Replacement to Zermelo's system had to do with doubts about the utility of Replacement, not with doubts about its truth. The fact that Replacement had no known applications outside of the theory of infinite ordinal and cardinal numbers but was nonetheless accepted as true shows that, whatever the basis on which Replacement was seen to be true, that basis must have been one distinctively concerned with the infinite.

In 1923 von Neumann worked out important consequences of Replacement [vN23], an axiom that he said [vN23, p. 347] "fills a substantial gap in Zermelo's axiomatization." When W is a well-ordered set, he defined a *numeration* of W to be a function f such that for all w in W $f(w) = \{f(u) : u < w\}$, that is, such that $f(w)$ is the range of f on the predecessors of w. The need to rely on the Axiom of Replacement to show that well-ordered sets have numerations is clear. He defined the range of f on W to be an *ordinal number* of the well-ordered set W. He then showed, still making strong use of the Axiom of Replacement, that every well-ordered set has a unique ordinal number and that similar well-ordered sets have the same ordinal number, which shows that the use of the term *ordinal number* is legitimate; he characterized the ordinal numbers in several different ways; he showed that the ordinal numbers are well-ordered by the membership relation; and he justified definition by induction on the ordinal numbers, which made it possible to introduce addition, multiplication, and exponentiation of ordinal numbers directly, without the use of auxiliary set-theoretic notions like that of ordered sets.

Once definition by induction on the ordinal numbers has been justified, it becomes possible to describe the von Neumann ordinal numbers by saying that each one is the set of its predecessors. Thus 0 is $\varnothing$, 1 is $\{\varnothing\}$, 2 is $\{\varnothing, \{\varnothing\}\}$, and so forth. Cantor had treated the ordinal numbers as separate objects that were obtained from well-ordered sets by "abstraction." Zermelo had axiomatized set theory without ordinal numbers. But von Neumann had shown how to introduce ordinal numbers as sets, making it possible to use them without leaving the domain of sets. The Axiom of Replacement is crucial. It is used

---

inated with the ordinal numbers, the ordinal numbers are rarely required outside of set theory, as just discussed in the text. Theorems concerning, for example, Cantor's derived sets seem to involve the ordinal numbers and hence Replacement, but they are a part of ordinary mathematics. The most convenient and natural formulations of such theorems do make use of the ordinal numbers, but they can be reformulated to avoid them and Replacement.

to show that every well-ordered set has an ordinal number and to justify definitions by induction on the ordinal numbers.

Though von Neumann apparently did not know it, Mirimanoff had already shown using Replacement that to every ordinal, and hence to every well-ordered set, there corresponds a unique von Neumann ordinal number and that similar well-ordered sets have the same von Neumann ordinal number; he had characterized the von Neumann ordinal numbers in several different ways; and he had shown that the von Neumann ordinal numbers are well-ordered by the membership relation [Mir17a, Mir17b]. He therefore, it seems to me, deserves some of the credit, and so I shall henceforth refer to the so-called von Neumann ordinal numbers as Mirimanoff–von Neumann ordinal numbers. But Mirimanoff did not propose identifying the ordinal numbers with the Mirimanoff–von Neumann ordinal numbers. He did not justify induction on the Mirimanoff–von Neumann ordinal numbers; indeed he made implicit use of induction on well-ordered sets to introduce them [Mir17a, p. 45]. He used Replacement to show that to every ordinal number—a notion taken to be antecedently understood—corresponds a Mirimanoff–von Neumann ordinal number [Mir17a, p. 49], instead of showing what his techniques sufficed to show, that to every well-ordered set corresponds a Mirimanoff–von Neumann ordinal number. He therefore had not shown that one could introduce Mirimanoff–von Neumann ordinal numbers initially as the sole notion of ordinal number. He had only shown that they could be used as a substitute for the ordinal numbers after the ordinal numbers had been used to introduce them.

§3. Definiteness and Skolem's Paradox

This section begins with a brief expository aside.

A *first-order* logic is one in which the quantifiers range only over the members of a domain. A *second-order logic* is one in which there are also quantifiers that range over things that determine relations and operations on a domain. Leopold Löwenheim [Löw15] usually receives credit for the distinction. See [Moo88a] for a careful history.

Second-order quantifiers might, for example, range over relations on a domain, operations on a domain, propositional functions on a domain, or collections of members of a domain. For simplicity, I shall just consider second-order quantification over relations.

The distinctive feature of second-order logic is that it presupposes that, given a domain, there is a fact of the matter about what the relations on it are, so that the range of the second-order quantifiers is fixed as soon as the domain is fixed. I shall not, in the end, make any essential use of second-order logic, and so I shall not endorse the presupposition. Nonetheless, it is important to see that the presupposition is a part of the use of second-order logic for many foundational purposes. There are a number of logical systems in which there are quantifiers that are second-order in that they range over something like relations over a domain, but in which the system is determined in an additional way—by giving a set of relations in addition to a domain or by axioms concerning the quantifiers. For present purposes—that is, with respect to Skolem's paradox—such systems may as well be first-order, and we do not include them in second-order logic. See [Sha91, Chapter 4] for an elementary exposition.

Most authors call all of the systems of logic with quantifiers over relations second-order—including those systems I just excluded. That is the natural thing to do if, for example, you think that a logic should be specified by the rules of its use rather than by truth conditions given in terms of domains and relations on domains. You might think that a logic should be specified by the rules of its use, for example, because it is not clear how domains and relations can be specified short of by giving the rules of reasoning about them. The logic that I am simply calling second-order is usually called "full" or "standard" second-order logic.

I do not wish to take a stand on the issue of which is the appropriate way to specify a logic. After all, I am arguing that logic has much less to do with set theory than is ordinarily supposed, and so the issue is not important here. But it is only full second-order logic that is relevant to Skolem's paradox and to Zermelo's work, and it is therefore a terminological convenience not to have to specify "full" or "standard" every time second-order logic is discussed.

Now, back to our history. Zermelo's second proof of the Well-Ordering Principle made use of a reduction of well-orderings to sets. That is what enabled him to carry out the proof on the basis of an axiomatization of a theory of sets and sets alone. But Zermelo's 1908 axiomatization in fact involved another sort of entity as well—definite propositional functions, which appeared in the statement of the Separation Axiom. That notion was criticized for lack of clarity by many. (See [Moo82, p. 260] for a list.)

In 1910 Hermann Weyl suggested that a property is definite if it can be obtained from $=$, $\in$, and members of the domain using a finite number of definition principles [Wey10, p. 304]. By 1917, he had arrived at a satisfactory list of definition principles: negation, identification of variables, conjunction, disjunction, substitution of constants, and existential quantification over the domain [Wey18, pp. 4–6, 36]. In modern terms, his idea was that a property is definite if it is definable in first-order logic with parameters.[21] That definition of definite presupposes the notion of finite iteration (of the definition principles) and hence the natural numbers. Thus, Weyl thought that the effort to found the natural numbers on set theory was misguided [Wey18, pp. 6–37].

The main purpose of the book in which Weyl discussed definiteness was the reconstruction of a portion of analysis in a theory that permitted quantification only over the natural numbers. The motivation was to give a predicative reconstruction of a portion of analysis: since real numbers are defined as sets of rational numbers, in effect as sets of natural numbers, a definition of a real number that involves quantification over real numbers is essentially a definition of a set of natural numbers in terms of (quantification over) the set of sets of natural numbers, and hence impredicative. To adopt a predicative program is to give up on standard set-theoretic analysis, and hence to give up on set theory. Perhaps that is why Weyl's suggestion had little impact on the development of set theory.[22]

In 1922 Skolem independently arrived at the same definition of definite as had Weyl [Sko23b, pp. 292–293]. The definite propositional functions of the Separation Axiom were, according to Skolem, just formulas of first-order logic with parameters. The axiom became a schema. The "functions" in Skolem's version of the Replacement Axiom correspondingly were functions definable in first-order logic with parameters. Thus, the Replacement Axiom is also a schema. Let me just give the details for Replacement. Separation is similar but simpler. When $\phi(x, y, u_1, \ldots, u_n)$ is a formula of the language of set theory, there is a corresponding instance of Replacement, which reads:

Fix sets $a_1, \ldots, a_n$. For any set S, if for every x in S there is a unique y such that $\phi(x, y, a_1, \ldots, a_n)$ (that is, if ϕ with parameters $a_1, \ldots, a_n$ defines a function on S), then there is a set T such that for all y, y is in

---

21. According to [Moo88a, p. 135], what Weyl had in mind was closer to ω *logic*: first-order logic with a built-in predicate for the natural numbers.

22. In 1946 Weyl gave a very readable brief account, in English, of his views and how they relate to those of others [Wey46, pp. 268–279]. Solomon Feferman is an eloquent modern advocate of predicative mathematics [Fef88].

T if and only if there is an x in S such that $\phi(x, y, a_1, \ldots, a_n)$. (That is, T is the range on S of the function defined by ϕ with parameters $a_1, \ldots, a_n$.)

That statement, naturally, is intended only as a relatively readable rendering of the official axiom, which is a sentence of first-order logic.

Like Weyl, Skolem [Sko23b, pp. 300–301] dismissed axiomatic set theory as clearly "not a satisfactory ultimate foundation of mathematics."[23] But, unlike Weyl, he dismissed it on the basis of a mathematical theorem about Zermelo's set theory, namely that if it is consistent, then it has a denumerable model. That means, as Skolem emphasized, that Zermelo's set theory has a model in which all of the "sets," even the supposedly nondenumerable ones like $\mathcal{P}(Z_0)$, are natural numbers. In fact the theorem, known as the Löwenheim–Skolem theorem, first proved by Löwenheim and strengthened and given a simpler proof by Skolem, shows that *any* finite or denumerable set of sentences of first-order logic that has an infinite model has a denumerable model. Thus it applies equally well to Zermelo's axioms with or without Replacement and also to every other first-order theory of sets.

There is no mystery about what has gone wrong: "Skolem's paradox" is not a formal contradiction. The natural number that is "$\mathcal{P}(Z_0)$" in a model of Zermelo's axioms with domain the natural numbers has only denumerably many "members," since there are only denumerably many natural numbers. But the theorem "$\mathcal{P}(Z_0)$ is nondenumerable" when applied in the model says only that there is no number in the model that within the model plays the role of a one-to-one correspondence between the number "$\mathcal{P}(Z_0)$" and the number "Z_0," where I have used scare quotes to emphasize that we are not talking about the real $\mathcal{P}(Z_0)$ and Z_0—if there are any such things—but the numbers that those descriptions pick out in our model with domain the natural numbers. Whatever correspondence we might use to show that "$\mathcal{P}(Z_0)$" is denumerable is just not in the model. If we allow (full) second-order logic, it is trivial to block the paradox: By *fiat* a second-order quantifier $\forall X$ includes every relation on the domain in its range, and hence every set of natural numbers if the natural numbers are in the domain. Thus, the second-order axiom

$$(\forall X)((\forall y)(X(y) \to y \in Z_0) \to (\exists x)(x \in \mathcal{P}(Z_0) \wedge (\forall y)(X(y) \leftrightarrow y \in x))),$$

23. He later changed his mind. See [Ben85, Geo85].

which says that if X holds only of numbers, then there is an x in $\mathcal{P}(Z_0)$ that is the set of numbers of which X holds, ensures that every set of natural numbers is in $\mathcal{P}(Z_0)$, hence in the domain, which is therefore uncountable. Skolem's paradox is blocked by stipulation. That is not much more helpful than just insisting that we intend our first-order interpretation to be full in the sense that its quantifiers range over all sets[24] and that the membership relation is the membership relation, which blocks the paradox directly.[25]

Skolem went on to point out that the notion "nondenumerable" is unavoidably "relative" in that a set that is nondenumerable in one model (for example, the $\mathcal{P}(Z_0)$ of the denumerable model above, which is nondenumerable in that model) may turn out to be denumerable in another (the model in which we carried out the construction of the denumerable model). He declared that "finite," "infinite," and other notions are similarly relative.[26] He concluded [Sko23b, p. 296] that on any consistent axiomatic basis the theorems of set theory "hold in a merely *verbal* sense." However [Sko23b, p. 300], "many mathematicians—indeed, I believe, most of them . . . do not have an axiomatic conception of set theory at all. They think of sets as given by specification of arbitrary collections."[27] One could, Skolem noted, introduce ab-

---

24. In fact all we need require is that the quantifiers range over a union of V_αs. See §4.

25. One might think it more natural to suppose that we know what the relations on a domain are given the domain than to suppose that we just know what sets there are, which makes second-order logic seem a bit more natural than the "full first-order set theory" just suggested in the text. But the point remains that the two block the paradox in essentially the same way.

26. Both he and, as we shall discuss below, von Neumann suspected that finitude would be relative like the other notions, but that did not follow from Skolem's results. It does, however, follow from Kurt Gödel's incompleteness theorem, announced in 1930 [Göd30] (using his 1929 completeness theorem too), as Gödel essentially pointed out in a review [Göd34] published in 1934 of [Sko33]. Skolem may not have recognized that, since he claimed to have derived that relativity in [Sko34] using essentially the techniques of [Sko33]. In fact, the result follows easily from Gödel's completeness theorem alone, but that does not seem to have been noticed until 1947 ([Hen47], see also [Kle88, p. 49]). Anatolii Ivanovich Maltsev gave arguments for the result similar to those based on the completeness theorem in 1936 and 1941 ([Mal36, Mal41], compare Robert L. Vaught's remarks [Vau86, p. 377].)

27. The quote is taken out of context—Skolem was making the point that the source of doubt about the Axiom of Choice is the "demand that every set be definable." But he did intend the point I am using the quote to make in the text, even if it was only in passing.

solutely nondenumerable collections on the basis of nondenumerably many axioms or on the basis of an axiom that yields nondenumerably many first-order consequences. But any such method would be circular.

Skolem raised two additional objections to the use of axiomatic set theory as a foundation for mathematics [Sko23b, p. 299]. First, in order to show any axiom system for set theory consistent, one must presuppose, outside of that set theory, a notion of a proof based on the axioms. But such a proof consists of "an *arbitrary finite number* of applications of the axioms." Thus, "the idea of the *arbitrary finite* is essential"; it must be presupposed, not introduced from within the axiomatic set theory. Skolem might have added, though he did not, that introducing the notion from within an axiomatic set theory to prove that theory consistent is not only circular, but that it is inadequate because the notion of arbitrary finite within the axiomatic system is a relative notion, while the one required of genuine proofs is not. In fact, as Gödel showed in 1930 [Göd30], if Zermelo's theory is consistent, then so is that theory plus a new axiom that says essentially that that very theory is *inconsistent*. Once more, there is no paradox. Models of the strange theory with the new axiom have a nonstandard notion of finite and hence a nonstandard notion of proof. The "proofs" of "inconsistencies in Zermelo's theory" that appear in such a model are not proofs in the ordinary, absolute sense: assuming Zermelo's theory is consistent, they are either infinite or not well founded (that is, they involve loops or infinite descending chains), as is easily seen from outside the model.

Skolem's remaining objection to taking axiomatic set theory as a foundation for mathematics was that it is absurd to define the natural numbers and then prove the induction principle on the basis of axiomatic set theory, since the natural numbers and induction are so much simpler and less open to question than any axiomatic set theory. That objection only shows that one cannot use a set-theoretic basis to justify the natural numbers or to render our theory of them more certain. It does not show that the theory of the natural numbers cannot be absorbed into set theory in a more technical sense. Indeed it can: the theory of the finite Mirimanoff–von Neumann ordinals, for example, provides a perfectly good mathematical substitute for the theory of the natural numbers. Note (though Skolem did not do so explicitly) that any axiomatic theory of the natural numbers will be subject to the same kinds of objections that axiomatic set theories are: the relativity of the notion of natural number, the need to presuppose a notion of finite to prove anything about the axiom system, and so forth.

We now turn to Fraenkel's theory of definiteness, published in 1925 [Fra25, p. 254]. Fraenkel simultaneously defined definite properties and a certain class of functions from the domain of all sets to sets, the Fraenkel functions, by induction.[28] Here is a slightly cleaned up version of the definition:

The base Fraenkel functions are the power set operation, the union operation, and the constant functions. The Fraenkel functions are closed under composition. When f and g are Fraenkel functions then so is the function of x that takes x to the set $\{f(x), g(x)\}$. When f and g are Fraenkel functions, the propositional functions $f(y) = g(y)$, $f(y) \neq g(y)$, $f(y) \in g(y)$, and $f(y) \notin g(y)$ are definite. When ϕ is a definite propositional function, the function that takes any x to $\{y \in x : \phi(y)\}$ is a Fraenkel function.

Fraenkel felt that his version of definiteness was superior to Skolem's since it did not require considerations of logic and stayed close to Zermelo's version [Fra25, p. 251]. There is, perhaps, some truth in that, since the Fraenkel functions are precisely the ones licensed by Zermelo's axioms. That is, Zermelo's axioms are logically equivalent to Extensionality, Choice, and Infinity, which do not assert that functions exist on the domain of all sets, plus the claim that the domain is closed under the Fraenkel functions. Fraenkel's version of Separation says: if m is a set and ϕ is a definite propositional function, then $\{y \in m : \phi(y)\}$ (the set of y in m such that ϕ holds of y) is a set. Fraenkel did not assume any Replacement Axiom.[29] Fraenkel showed [Fra25, Fra26,

---

28. The term *Fraenkel function* is due to Hallett [Hal84, p. 283].

29. One corresponding Replacement Axiom would now be: if m is a set and f is a Fraenkel function, then $\{f(y) : y \in m\}$ is a set. That version is suggested by [Fra26, p. 134]. Von Neumann showed in 1928 that that version of Replacement is derivable from Fraenkel's other axioms [vN28, p. 324]. It therefore does not do the required job. There is a different version of Replacement that is more in line with Fraenkel's procedure: since the Replacement Axiom asserts the existence of a function, that function should be added to the definition of Fraenkel functions. Define Fraenkel* functions by adding the following condition to the definition of Fraenkel functions: If f is a Fraenkel* function, then so is the function that takes x to $\{f(y) : y \in x\}$. Fraenkel did at one time suggest that adding Replacement would require widening the notion of Fraenkel function [Fra25, p. 271]. Unfortunately, it is a consequence of von Neumann's result that the new version is no better than the old one. Indeed, it follows easily from von Neumann's work that the Fraenkel* and Fraenkel functions coincide. Von Neumann [vN28, p. 323] suggested a different addition to the definition of Fraenkel function that does the job: Let $\phi(x, y)$ be

Fra32] that his version of Zermelo's system sufficed for the development of much of set theory. As we noted in the previous section, he explicitly admitted [Fra25, p. 251] that von Neumann [vN23] had shown that the system without Replacement does not suffice for the "special" theory of ordinals and cardinals.

In 1925, von Neumann published an axiomatization of set theory [vN25]. He had actually developed it two or three years earlier [Hal84, p. 283]. He credited Fraenkel with the Replacement Axiom [vN25, p. 398], ignoring Skolem's contribution, though he cited Skolem's paper in which Replacement appeared [Sko23b] concerning Skolem's paradox.

Von Neumann axiomatized, in effect,[30] a theory of what he called *classes*. Among the classes, certain ones are members of other classes. Such classes are *sets*. Thus, much like Cantor, he allowed the class of all sets, the class of all Mirimanoff–von Neumann ordinal numbers, and so forth. As he knew, such classes would lead to paradoxes if they were allowed to be members of other classes—hence the distinction between sets and classes, which is remarkably similar to Cantor's distinction between sets and absolutely inconsistent multiplicities.

Von Neumann's new system included the following distinctive axiom:

AXIOM 3.1 (LIMITATION OF SIZE). *A class is of the same power as the universe of sets if and only if it is not a set.*

To say that a class S is of the same power as the universe of sets means that there is a function from the class onto the universe of sets, that is, a class F of ordered pairs such that each member of the class S is the first component of exactly one pair in the class F and such that every set is the second component of at least one pair in the class F. In effect, the axiom says that the only way a class can fail to be a set is by being as large as possible—as large as the class of all sets. Von Neumann argued for that by saying [vN25,

---

a definite propositional function. If for each x there is a set that has as members exactly those y such that $\phi(x, y)$, then the function that takes each x to the set of y such that $\phi(x, y)$ is a Fraenkel function.

30. Von Neumann took functions to be primary, not sets. He introduced sets as characteristic functions, that is, as functions from the domain to two objects, functions like the ones Cantor used in his diagonal argument. No one has followed von Neumann in taking functions to be primary, and so I shall ignore that, acting as if he had taken sets to be primary. In the transformed picture, von Neumann's functions become classes of ordered pairs such that every set is the first element of exactly one member of the class.

p. 402] that it clarified existing confusion, is extraordinarily powerful, and "enlarges rather than restricts the domain of set theory." That limitation of size picture is extremely close to Cantor's, discussed in §IV.2.[31] The Axiom of Replacement—the range of a function on a set is a set—is immediate from von Neumann's axiom, since the range can be no larger than the original set. Cantor's argument from the letter to Dedekind shows that the class of ordinal numbers is not a set. Hence, it follows from von Neumann's axiom that there is a function from the class of ordinal numbers onto the class of all sets. Thus, the class of sets can be well-ordered, and so Choice follows from the axiom as well.

Technical Remark. In fact, a strong form of Choice follows from Limitation of Size: it says that for every class S there is a class T such that for every x, if there is an ordered pair in S whose first element is x, then there is exactly one such pair in T. Von Neumann showed [vN29, pp. 506–508] that the system with Limitation of Size is consistent if and only if the system with the strong form of Choice just introduced plus Replacement is consistent. He also showed by assuming Foundation (see §4) plus his other axioms that Limitation of Size is equivalent to those two axioms. (The consistency result is immediate from the other result plus his theorem [vN29, pp. 494–508] that the system obtained from his by replacing Limitation of Size by Strong Choice and Replacement is consistent without Foundation if and only if it is consistent when Foundation is added.) Since we shall consider Limitation of Size without Foundation in subsequent sections, let me mention that it is straightforward to check that all von Neumann's proof requires is that there be a function from the ordinals to sets such that every set is a member of some member of the range. The equivalence therefore also holds in models of the Anti-Foundation Axiom (see §4).

In von Neumann's system, the Separation Axiom is just the fact that the intersection of a set and a class is a set. Von Neumann thus, in effect, identified

---

31. Von Neumann believed that Cantor's set theory was the "naive" set theory that is in fact due to Russell. He attributed the basic limitation-of-size idea to Zermelo [vN25, p. 397]. In so far as the Separation Axiom only introduces subsets of an already given set, there is a kind of limitation-of-size principle to be found in Zermelo's system—a limitation to subsets—but it is not clear how it relates to von Neumann's proposed limitation on the *cardinality* of sets, a limitation that, as we have seen, results from Cantor's conception. See §5.

the definite propositional functions of Zermelo's Separation Axiom with the classes. To ensure that there are enough classes, von Neumann used axioms that entailed that any collection of sets is a class if it is first-order definable (with class parameters). Since first-order formulas are built up from atomic formulas using, say, negation, disjunction, and existential quantification, it is enough to ensure that the complement of any class is a class, that the union of two classes is a class, that the domain of any binary relation is a class,[32] and so forth. Von Neumann thus obtained a theory in which all of the relevant notions are axiomatized as members of the domain. In contrast, Skolem had required the auxiliary notion of a first-order formula, and Fraenkel had required that of a Fraenkel function. Von Neumann's theory is finitely axiomatized. No schemas are required.

Von Neumann said [vN25, p. 395] his work was in what would today be called a "formalistic" spirit: "one understands by 'set' nothing but an object of which one knows no more and wants to know no more than what follows about it from the postulates." Nonetheless, he carefully observed [vN25, p. 403] that his axioms "are nothing but trivial facts of naive set theory." That observation was important to him because it showed that the axioms, in the specific sense indicated, do not require too much [vN25, p. 403]. Thus, it was important to von Neumann that the axioms be "evident and reasonable," a constraint that is not formalistic. (Elsewhere [vN25, p. 402] he qualified his claim with respect to his main axiom, which he admitted is stronger than "what was up to now regarded as evident and reasonable.")

Von Neumann went on to sharpen some of Skolem's arguments against axiomatic set theory. He noted that probably no theory that has infinite models is *categorical*—that is, is such that all of its models are isomorphic—and so no theory of any infinite mathematical system can characterize that system. He concluded [vN25, p. 412], "This circumstance seems to me to be an argument for intuitionism."[33] He noted that the notion of "well ordering" is subject to Skolem's relativity. About the relativity of the notion of finitude, he said that it is difficult to say whether this militates more strongly against its intuitive character or its set-theoretic formalization. It counts against both, since it shows [vN25, p. 413] that we lack "any foothold that would enable us to make the definition of 'finite' determinate."

---

32. A binary relation is a class of ordered pairs. The domain of a binary relation is the class of all x for which there is a y such that the ordered pair $\langle x, y \rangle$ is in the relation.

33. Intuitionism, a philosophy of mathematics that rejects set-theoretic foundations, is discussed in §VI.2.

When combined with Skolem's arguments, those of von Neumann amount to a devastating criticism of our present-day axiomatic foundations of mathematics on the basis of set theory: Those foundations rely on a notion of proof, which requires a notion of finitude for its definition. But, once we specify the notion of definiteness, our axioms enable us to show that the definition of finitude they provide is an inadequate foundation for the notion of proof. To be sure, every theorem of mathematics has a counterpart within set theory— including the whole theory of finitude, based on the "finite" Mirimanoff–von Neumann ordinals. But that theory cannot serve as a basis for the notion of proof, and hence set theory cannot serve as a basis for an axiomatic mathematics, even if concerns about the certainty of the basis, like consistency or self-evidence, are not at issue. As Skolem said, we cannot make sense of what we are doing without presupposing the notion of "arbitrary finite number." (None of that provides an argument against a realist view of set-theoretic mathematics, on which axioms and proofs play only an incidental role.) The criticism is not directed at the practice of using proofs, a practice that we can certainly acquire without having a theory of proofs or an adequate characterization of finitude. In practice all we need in the way of a theory of finitude is the recognition that any proof we actually encounter in completed form is finite, which is very far from a complete characterization of what it is to be finite. The criticism is directed at a certain kind of attempt to characterize or define what the practice allows in the way of proofs. The attempt fails.

In 1929 Zermelo clarified his own view of definiteness. He began by discussing the various attempts by others at defining definiteness. The difficulty, he said, with eliminating the notion of definiteness in favor of general logic is that there is no widely accepted general logic [Zer29, pp. 339–340]. He clearly had in mind proposals like that of Weyl and Skolem. He criticized Fraenkel for introducing the notion of what we have called Fraenkel functions via a construction, since the construction depends on the notion of the finite numbers, the clarification of which is an important job for set theory [Zer29, p. 340]. He therefore preferred von Neumann's purely axiomatic approach, though he thought that the use of functions made von Neumann's foundation intricate and hard to understand [Zer29, p. 340].

Since we have avoided von Neumann's use of functions, our version of von Neumann's system bears some resemblance to the kind of system Zermelo seems to have had in mind. There are two differences worth emphasizing: Zermelo allowed something like quantification over classes [Zer29, p. 343], and he introduced a restrictive axiom, which said essentially that no proper part of the collection of all classes satisfies the axioms for classes. That axiom was intended to have the effect of ensuring that there were no more classes than those required by the axioms [Zer29, p. 344].

Skolem replied to Zermelo promptly. For Skolem, the question whether a notion of definiteness is presented axiomatically or by means of a construction was merely a matter of formulation [Sko30, pp. 337–338]. He noted that Zermelo's proposal was quite similar to what his own would be, were it to be presented axiomatically.

The main differences between the proposals of Zermelo and Skolem were the restrictive axiom and the quantification over classes. About the restrictive axiom, Skolem asked [Sko30, p. 338]: Since Zermelo did not want to use the notion of finite number, why did he use the notion of proper part? Isn't that also a notion that is to be fixed by set theory? The quantification over classes stood in need of further clarification. Did Zermelo intend to clarify by means of a construction, or further axioms? Most important, if classes were introduced initially without quantification over classes, then allowing quantification over them does not make new sets or classes possible [Sko30, pp. 339–340] and is therefore superfluous. Finally, Skolem noted, much as before, that axioms for set theory will not specify a single model, since they will always have a denumerable model because of the Löwenheim–Skolem theorem [Sko30, p. 340–341].

Zermelo was implicitly relying on what we would today analyze as second-order notions when he quantified over classes and employed parts of the domain. Skolem's request for clarification was a request for a first-order version of those notions, to be supplied by the use of a construction or axioms.

Any first-order theory (that is, any theory formulated in a first-order logic) is subject to Skolem's argument for relativism. Second-order theories are not, but Skolem's request for clarification seems legitimate. Thus, the stand-off between Skolem and Zermelo.[34]

§4. Zermelo

In 1930, Zermelo proposed a new axiomatization of set theory [Zer30, p. 30]: an Axiom of Pairing (any two objects compose a set) much like that of Fraenkel [Fra25, p. 254] replaced his old Axiom of Elementary Sets. The Axiom of Infinity was dropped, on the grounds that it does not belong to general

---

34. For a modern discussion, and access to the extensive modern literature on the stand-off, see [Sha90]. We return to the subject in §VII.4.

set theory. The Axiom of Choice was not expressly part of the system, since Zermelo thought of it as part of the background logic. The Axiom of Separation took a quite general form: one can separate out of any given set the subset of elements for which a given propositional function holds, where the propositional function can be any whatever, without restriction. As he also put it, any part of a set is itself a set. He mentioned his 1929 article summarized in §3 and Skolem's criticism of it, but simply reserved the right to add to that discussion. He added a strong version of the Axiom of Replacement analogous to his Axiom of Separation and kept the old Axioms of Extensionality, Power Set, and Union. Finally, he added a new axiom [Zer30, p. 31]:

AXIOM 4.1 (FOUNDATION). *Every (descending) chain in which each element is a member of the previous one is of finite length.*

He claimed that that is equivalent to the following: every nonempty part P of the domain has a member that has no members in P.[35] The Axiom of Foundation forbids circles of membership and ungrounded sets [Zer30, pp. 29, 31]. Foundation had a different status from the other axioms, as is indicated by the fact that Zermelo referred to the proposed system as supplemented ZF or ZF′, where Foundation is the supplement. The difference is an important one—Zermelo did not believe that Foundation is true. He commented that the axiom had been satisfied in all useful applications of set theory up to that time, and thus it, provisionally, imposed no essential restriction on the theory [Zer30, p. 31]. He thus apparently believed that while there are non-well-founded sets (that is, sets that are at the top of infinite descending chains) they are of little importance in known applications of set theory. Zermelo used the restriction to well-founded sets to great effect to investigate the models of the supplemented theory.

Zermelo's axiom system is much like so-called ZFC, the axiomatization of set theory that is in most common use today. The two main differences are these: First, he allowed what he called *urelements,* objects that are not sets and have no members, in the domain, while it is conventional to exclude them today. Fraenkel was the first to propose excluding urelements [Fra22b, p. 234], as part of an attempt to give categorical axioms. Skolem mentioned [Sko23b, p. 298] that axioms that do not exclude urelements have

---

35. They *are* equivalent given a sufficiently strong form of Choice. Details are omitted, since we never consider dropping Choice in this book.

models both with urelements and without, and he sketched a proof. In much the same way that Zermelo argued for Foundation, Fraenkel argued that urelements serve no mathematical purpose and that eliminating them simplified matters.[36]

Second, Zermelo's Axioms of Separation, Replacement, and Foundation are based on a strong second-order understanding of the notion of definiteness: he considered *all* propositional functions, functions, chains, and parts of the domain, while today we allow only first-order definable (with parameters) propositional functions and functions, and only chains and parts of the domain that are themselves members of the domain. The system Zermelo proposed is essentially what we would today call second-order ZFC, though without Infinity.[37] The first to use a first-order system much like the one so universally adopted today was von Neumann, in 1928 [vN28, pp. 321, 323], who named it Zermelo–Fraenkel set theory.[38]

Though the Axiom of Foundation had several precursors, Zermelo did not discuss them. Mirimanoff [Mir17a, p. 42] was the first to distinguish the well-founded sets, which he called "ordinary sets," from the non-well-founded, "extraordinary" ones. Though Mirimanoff never considered the Axiom of Foundation, which would in his terminology have been that all sets are ordinary, he did the next best thing: he restricted his attention to the study exclusively of ordinary sets [Mir17a, p. 39]. That is not far from Zermelo's attitude. After all, Zermelo did not present Foundation as a new truth about sets, as he had presented Choice. He used Foundation to restrict his considerations to well-founded sets, just noting that the other sets have no use. Mirimanoff defined the *rank* of an ordinary set as follows [Mir17a, p. 51]: The rank of an urelement or the empty set is 0, while the rank of any other set is the least ordinal number greater than the ranks of any of its members. He showed, us-

---

36. It is ironic that Fraenkel was the first advocate of excluding urelements, since he was also the first to put them to serious mathematical use: in 1922, he proved that Choice is not a consequence of the other axioms if one allows urelements [Fra22a] (assuming the other axioms are consistent). That was not shown without urelements until 1963, by Cohen, see [Jec78, p. 184].

37. Today one drops Separation, since it is a consequence of Replacement, and one uses a first-order version of Foundation, since the second-order versions follow from that version and second-order Replacement.

38. The system is Zermelo's, using Fraenkel functions to specify what definiteness means, supplemented by Replacement amended as discussed in a note to §3. Urelements are allowed, and Foundation is not added.

ing Replacement, that every ordinary set had, according to that definition, a unique rank.

Skolem [Sko23b, p. 298] mentioned that for any domain that is a model of Zermelo's axioms, the elements of the domain that are not at the beginning of an infinite descending chain also form a model. He was not advocating a restriction, merely noting that Zermelo's 1908 axioms apparently do not determine whether such chains exist.

In 1925 von Neumann [vN25, pp. 404, 411–412] mentioned the possibility of an axiom that there is no function f with domain the natural numbers such that for all n, $f(n + 1) \in f(n)$, that is, that there is no infinite descending chain. Zermelo [Zer30] cited that paper for another purpose. Von Neumann did not have his version of Foundation in his basic system. He mentioned it in the second part of the paper, "Investigation of the axioms," primarily to note that it probably did not help to ensure that the theory is categorical, because of the relativity of the notion of descending chain. He did observe [vN25, p. 412] that adding the Axiom of Foundation would not lead to contradictions if there were none already (on the basis of his axioms) and that it would have the "desirable" effect of excluding "superfluous" non-well-founded sets. He published the proof that adding Foundation would not lead to contradictions in 1929 ([vN29, pp. 494–508]; see [Vau85] for an elementary presentation) in a paper in which he also gave (p. 498) the alternative form of the Axiom mentioned by Zermelo and (p. 503) the definition of rank, apparently independently of Mirimanoff.

Zermelo developed the use of rank to understand the structure of the well-founded sets. He went far beyond Mirimanoff. He observed that ranks stratified the well-founded sets: at the zeroth layer is the set of all objects with no members—the empty set[39] and all the urelements. Every *layer* (indexed by an ordinal)[40] is the set of all sets that are made up of objects that occur in previous layers. Thus, if we let $V_0(U)$ have as members the empty set plus the members of U, where U is a (possibly empty) set of urelements, and let $V_\alpha(U)$ otherwise be the union of $\bigcup_{\beta < \alpha} V_\beta(U)$ and its power set, then every well-founded set (over the urelements in U) is a member of some $V_\alpha(U)$, and the rank of an object a is just the least α such that a is in $V_\alpha(U)$.[41] The members of each well-founded set lie in preceding layers, and each well-founded

---

39. Zermelo took the empty set to be an arbitrarily selected urelement.

40. Zermelo distinguished between the ordinal numbers and their substitutes in his domains, the Mirimanoff–von Neumann ordinals. I shall suppress the details.

41. Here and below, I have modernized Zermelo's notation, and I have rephrased his

set serves as material for following ones [Zer30, pp. 29–30]. Zermelo did not emphasize that picture. After all, it did not apply to all of the sets, and he did not view the sets as being in any important sense constructed or built up. He merely said that it helped with his investigation of models of the axioms, and he introduced it as an aid on a par with that of using the Mirimanoff–von Neumann ordinals. The layers are not even mentioned in the final, concluding section of the paper.

Zermelo showed that his axioms serve to guarantee that every model is isomorphic to one of the form $\bigcup_{\beta<\kappa} V_\beta(U)$. Zermelo called models of that form *normal domains*. Normal domains are the subject of Zermelo's investigation. The stratification of sets within each normal domain plays only an auxiliary role. Since every model is isomorphic to a normal domain, we can afford to ignore all other models—results about them will follow immediately from the ones about the normal domains. Moreover, when we disallow urelements and abbreviate $V_\beta(\varnothing)$ (the $V_\beta(U)$ with the empty set of urelements) by V_β, if $\bigcup_{\beta<\kappa} V_\beta$ is a normal domain—a model of Zermelo's axioms—then V_κ is a model of von Neumann's axioms—if we take the sets to be exactly the members of $\bigcup_{\beta<\kappa} V_\beta$. Zermelo showed that a normal domain is characterized up to isomorphism by just two numbers—the cardinality of the set of urelements, which can be any cardinal number, and the least Mirimanoff–von Neumann ordinal number κ not in the model, which can be any strongly inaccessible initial ordinal.[42] Zermelo argued that if there is any normal domain at all that contains infinite members, then strongly inaccessible initial ordinals must certainly exist. Naturally one can assume otherwise, just as one can assume that there are no urelements, but only at the cost of generality [Zer30, pp. 44–45].

Distinct normal domains stack neatly. Suppose, for example, that two non-isomorphic normal domains $\mathfrak{V}$ (German capital V) and $\mathfrak{V}'$ have sets of urele-

---

results to deal only with the cumulative $V_\alpha(U)$s. He often considered the set of all sets of rank α, that is, in our notation, the set $V_\alpha(U) - \bigcup_{\beta<\alpha} V_\beta(U)$, not $V_\alpha(U)$.

42. A *strongly inaccessible initial ordinal* κ is an ordinal κ such that κ is of greater cardinality than any of its predecessors, κ is not the least upper bound of a set of cardinality less than that of κ of ordinals that are less than κ, and the power set of any ordinal less than κ has cardinality less than that of κ. One usually requires in addition that κ be greater than ω, but Zermelo allowed ω as a strongly inaccessible initial ordinal since he omitted the Axiom of Infinity from his set theory. With the definition I have given of $V_\alpha(U)$, not every strongly inaccessible initial ordinal gives rise to a model when the set of urelements is large. Zermelo showed that that can be patched up [Zer30, pp. 38–39]. I omit the details.

ments of the same size. Then they are characterized by distinct ordinals, say κ and κ' with, say, $\kappa < \kappa'$. Zermelo showed that $\mathfrak{V}$ is isomorphic to a substructure of $\mathfrak{V}'$ and indeed that κ is in $\mathfrak{V}'$ and that $\mathfrak{V}$ is isomorphic to the set $\bigcup_{\alpha < \kappa} V_\alpha(U)$ as defined within $\mathfrak{V}'$.[43] That is, in particular, the smaller domain is isomorphic to a set in the larger domain.

The above results make strong use of the second-order character of Zermelo's axioms: The use of *all* chains in the Foundation Axiom, instead of just those in the domain, guarantees that the ordinals of a domain really are well-ordered, which a first-order version of the axiom could not have done, and so the ordinals of any model are guaranteed to be (order) isomorphic to "genuine" ordinals. The use of *all* propositional functions in the Separation axiom guarantees that the $V_\alpha(U)$s of different normal domains with the same U are the same when α is in both domains: Suppose not. Then there is a least α in the domains at which the domains differ. Since they differ, one of them must include a set C in $V_\alpha(U)$ that the other omits. But every member of the set C is in both domains, since α is minimal. Separation guarantees that that set is in the other domain, contrary to our assumption, since C is the subclass of $\bigcup_{\beta < \alpha} V_\beta(U)$ determined by the propositional function that holds of the members of C and nothing else. The proof would not go through with Skolem's first-order version of Separation, since the requisite propositional function (x is in C) might not be given by a formula.

Zermelo accepted that there are many normal domains. They are nested as described above. That enabled him to make sense of his use of all propositional functions. Let me speak in terms of the collections that the propositional functions define, the parts of the domain, instead of in terms of the propositional functions themselves. That will serve to put Zermelo's position in the terms of our present-day view, and it will make clear what Zermelo's implicit reply to Skolem was.[44] Zermelo's Separation Axiom says that every part of a set is a set. Skolem's question had been, Isn't the notion of "part" to be fixed by set theory? Zermelo could now reply that it is: A model of set theory is a set in a higher model (that is, a model containing all of the urelements of the first model that has a larger characteristic ordinal), and so the notion of a part of a given model can now be explained set theoretically: a part of the

---

43. None of the results requires the Axiom of Infinity. Without it, $V_\omega(U)$, the set of hereditarily finite sets, is a normal domain, and hence included in the analysis. That is probably why Zermelo omitted Infinity.

44. It is no accident that putting things in terms more congenial to Skolem helps to make contact with today's views. Today's views are direct descendants of those of Skolem.

model is nothing more than a subset of the model, where the term *subset* is used in the sense of a higher model in which the original one is a set. Zermelo assumed that there always is a model higher than any given one [Zer30, p. 46]. The parts of a normal domain, its proper classes, are ordinary sets in a higher domain. Thus was the failure of the axiom system to be categorical turned into a virtue.

Note that all that is required for Zermelo's reply to Skolem is that every model of the axioms is a set in a higher model. If one could show that a similar result held in some system without the Foundation Axiom, then Zermelo's basic philosophical attitude would go over to that system unchanged. That is no idle speculation—it applies to a system of non-well-founded set theory that has some advocates today, ZFC minus the Foundation Axiom plus the "Anti-Foundation Axiom" (ZFC⁻+AFA). See [Acz88] for an exact statement of the axiom.[45] Zermelo assumed that there are as many strongly inaccessible initial ordinals as ordinals, and hence that there is a model of set theory for each ordinal. The same result follows for ZFC⁻+AFA. Alfred Tarski made a proposal equivalent to that of Zermelo but as an axiom of set theory instead of an assumption about set theory, eight years later [Tar38]. The succession of models also clears up the paradoxes: the proper classes of one model are sets in all higher models [Zer30, pp. 29, 47].

Zermelo went on [Zer32] to publish a strong attack on Skolem's "assumption that all mathematical concepts and theorems must be representable by a *fixed finite system of signs.*" Zermelo thought the resulting relativity of set-theoretic notions ought to convince anyone to abandon Skolem's "prejudice." He went on to say that "our system of signs is always an *incomplete device,* shifting from case to case. It reflects our *finite* understanding of the infinite, which we cannot *immediately* and *intuitively* 'survey' or comprehend, though at least we can approach mastery step by step." (See [Zer32, p. 85] as translated by Gregory H. Moore [Moo80, p. 126].) He went on to propose an infinitary logic, with conjunctions and disjunctions of any set of propositions and well-founded infinitary proofs ([Zer32, pp. 86–88], see also [Zer35]).

---

45. The construction of a model of ZFC⁻+AFA given in [Acz88, Chapter 3] yields a separate model of the axioms for each pure (that is, urelement-free) normal domain. Those models of ZFC⁻+AFA are uniquely determined up to isomorphism by Zermelo's parameter (the least ordinal not in the model), and they are nested as required: each model is a set in any higher model, and any higher model contains all the parts of the original model.

Within that logic, arbitrary propositional functions can be used to establish
Zermelo's structure theory described above, and, moreover, Zermelo noted
that that logic is not subject to the Gödel incompleteness results.

Zermelo summarized his view eloquently in an unpublished manuscript:

> How must a 'domain' of 'sets' and 'urelements' be constituted, to satisfy
> the 'general' axioms of set theory? Is our axiom system 'categorical,' or
> does it give a multiplicity of essentially distinct 'set-theoretic models'?
> Is the idea of a 'set' in contrast to a pure 'class' absolute, determined
> by logical characteristics or only relative, dependent on the set-theoretic
> model suitable as a basis at a time? . . .
>
> Every 'normal domain' is a 'closed domain' and therefore can in a
> higher one also be interpreted as a 'set.' . . . No (closed) normal domain
> can represent the whole of set theory . . . The whole of set theory is only
> represented by the 'open' totality of all normal domains. ([Moo80,
> pp. 131–133], my translation)

He also said that mathematics begins with the infinitary logical assimilation
of intuitively given material; it cannot be based on intuition [Moo80, pp. 134,
135]. He did not explain much more. I mention that only because the statement may seem to conflict with the kind of idea of self-evidence that I attributed to Zermelo earlier. It does not. All he means to deny by denying that
mathematics is founded on intuition is that mathematics is about space, time,
or the like. The notion of assimilation of intuitively given material leaves ample room for self-evidence.

§5. Go Forward, and Faith Will Come to You

There are several important aspects of today's approach to set theory that have
not come up in our historical account so far. Chief among them are the idea
that axiomatic set theory concerns a single intended domain consisting of all
the sets, the iterative conception of sets, and the primacy of the first-order
versions of the axioms.

Let me begin by briefly discussing the primacy of first-order versions of
the axioms, since there is at least some reason for that in what has come before, even though it has not been emphasized. The second-order quantifiers
in second-order axiomatizations of set theory essentially range over classes
in von Neumann's sense of class—a collection of sets. Since von Neumann's
axioms take the classes to actually be in the domain, they are first order. With

the classes in the domain, the temptation to use second-order quantification over them has been removed. Zermelo said (see §4) that he preferred von Neumann's approach to definiteness since it is purely axiomatic, not relying on a construction that is described outside the set-theoretic axioms. He applied that criticism to Fraenkel's proposal, but it applies equally well to the proposal of Weyl and Skolem, which relies on a logic specified outside the set-theoretic axioms. A preference for von Neumann's axioms on whatever grounds is *ipso facto* a preference for first-order axioms.[46]

Skolem's criticisms apparently led Zermelo to change the type of axiom system he favored, but his old reasons for preferring von Neumann's system were adopted by others. In addition, Skolem's insistence that second-order systems required further explanation had the effect of discouraging their use. See [Moo88a] for a discussion of how influences not particularly connected to set theory led to a general preference for first-order logic. Now on to the main business of this section.

Cantor thought of himself as studying sets, not some limited partial domain of sets, and certainly not the formal consequences of axioms. Thus, it is not unreasonable to say that Cantor believed that set theory concerns a single intended model. Since Zermelo just presupposed Cantorian set theory in his 1904 proof of the Well-Ordering Principle from the Axiom of Choice, he may be thought to have taken a similar view—his remarks are so abbreviated that one cannot rule that out.

By 1908, the situation had subtly changed: Since Zermelo argued for the truth of Choice and his other axioms, he was certainly committed to the idea that there are sets, and that they are what is being studied. That also becomes clear in his argument [Zer08a, p. 191] that impredicative definitions are acceptable—since, "after all, an object is not created through such a 'determination.'" On the other hand, in reply to those who had argued that the 1904 proof used principles that let to paradoxes, he said [Zer08a, p. 195], "it is not permissible to treat the extension of every arbitrary notion as a set . . . But if in set theory we confine ourselves to a number of established principles . . . —principles that enable us to form initial sets and to derive new sets from given ones—then all such contradictions can be avoided." He was proposing to abandon the general study of all sets, concentrating on those that could be shown to exist by means of a few principles. Since, as we have seen,

---

46. Zermelo proposed second-order axioms anyhow, but, as we have seen, Skolem showed that allowing second-order quantification in a system like von Neumann's did not make it possible to define new classes.

his principles were chosen in an *ad hoc* way, there was no reason to suspect that they were strong enough to generate all sets. Indeed they were not strong enough, as the later discovery of the need for Replacement showed.

As we have just seen, Zermelo proposed abandoning the general study of sets when he introduced axioms for set theory in 1908. In the period we have discussed after 1908 no one investigating the foundations of set theory concerned themselves with a single intended model of all sets. Skolem and von Neumann argued against the possibility of determining such a model. Fraenkel followed Zermelo's hint: He abandoned the general study of sets, trying to see instead how few sets were needed to develop the needed theory, rejecting Replacement, introducing an axiom of restriction (which said that no sets existed other than those required by Zermelo's axioms [Fra23, p. 219]), and giving a first-order version of definiteness. He could therefore avoid questions about all sets. Zermelo later (after 1929) did try to clarify what sets are instead of restricting his investigations to those sets whose existence is guaranteed by some principles, but he denied that the sets form a single domain.

Gödel introduced a very different point of view in 1947 [Göd47, vol. 2, p. 180]:

It might at first seem that the set-theoretical paradoxes would stand in the way of such an undertaking, but closer examination shows that they cause no trouble at all. They are a very serious problem, but not for Cantor's set theory. As far as sets occur and are necessary in mathematics (at least in the mathematics of today, including all of Cantor's set theory), they are sets of integers, or of rational numbers (i.e., of pairs of integers), or of real numbers (i.e., of sets of rational numbers), or of functions of real numbers (i.e., of sets of pairs of real numbers), etc.; when theorems about all sets (or the existence of sets) in general are asserted, they can always be interpreted without any difficulty to mean that they hold for sets of integers as well as for sets of real numbers, etc. (respectively, that there exist either sets of integers, or sets of real numbers, or . . . etc., which have the asserted property). This concept of set, however, according to which a set is anything obtainable from the integers (or some other well-defined objects) by iterated application[a] of the operation "set of",[b] and not something obtained by dividing the totality of all existing things into two categories, has never led to any antinomy whatsoever; that is, the perfectly "naïve" and uncritical working with this concept of set has so far proved completely self-consistent.[c]

But, furthermore, the axioms underlying the unrestricted use of this concept of set, or, at least, a portion of them which suffices for all mathematical proofs ever produced up to now, have been so precisely formulated in axiomatic set theory[d] . . .

a. This phrase is to be understood so as to include also transfinite iteration, the totality of sets obtained by finite iteration forming again a set and a basis for a further application of the operation "set of".

b. The operation "set of x's" cannot be defined satisfactorily (at least in the present state of knowledge), but only be paraphrased by other expressions involving again the concept of set, such as: "multitude of x's", "combination of any number of x's", "part of the totality of x's"; but as opposed to the concept of set in general (if considered as primitive) we have a clear notion of this operation.

c. It follows at once from this explanation of the term "set" that a set of all sets or other sets of a similar extension cannot exist, since every set obtained in this way immediately gives rise to further application of the operation "set of" and, therefore, to the existence of larger sets.

d. [At this point, Gödel referred in a note to the axiomatization of von Neumann and to those of Paul Bernays and himself, which are based on von Neumann's. All are first order.]

In that brief passage, Gödel introduced the idea that axiomatic set theory is the study of a single intended domain of all sets. He introduced the *iterative conception*—the idea that sets are to be conceived as just the objects obtained by iteration of the "set of" operation, in other words, the idea that sets are to be conceived as just the objects in the $V_\alpha(U)$s.[47] And he gave his support to the idea that the axioms of first-order axiomatic set theory "underlie" the concept of set. Each of those ideas is tremendously influential today.

The iterative conception gives the Axiom of Foundation center stage: as Zermelo showed, that axiom ensures precisely that each set is a "set of" sets that occurred at previous "layers" or iterations of the "set of" operation. The axiom guarantees that all sets are iterative sets, and the iterative conception makes the axiom obvious.

The iterative conception of set was not, as we have seen, present in the de-

---

47. The term *iterative* was used earlier by Bernays in connection with set theory in a passage that is in some ways very similar to the one just quoted [Ber35b, p. 260]. Gödel briefly mentioned the iterative conception in print in 1944 [Göd44, p. 462]. The mathematical background for the iterative conception was, as we saw in §4, developed in [Zer30]. But the idea that it is constitutive of what sets are that each one is in some $V_\alpha(U)$, which I take to be the essential component of the iterative conception, was *not* in [Zer30], as we have seen.

velopment of set theory up until at least 1930, at least not so far as Zermelo knew. For otherwise he would have introduced the Axiom of Foundation, which is characteristic of the iterative conception, not merely as a provisional restriction for mathematical convenience but as a provisional further specification of what is meant by "set."

Gödel apparently knew of no precedent for the iterative conception either: in the 1960s, when revising his article, he considered crediting Zermelo with "substantially the same solution of the paradoxes,"[48] citing [Zer30], the 1930 article discussed in §4. In the end, quite rightly, he did not. (See Moore's discussion of Gödel's article [Moo86, p. 167].) It was an extraordinarily bold move on Gödel's part to introduce the iterative conception as fundamental so late in the development of set theory.

Someone will surely object that the iterative conception was already implicitly present in Cantor's definitions of a set as "bound up" or "collected." But, as we have seen, Cantor's theory can be spelled out without Foundation. Cantor never, so far as I know, commented on whether a set can be a member of itself. Given the impredicativity of Cantor's theory, there seems to me to be no reason why an enumeration of elements cannot, after the fact, turn out to be such that its range is one of the elements, in which case a non-well-founded set would be a Cantorian set.[49] Moreover, none of Cantor's successors saw such an idea in his work, at least not until after 1947.

Can there be a mathematical argument for the iterative conception? First of all, there can be no proof that the Axiom of Foundation is either true or false. Von Neumann's proof that Foundation cannot lead to contradictions (mentioned in §4) shows that in any model[50] of any of the usual axioms for set theory (without Foundation) the well-founded sets form a model of those axioms plus Foundation. Thus, even when we allow non-well-founded sets it will remain possible to consistently add Foundation by restricting our attention to the well-founded sets. On the other side, results concerning Anti-Foundation mentioned in §4 in fact establish that for any given model[51] of any of the usual axioms plus Foundation, there is a model of Anti-Foundation extending it and such that the given model is the well-founded part of the ex-

---

48. No solution of the paradoxes was in fact proposed by Gödel. What was proposed instead is that the paradoxes are not relevant to set theory.

49. A collection whose only member is itself provides a cheap example: the function that takes 0 to that object witnesses that it is a set.

50. The result even applies to class models, which will be introduced in §VII.4.

51. Once again, the result even applies to class models.

tension. Thus, even if we exclude non-well-founded sets it will remain possible to consistently add Anti-Foundation by embedding the well-founded sets into a model of it. The combined effect of these facts is to show that, so far as our present knowledge is concerned, non-well-founded sets are perfectly good mathematical objects and that we should not expect to settle the question whether or not there are non-well-founded sets by proving or disproving Foundation on the basis of some new independently supported axioms.

The only other way to settle the question whether Foundation is true on internal mathematical grounds is in terms of its usefulness, but, as the historical development of set theory makes clear, Foundation has no mathematical use.[52] As Azriel Levy noted in a work that adopted the iterative conception [FBHL73, p. 87], each of the other axioms "was taken up because of its essential role in developing set theory and mathematics in general; if any single axiom were left out we would have to give up some important fields of set theory and mathematics . . . The case of the axiom of foundation is, however, different; its omission will not incapacitate any field of mathematics."[53] It must be granted, however, that it is equally true that the inclusion of Foundation will not incapacitate any field of mathematics: Every structure is isomorphic to a well-founded one, so that when one works only up to isomorphism, as is usual in mathematics, there is no loss in excluding the non-well-founded sets. That provides an excellent justification for Zermelo's policy of adopting Foundation as a simplifying assumption, but it provides only the weakest kind of support for the iterative conception as a conception. The iterative conception entails that it is part of the very idea of what a set is that all sets are well founded. The fact that for many purposes we can live without the non-well-founded sets hardly shows that the very idea of a non-well-founded set is incoherent. (Indeed, it *is* coherent as the arguments mentioned above based on Anti-Foundation show. That is already enough to show in a certain weak sense that the iterative conception is false.)

---

52. It has sometimes been claimed that Foundation is needed to obtain an adequate theory of cardinality. Levy's theorem concerning the definability of cardinal numbers is cited [Lev69]. But what Levy showed is that either Foundation or Choice is required to obtain an adequate theory of cardinality. Since for our purposes in this work Choice is always assumed, Foundation is not needed for an adequate theory of cardinality. The most common theory of cardinality is in fact the one we rely on here, the one based on Choice, not Foundation. See, for example, [Vau85] for an elementary treatment.

53. Levy parenthetically expressed doubts about the necessity of the Axiom of Extensionality.

Whatever defense the iterative conception has must be philosophical. To the extent that the conception integrates and provides a motivation for adopting the axioms of set theory as true of all sets, it is worth adopting. If it showed how to integrate the axioms into a coherent picture of what sets are, the loss of generality occasioned by restricting our attention to well-founded sets might be worth it.

Because the critical issue is the extent to which the iterative conception integrates the theory of sets, I have characterized the iterative conception very narrowly. Many advocates of the iterative conception include in it not only the idea that the sets are built up by iteration of the power-set operation, but also some aspects of what has here been called the combinatorial conception. (As Maddy kindly pointed out to me, Gödel may have been one of those who advocated such a combined picture—that may be the force of "underlying" in the opening sentence of the second paragraph of the passage quoted earlier.) But advocates of the kind of mixed view just described cannot claim that it provides an integrated theory of sets without telling a story about how the two aspects of their view come naturally together—a mere conjunction is not enough. No such story has been provided. I therefore adopt a narrow characterization of the iterative conception, which makes it possible to carefully analyze how that conception fits in with ideas drawn from the combinatorial conception.

So let us see how much the iterative conception helps us to integrate and motivate the axioms of set theory.[54] "Sets of" objects are certainly sets in some antecedent sense. We must therefore adopt the Axiom of Extensionality, not motivated by the iterative conception but prior to it. We are instructed by the iterative conception to take all sets of sets obtained by iteration, but the instruction "all sets" is of no help without some prior understanding of what sets there may be, an understanding that requires, at least, the Axiom of Separation. The Axiom of Separation does not follow from the iterative conception either. Like Extensionality, it must be part of a prior concept of set.

The Axiom of Choice also depends on the prior concept of set. If Choice is true of sets, then it will, on the iterative conception, be true of sets obtained by iterating the "set of" operation: Let S be a nonempty iterative set of pairwise disjoint iterative sets. If the Axiom of Choice is true, then there will be a set T that contains exactly one member of each member of S. Since each member

---

54. Many expositions of the iterative conception have appeared. For critical discussions of some of the most important ones, see [Hal84, especially pp. 214–223] and [Par77].

of a member of an iterative set is itself an iterative set, and since all of the members of members of S appeared at some stage before S, the set T will be an iterative set that appears by the stage at which S does. Thus, if the Axiom of Choice is true of sets, it will also be true of iterative sets. But that is of no help in determining whether or not the Axiom of Choice *is* true. To the extent that the iterative conception is an autonomous conception, it fails to help in settling whether or not Choice holds. Choice must be taken as an additional assumption, as some advocates of the iterative conception have noted.[55]

The iterative conception does not tell us how far to iterate (see Gödel's note *b*), and so we must also start with an Axiom of Infinity. In addition, for the same reason, the iterative conception presupposes the notion of "transfinite iteration." In effect, the ordinal numbers are supposed given in advance.[56] One of the symptoms of the need to start with whatever ordinal numbers are used is that the Axiom of Replacement, which, as we have seen, is intimately associated with the ordinal numbers, is not a consequence of the iterative conception.

The remaining axioms do follow from the iterative conception:

Pairing. If s and t are two iterative sets, then the set of s and t is an iterative set formed at the first stage after both s and t are.

Union. If S is an iterative set, then its members are all iterative sets formed before S and their members are all iterative sets formed before they

---

55. Hao Wang [Wan74] called his concept of set the iterative conception. In my terminology, he—like others—combined a combinatorial conception of set, the iterative conception, and some elements of limitation of size to motivate the axioms of set theory, including Choice and Separation. But Choice and Separation follow from the combinatorial aspect of his conception, not the, in my terms, iterative one. Wang described intuitions behind the axioms of set theory in a way that does capture, I believe, a lot of the picture that goes with the present widespread acceptance of those axioms, but the iterative aspect is not the crucial one.

56. George Boolos [Boo71] and Dana Scott [Sco74] have given variants of the iterative conception that avoid the need for that. Boolos assumed in his formal theory that one can prove things about the stages by induction instead of explicitly assuming that the stages are constructed in a sequence indexed by ordinal numbers. But the assumption about induction is motivated, as Boolos clearly stated, by a "rough description" that does involve the ordinals. Scott employed a reflection principle. While his axiomatization of set theory is interesting in its own right, the use of reflection principles is not a part of the notion of "set of," and so his axiomatization is not relevant to an evaluation of the basic iterative conception.

were, and hence before S. We can therefore form the set of members of members of S at the same stage we form S.

Power Set. Any subset of an iterative set S will be formed by the stage at which S is. We can therefore form the set of all of them at a stage after S is formed.

The reader should by now be uncomfortable with my talk of "forming" some sets "before" or "after" others. It is crucial to maintaining the full impredicative forms of Separation and Replacement that sets be construed realistically, not as being constructed by us, as Zermelo essentially argued in 1908 (§1). So whatever notion of priority is being invoked by the iterative conception cannot be temporal, and whatever notion of formation is being invoked cannot be construction. Moreover, as Charles Parsons pointed out [Par77, p. 507], the temporal metaphor of a mind collecting objects already constructed breaks down for nondenumerable iterations and collections: "It is hard to see what the conception of an idealized mind is that would fit here; it would differ not only from finite minds but also from the divine mind as conceived in philosophical theology, for the latter is thought of either as in time, . . . or its eternity is interpreted as complete liberation from succession." Even a Cantorian appeal to God's powers proves to be inadequate here!

Without the temporal notion of constructing some sets after others the iterative conception loses much of its appeal. Parsons has suggested that a modal interpretation of the iterative conception be given to avoid the reliance on time: a multiplicity of actual sets is a possible set ([Par77, pp. 509, 515], [Par83c]). That is a fascinating suggestion well deserving of further exploration. It is not yet, however, clear how to interpret the necessary notion of possibility without relying on metaphors of time and construction.

At least at present the advantages of the iterative conception do not suffice to justify adopting it: It does not provide a conception that unifies the axioms of set theory. It is based on a picture of construction that does not seem to have a coherent interpretation. And, most damning of all, even though the iterative conception has been widely embraced in recent times, it has had very little impact on what theorems can be proved—no essential mathematical use has been found for Foundation.[57]

---

57. Given the widespread employment of Foundation today it remains possible that some use will yet be found. In that case, I would reverse my position. Caution therefore dictates that I not make my other arguments depend on a denial of the iterative conception, and indeed none of them do depend in any way on such a denial.

There is a final advantage of the iterative conception that is best brought out by contrasting it with the limitation-of-size conception. The limitation-of-size conception, recall, has it that if a collection is the same size as a set, then it is a set. Just as the Axiom of Foundation is characteristic of the iterative conception, the Axiom of Replacement is characteristic of limitation of size. But note that limitation of size tells which among the collections are sets—those that are small enough. Unless one just takes Replacement to express the limitation-of-size idea, limitation of size suggests, on pain of vacuity, that there are some collections that are not sets. Thus, for example, it is immediate from von Neumann's Limitation of Size Axiom that the universe is a class that is not a set. Though "set of" requires an antecedent notion of set, it does not require that there turn out to be collections that are not sets. If, like Gödel, one wants a domain of all sets, then one based on limitation of size will plausibly involve collections that are not sets,[58] while one based on the iterative conception could be construed as a domain of all collections as well as all sets. That is, the iterative conception leaves open the possibility of claiming that there is no collection of all sets and that every collection is a set.

If one is willing to give up on a single domain of all sets, as Zermelo was, then limitation of size is compatible with the claim that all collections are sets. It is just not compatible with the claim that all collections are sets in a single domain. Nonetheless, we can see immediately that limitation of size will not serve as an overall guiding principle for our set theory any more than the iterative conception does: limitation of size does not justify the Power Set or Union Axioms. It is not clear that it justifies Choice either: Given a set S of disjoint nonempty sets, a collection T consisting of exactly one object from each of them will be the same size as S and will therefore be a set—if it exists. The truth of Choice depends on our antecedent theory of collections, which is not given by limitation of size. There is, however, as we have seen

---

58. Given suitable background assumptions, one *can* formulate limitation of size without classes. One version, suggested to me by Vann McGee, is that given a well-ordering of the universe, limitation of size can be formulated as follows: If there is an x such that every y such that $\phi(y)$ is less than x, then there is a set of all ys such that $\phi(y)$. (Naturally, I have in mind a schema in which formulas with no free occurrences of x may be substituted for ϕ.) I do not know of a proposal along such lines that seems sufficiently natural to merit serious consideration, and so in the text I have pretty much ignored the possibility of formulating limitation of size without classes.

in §3, an extension of limitation of size from which Choice does follow: von Neumann's Limitation of Size Axiom.

There is another limitation theory besides limitation of size: Fraenkel's theory of limitation of comprehensiveness, suggested to him by Zermelo's Separation Axiom. Since it played no role in the development of the axioms, I have not discussed it before. But Fraenkel used various versions of it to justify the axioms from 1924 through at least 1958. See [Hal84, pp. 200–202]. The basic idea is that, starting with given sets, we only form sets from them that are somehow connected to them (how is never exactly clear), so that the sets formed are not absolutely comprehensive like the sets of the paradoxes. Here is a version—my own, not Fraenkel's: One way to guarantee that a collection is not too comprehensive is to require that its members already be bounded by a set. That idea suggests the following axiom: every subcollection of a set is a set. Given a reasonable notion of collection that yields the Separation Axiom. But it doesn't get us much else. However, it seems a reasonable extension of the idea to allow that a collection is not too comprehensive, and hence forms a set, if the members of its members are bounded by a set. That suggests the axiom: a collection is a set if its union is included in a set. That, in combination with Separation, which we have just seen follows from limitation of comprehensiveness, yields Power Set. It should be fairly clear from the above that limitation of comprehensiveness shares some similarities with the iterative conception. Gödel's note c in the quote above is, as Hallett suggested [Hal84, p. 202], reminiscent of the limitation-of-comprehensiveness.

Just as in the case of limitation of size, assuming the converse of the limitation of comprehensiveness seems like a reasonable extension: the sets are exactly the collections that are not too comprehensive. The converse of the limitation-of-comprehensiveness principle that yielded Power Set yields Union.

We can combine the two limitation theories to obtain an axiomatization of set theory that is as well motivated as any. Since both theories start with collections and delimit the sets among them, the theory will—like von Neumann's in the usual presentation—be a theory of classes with axioms to tell which classes are sets.

As in the modified von Neumann theory, a set, by definition, is a class that is a member of a class. The first axiom is Extensionality for classes.

AXIOM 5.1. *Classes with the same members are equal.*

We shall also need to have axioms to ensure that there are enough classes.

Von Neumann's axioms on classes would do, as would those of Bernays [Ber76, p. 5] or Gödel's based on them [Göd40, p. 37], but those axioms all assume the existence of ordered pairs of sets that are sets, and so I prefer the following axiom schema, in which $\phi(x)$ is any definite formula:[59]

AXIOM 5.2. *There is a class with members exactly the sets x such that* $\phi(x)$.

I used the vague term *definite* because there are several ways of spelling it out. There is the first-order version of the axiom in which a definite ϕ is simply a first-order formula in the language of the theory with quantification only over sets. That version is the "class theorem" of Bernays [Ber76, pp. 12–13]. There is Zermelo's second-order version, in which the domain of a model is a set in a higher model. There, the definite formulas will also include ones of the form $x \in S$, where S is an arbitrary subset (in the sense of a higher model) of the domain. There will be a first-order version that allows quantification over classes, a first-order version for any expanded language, and there is at least one other version as well, to be discussed in §VII.4. In any event, the axiom will serve to guarantee that the classes are closed under simple operations. In particular, if S is a class, then so is its union, the class of members of members of S. We need an Axiom of Infinity. I like this one:

AXIOM 5.3. *There is a nonempty set on which the membership relation is a discrete linear order with no last element.*

Next, we have von Neumann's Limitation of Size Axiom 3.1. Last of all, we have the following:

AXIOM 5.4 (LIMITATION OF COMPREHENSIVENESS). *A class is a set if and only if its union is a set.*

Technical Remark. The above axiom system (in its first-order version) is equivalent to von Neumann's, reformulated as usual, as is easily seen: Limitation of Comprehensiveness yields Union and Power Set, and conversely. The version of the axiom system along Zermelo's lines has as

---

59. I have excluded urelements to simplify the axioms. A fully general version should allow them. As usual, there is no essential problem in doing so, but some minor increases in complexity result. I leave the needed changes to the reader.

models those Zermelo noted were models of von Neumann's system: domains of the form V_κ, where κ is strongly inaccessible and the sets are exactly the members of $\bigcup_{\beta<\kappa} V_\beta$.

The "only if" of Limitation of Comprehensiveness is redundant. That direction is the Union Axiom; the remaining direction is what gave Power Set. But Levy showed that Limitation of Size plus Power Set yields Union [Lev68].

First order or second order? One intended domain or many? Are all collections sets? Do the iterative conception, limitation of size, and limitation of comprehensiveness fit into a coherent conception of sets? The last question is the heir of Cantor's quandary, How does Power Set fit into a combinatorial conception of sets? Our understanding of the foundations of set theory is not much better than d'Alembert's understanding of the foundations of analysis was in the latter half of the eighteenth century.[60]

---

60. Fraenkel expressed a similar sentiment in 1927 [Fra27, p. 61]. The situation has not changed all that much since.

Knowing the Infinite

§1. What Do We Know?

The theme of this section is that Mathematics Is Infinite: contemporary mathematics has a strong commitment to the mathematical existence of infinite mathematical objects with rather specific properties. There is nothing mysterious to that claim. It means nothing more than that mathematicians prove theorems that assert the existence of infinite objects, study infinite objects, teach their students about infinite objects, and so forth. Before discussing the details, let us consider why a philosopher of mathematics should care what contemporary mathematics is like.

There is ample precedent for arriving at norms for mathematics on the basis of philosophical considerations and then telling the mathematicians that they must conform to those norms, changing mathematics if need be. Luitzen Egbertus Jan Brouwer is frequently taken to provide an example of someone who made such a drastic recommendation. Since he failed to get most mathematicians to change, the example is a cautionary one.

A philosophy of mathematics must make at least some contact with the pre-existing practices of mathematicians—one that doesn't has almost certainly missed the mark. I am using the term *practice* here broadly: I mean to include not only the activities and procedures mathematicians employ but also their beliefs, heuristics, standards, and theories. What aspects of practice need to be taken seriously and explained—or explained away—and what aspects can be ignored? Some cases are clear: a philosopher of mathematics can ignore the coffee consumption of mathematicians but not their axiomatic method. Other cases are much less so: Certain axioms are said to be self-evident. Do we therefore need a corresponding theory of mathematical evidence? Many contemporary mathematicians would describe themselves as formalists. Does that count in favor of a formalist philosophy of mathematics?

My imprecise and partial answer is this: as philosophers, we must take account of certain core practices universally agreed to be mathematical. The practice of proving theorems about the ring of polynomials over the natural numbers provides an example. We need to take account of more peripheral practices to the extent that they are tied to the core. A practice that is widespread or influential is more likely to be worth taking seriously than one that is not, but we may wish to dismiss it. All else being equal, a theory that explains a practice will be superior to one that dismisses it, and a theory that promises to lead to fruitful developments within mathematics will be superior to one that does not. That will all sound familiar, since it parallels standard methodological recommendations for science, here taking the philosopher of mathematics to be like a scientist, with the practices of mathematicians playing the role of empirical data.

As I shall use the term, "explaining a practice" is not at all the same thing as justifying it. Scientists do not justify their data. To show that certain assumptions and procedures lead to a practice is to explain why that practice is adopted, just as deducing the outcome of an experiment from a law explains the outcome. Such scientific explanation does not "justify" the underlying laws, and neither does an explanation of how a practice comes out of certain assumptions and procedures justify those assumptions and procedures. For example, the law of the excluded middle (which says that P or not P holds for any statement P) is (part of) the foundation of the mathematical practice of employing proofs by contradiction, and the fact that many mathematicians assume the law of the excluded middle helps to explain the practice of proof by contradiction, though it does not justify it to anyone who has doubts about the law.

My reason for taking the stance I do is anything but a skeptical one. I believe that the reasons mathematicians have provided for their practices are, by and large, eminently reasonable and that the mathematicians have, by and large, gotten things right. The reason for my stance is nearly the opposite of skeptical: Most philosophers of mathematics have felt it necessary to provide a justification for the practices of mathematicians that contradicts the reasons provided by the mathematicians themselves. By getting out of the way, I hope to let the excellent reasons of the mathematicians be clearly heard.

There is an important kind of reason that mathematicians have given for one or another belief—namely, that the belief is self-evident or intuitively obvious—that has not been taken seriously by any philosopher of mathematics, with the notable exception of Gödel, for a long time. (Some have been willing to accept such claims for trivial principles—for example, the Union Axiom V.1.4—but not for the central cases like Choice and Replacement.)

The arguments against accepting claims of self-evidence have seemed overwhelming. There are basically two such arguments. The first is the one we have spent the last five chapters disposing of. It was the argument based on the supposed historical facts that there have been strong disagreements about what is self-evident and that supposedly self-evident principles have led to contradictions. The argument would indeed have been devastating for self-evidence, but—as we have seen—it was based on flawed historical accounts.

The second argument will be a dominant theme for the rest of the book. It is that some of what has been claimed to be self-evident—including Choice and Replacement—could not possibly be evident to any human being, since such self-evidence would require some sort of access to infinitary objects, and that is clearly impossible. Gödel, who accepted a form of self-evidence, seems to have accepted the conclusion that we have access to the infinite. Once one has accepted the arguments against self-evidence only two sorts of reactions are possible: one can reject the supposedly self-evident assertions along with the claim that they are self-evident, or one can look for new justifications for them.

In this chapter I shall give a list of important principles that have been held to be self-evident, discuss the nature of the second argument against self-evidence, and discuss the first sort of reaction. In the next chapter I shall discuss the second sort of reaction. In the final chapters I shall rehabilitate self-evidence—at least in the context of the self-evidence of set-theoretic principles—by overcoming the second objection to it.

Once self-evidence has been rehabilitated it will be possible to view my account of the reasons mathematicians have given for endorsing set theory as an account of the compelling reasons we actually have for endorsing set theory. No further external justification is required, and the old reasons for thinking one might be have been removed.

As we have seen, mathematical practice was transformed in the nineteenth century. For the first time, a proof from explicit assumptions became a sufficient condition for the truth of a theorem. Before, that had only been true for elementary number theory. In Euclid's geometry there were axioms and proofs, but the correctness of the proofs relied on the use of diagrams. In the analysis of the eighteenth and early nineteenth centuries, the proof techniques were well known to be inadequate, and so to be convincing a proof had to be supplemented by a plausibility argument typically based on physical considerations—as we saw in the dispute about whether every function has a trigonometric-series expansion. Some readers will undoubtedly rise to the defense of the mathematicians of those years and claim that in some sense they—at least the better ones—knew exactly what they were doing. Let me just say that if that is correct then their knowledge did not become fully explicit until the work of Cauchy and Weierstrass.

The rigorization of analysis led to remarkable further changes, not least that it became possible for mathematicians to drop any reference of their ideas to intended extramathematical interpretations. That possibility has culminated in modern mathematics, in which virtually all work is only up to isomorphism and no one cares what the numbers or points and lines "really are." For the purposes of this book the most important effect of the rigorization of analysis is that the changes wrought by Cauchy and Weierstrass led inevitably to our present strong commitment to actually infinite sets, which have become an essential part of the way we understand even the most central parts of mathematics.

What do we know? More cautiously, What are the commitments of contemporary mathematical practice? Here is a list of some commitments that concern the infinite: There are actually infinite combinatorial collections, most centrally some set that will serve as the natural numbers and some set that will serve as the power set of that one, which gives us the real numbers. Those collections exist in some sense that licenses reasoning about them impredicatively. They all have power sets. We have at least some clear and apparently immediate "self-evident" knowledge about the properties of actually infinite combinatorial collections, most strikingly that they obey the Axioms of Choice and Replacement.[1] And we have a notion of finiteness, at least the finiteness of proofs, that is prior to the notion of finiteness given within axiomatizations of set theory.

Before I discuss the items on my list, let me defend an omission from it and issue a disclaimer. The omission: It may well be that the vast majority of those mathematicians who accept the items I have mentioned also accept the Axiom

---

1. As we discussed in §V.1, extrinsic considerations motivate focusing exclusively on combinatorial collections. Given that we are interested in combinatorial collections, Choice is seen to be self-evident.

of Foundation. That is a datum I cannot completely dismiss, and so I shall make sure that everything I claim below is compatible with accepting Foundation. I *shall* dismiss it to the extent that I shall not attempt to account for the acceptance of Foundation, much less attempt to account for the claim, made by proponents of the iterative conception, that Foundation is central for the understanding of sets. I am relying on the fact that Foundation is much more peripheral to our practice than the other things that I have mentioned. After all, not a single theorem of analysis depends on Foundation. Moreover, the use of Foundation as a principle constitutive of what mathematicians mean by "set" was not introduced until well after the main ideas and methods of set theory were well in place. That includes the introduction of everything on my list. Admittedly, Foundation was in some sense "in the air" from as early as 1917 (when Mirimanoff restricted his discussion to "ordinary" well-founded sets). But Zermelo denied its general truth as late as 1930, and no one took it to be a general truth about all sets until Gödel did in the 1940s.

The disclaimer: Some will feel slighted by my bald claim that the items on the list are part of "contemporary mathematical practice." There is, after all, a tradition of constructive mathematics thriving today whose practitioners reject much of the list. I discuss that tradition below. Different constructivists understand the infinite in different ways, and some accept more Cantorian set theory than others. One therefore cannot give a unified account of the constructive understanding of the infinite. But giving such an account is not part of the purpose of this book. That purpose is explaining the nature of the infinite in the dominant tradition in mathematics today, the tradition which accepts Cantor–Zermelo set theory and makes frequent use of set-theoretic ideas. One might say with some justice that the title of the book should have been *Understanding the Transfinite,* not *Understanding the Infinite.*

Now back to the list. We have seen in some detail how constructions of the real numbers led to the use of infinite sets of rational numbers and how the need to classify the points in an arbitrary set of real numbers into isolated points, limit points, limits of limit points, and so on into the transfinite led to the transfinite ordinals—including ω. We have also seen (in §III.3) how the concepts that came out of all that led to a redefinition of the integral. Mathematics is committed to infinite sets.

Why must we take the infinite sets of contemporary mathematics to be completed actually infinite sets, not just *potentially infinite* ones—that is, incomplete objects that are unboundedly increasable? One reason is that the sets obey Extensionality: if sets were potentially infinite, and hence did not already have all of their members, then even sets that have the same members so far might come to differ, and there might be no present fact of the matter concerning whether they (will) differ. The temporal language used here, or a modal replacement for it, is essential when one speaks of potential infinities. One might even argue that the infinite sets of contemporary mathematics are actually, not merely potentially, infinite just on the basis of the independence of their characterization from temporal or modal considerations [You91, p. 129]. Sets either differ or not, determinately, and there is no temporal or modal component to the matter.

Another reason that we must take the infinite sets to be actually infinite—the most important reason—is that we can define sets impredicatively: one can, for example, define a real number as the least upper bound of a bounded set of real numbers. But the least upper bound of a set of real numbers may very well be a member of the set. Thus, the usual definition of the least upper bound of a set would be inadequate if the set were potentially infinite: as Zermelo observed (§V.1), an impredicative definition of a member of a set makes sense only if the definition is picking out that member from a determinate collection of previously existing objects, that is from a (finite or) actually infinite set. The example of the least upper bound is taken from analysis to help emphasize the close ties between our set-theoretic principles and analysis. Nonetheless, the principle that infinite sets are actually infinite must apply to sets generally, not just sets of real numbers, if we are to allow free use of unrestricted quantification: in a predicative theory quantification must always be restricted to the entities that have been defined so far. Moreover, the set-theoretic proof that every bounded set of real numbers has a least upper bound is an example of an application of Separation, a general impredicative principle that applies to all sets, not just the real numbers. Its truth requires that all infinite sets be actually infinite. I should now discuss how deeply Separation is embedded in mathematical practice, but it strikes me as clear that mathematicians are specifying subcollections of domains of interest in impredicative ways all the time, and then taking the resulting subcollections to be perfectly good collections—the example just given of least upper bounds will do. Besides, Separation is a consequence of Replacement, which is discussed two paragraphs below.

As we have seen, it came as a rude shock to Cantor that sets have power sets. So why do we suppose it? Cantor's version of Power Set was that if S is a set, then so is the collection of all functions from S to a set with two members. That is a special case of the general principle that if S and T are sets, then so is the collection of all functions from S to T, and the special case (given other set-theoretic assumptions) entails the general principle. The principle is used throughout mathematics. Here are some examples: When we define the real numbers in terms of certain functions from the natural numbers to the rational numbers (that is, in terms of Cauchy sequences) we only need each such function, not the collection of them. But when we define a Hilbert space with domain the set of square-integrable functions from the real numbers to the real numbers, we take the real numbers to form a collection suitable to act as the domain of the square-integrable functions. When we consider the group of automorphisms on an algebraic structure *as* a group, the domain of the group is the collection of all such automorphisms, which are functions from the algebraic structure to itself. In either case we select a set of functions of interest from the set of all functions from the domain of interest to the range of interest. Power Set is just the codification of the fact that the collection of functions from a mathematical collection to a mathematical collection is itself a mathematical collection that can serve as a domain of mathematical study.

Choice and Replacement are both self-evident principles about sets, and that self-evidence is an important part of our grounds for accepting them. It is often claimed that those axioms have a contentious history and that that history puts them in doubt, showing that they are not self-evident. But the claim is false, as we saw in Chapter V. No one has ever had any serious doubts about whether Choice and Replacement are true of combinatorial collections. What doubts there were took the quite different forms of doubts concerning whether we should employ a theory of combinatorial collections in mathematics, whether the notion of a combinatorial collection is coherent,[2] whether some other notion (for example, that of a logical collection) wouldn't be preferable for mathematics, whether we could live with a weaker theory than that of everything true about combinatorial collections, whether Choice and Replacement hold for something other than sets (in particular, logical collections), and so forth. Today, combinatorial collections play a central role in mathematics, and doubts about them have been dismissed long since. It is a clear and striking fact that crucial assumptions about the combinatorial infinite—Choice and Replacement—enter as self-evident truths suitable to take as axioms in a sense that makes them far more than arbitrary or convenient assumptions. We seem to have nontrivial intuitions concerning the

---

2. Even if the notion of a combinatorial collection were vacuous, that would not put Choice and Replacement in doubt—it would make them vacuously true!

infinite going far beyond simple things like Extensionality, Pairing, or even Power Set.

The final item on the list was a notion of finiteness that is prior to first-order axiomatic set theory. Proofs have always played an important part in reasoning within mathematics. They have only become more important with the advent of modern rigor. As a result, reasoning about proofs is an important part of reasoning about mathematics. To reason mathematically about proofs, one must be able to give a mathematically adequate definition of proof. Modern logic has provided us with such a definition, but the notion of finiteness plays a crucial part: a proof is a *finite* tree of sentences, each of which is an axiom or follows from sentences before it in the tree according to a rule.[3] As we saw in §V.3, the notion of finite that is defined within a first-order theory of sets is not adequate for present purposes: it is relative to the choice of model, and not only do new "sentences" get added to the genuine[4] ones in models with a nonstandard notion of finiteness, but, even worse, new "proofs" of formerly unprovable genuine sentences may appear, and so more genuine sentences may be "provable" in a nonstandard model than in a standard one in which all the "proofs" are genuine proofs. The very fact that we are committed to the distinction between genuine sentences and proofs and nonstandard ones shows that we are committed to a notion of finiteness that goes beyond the internal one of first-order axiomatic set theory—and we must be committed to that distinction if we are to be able to give an adequate (meta)mathematical characterization of what we are doing when we give a proof.

When, in defining proof, we say "finite," we mean really finite, not just finite in a formal sense that allows calling any collection of sets the finite ones

---

3. I am not ruling out the interest of infinitary logics or languages in a dogmatic way. I am simply arguing that we are committed to a notion of finiteness that goes beyond the one definable in first-order set theory because such a notion of finiteness is required to understand our practice concerning proofs. Most of the time infinitary logics are introduced within a background set theory that has an underlying axiomatic basis that depends on a notion of proof of the sort discussed in the text, and so such a notion is required for metamathematical purposes. But even if we should avoid ordinary logic by granting a direct commitment to any reasonable infinitary logic, the requisite notion of finiteness will be needed in the metatheory of that logic, since the logic will allow only n-placed relation and function symbols for *finite n*, and the like. Moreover, finiteness is outright definable in a sufficiently strong logic. Infinitary logics and languages, like ordinary first-order logic, need a strong notion of finiteness. As a result, they do not require any more consideration for present argumentative purposes.

4. By a "genuine sentence" I mean one that really is a sentence.

so long as the collection meets certain axiomatic conditions. If that seems no more informative than just proclaiming loudly that I mean *REALLY FINITE*, that is absolutely correct. The point here is not that we have some clear means of establishing the notion.[5] It is rather that we do presuppose such a notion in our theorizing about mathematics, at least insofar as mathematics has an axiomatic component, like it or not.

We are committed to quite a lot about set theory, and each of the commitments taken singly seems to have a lot of good reason or intuitive force behind it. At the end of Chapter V, I emphasized that we are nearly as confused about the foundations of set theory as pre-Weierstrassian analysts were about the foundations of analysis. Let me balance that here by noting that just as those analysts knew quite a lot about analysis despite the foundational problems, we know quite a lot about set theory and about the infinite.

§2. What Can We Know?

The theme of this section is that We Are Finite: while our knowledge of the number 3 is intimately associated with experiences of some kind or other of triples of one or another sort, surely nothing like that is the case for our knowledge of the number ω or, for that matter, the number $10^{10^{10^{10}}}$.

To avoid misunderstanding, let me emphasize that the issue being addressed is not "How can we have knowledge of mathematical objects despite their abstractness?" but rather "How can we have knowledge of infinite mathematical objects despite their huge size and consequent remoteness from experience?" Here is what I mean by remoteness: Not only do we not have experience of infinite mathematical objects, but we do not have any experience of anything suitably like them. The number 2 bears some relation to pairs of objects. The points and lines of Euclidean geometry bear some relation to pencil points and lines. But there does not seem to be anything suitable to play any analogous role for infinite mathematical objects.

Of course infinite mathematical objects are abstract. The issue of remoteness is an addition to that of abstractness. Both are apparent epistemological difficulties caused by the distance of mathematical objects from experience. But if we divide the problem of the abstract into two parts, the problem of the finite and the problem of the infinite, it becomes clear that the two have quite distinct features.

---

5. But see §VII.4.

The problem of the abstract, in Paul Benacerraf's words, is this [Ben73, p. 409]: "the concept of mathematical truth, as explicated, must fit into an over-all account of knowledge in a way that makes it intelligible how we have the mathematical knowledge that we have. An acceptable semantics for mathematics must fit an acceptable epistemology." For Benacerraf, an acceptable semantics is a Platonist one, and so his problem is that of the difficulties involved in giving an acceptable account of knowledge of abstract objects. Benacerraf concentrated on a weaker version of the problem: not on accounting for the mathematical knowledge that we have but on accounting for how we can have any mathematical knowledge whatever. "The minimal requirement, then, is that a satisfactory account of mathematical truth must be consistent with the possibility that some such truths be knowable" [Ben73, p. 409]. Philosophers have tended to concentrate on the minimal requirement and to offer up solutions to the problem of the finite to solve it.

Even those skeptical about the existence of any abstract mathematical objects at all want to endorse the truism '$2 + 2 = 4$' in some way or other that acknowledges that it is better than '$2 + 2 = 5$,' even though they may not grant that it is true. Even without an abstract number 2, one is still faced with explaining the general fact that the members of two nonoverlapping pairs form a quadruple.

Whatever one's views about mathematical objects, it is necessary to make sense of our counting, computing, and bookkeeping activities. Skepticism about small finite mathematical objects—in particular small natural numbers—is just not doubt about the acceptability in some form or other of many of the putative facts about them. Moreover, there are many stories one could tell about a source for genuine knowledge concerning some finite mathematical objects, including, as the most trivial special case, knowledge of their existence. Take, for example, small natural numbers. Various explanations of our knowledge about them might invoke the experience of time, the experience of bunches of physical objects or of patterns exhibited by them, or the sequencing of words in sentences. Other explanations might rely on the exigencies of the construction of theories of the physical world. We are faced with too many ways of accounting for our knowledge of small finite mathematical objects, not too few. For example, Parsons [Par80] showed how to account for such knowledge on the basis of our linguistic capacities, while Maddy [Mad90] showed how to do it on the basis of experiences with medium-sized physical objects.

Benacerraf's minimal requirement can surely be met with an acceptable solution to the problem of the finite—though I make no claim to know what the actual solution is. That is a question whose answer involves detailed psychological information about how people typically actually acquire knowledge of small finite mathematical objects, and we do not yet have sufficiently detailed information to answer it. Acquisition of the number concept does, however, involve both linguistic components—learning to count aloud—and experience of medium-sized physical objects—counting them using the spoken number sequence. It is therefore likely that the actual solution involves components of both the one proposed by Parsons and the one proposed by Maddy—and probably other components as well.

In sharp contrast to the situation about '2 + 2 = 4,' many of those who are skeptical about the existence of infinite combinatorial collections would want to doubt or deny the Axiom of Choice—not only its truth, but its acceptability in any form whatever. General facts about the infinite are not robust in the same way that the facts of counting, computing, and bookkeeping are. Moreover, it is not at all clear what we can fall back on as a source of mathematical knowledge concerning the infinite—what can play the role that bunches and sequences of moments, objects, or words seem so well suited to play for small finite mathematical objects. It is that lack that raises the problem posed by the remoteness of the infinite: it seems that we cannot have grounds to know what we find we actually do know about the infinite.

In Chapter VIII I shall show that as a matter of fact the combinatorial infinite is not remote—it has pretty much the same kinds of ties to experience as do small natural numbers. (That may be a bit misleading—see Chapter VIII for a more careful formulation.) That solves the problem of the remoteness of the infinite—philosophical problems concerning infinite mathematical objects become just like the familiar ones concerning finite mathematical objects. That is important because the problems concerning finite mathematical objects are not skeptical ones—the genuine doubts about the acceptability of our theory of the infinite are refuted. It also provides the essential missing ingredient for an explanation of the grounds on which mathematicians are entitled to make claims of self-evidence. But before presenting the solution, it is necessary to become clearer on the nature of the problem.

The two Benacerrafian problems—of the finite and the infinite—are both important, and every adequate philosophy of mathematics must be compatible with solutions to them. Nonetheless, the problem of the infinite deserves special emphasis because it is in danger of being lost as the result of the huge amount of attention being devoted to the Benacerrafian problem of the abstract in its simplified guise as the problem of the finite. That loss would be most unfortunate. The problem concerning the infinite was a primary concern of the philosophy of mathematics for many years—as may be seen in the work of Brouwer and that of several philosophers discussed by Benacerraf, such as Hilbert, Gödel, and Quine.

One might think that a solution of the problem about small mathematical objects is a prerequisite for the solution of the problem about infinite ones or that a common solution is required. But outlined solutions of the first problem have not generally helped with the second. In addition, proposals concerning the problem of the infinite have frequently turned out to be pretty much independent of the details of any assumed solution to the problem of the finite, as we shall see in the discussion of various proposals throughout the rest of this chapter.

In the rest of this section, I shall try to bring out a common—though by no means universal—strand of epistemological concern within the philosophy of mathematics about how mathematical knowledge is connected to human experience, and so my initial formulation will intentionally be left fairly broad. I then go on to discuss various forms of finitism, constructivism, and intuitionism—philosophies of mathematics that often make epistemological concerns central.

The knowledge that $3 \times 3 > 3 + 3$, for some epistemologists, may be knowledge about the number 3 itself in some *de re* sense that requires that the number exist, while for others it might be *de dicto* declarative knowledge that something is true without any requirement that that truth be about anything in particular. Indeed the supposed knowledge concerning a number might, on some accounts, turn out to be misdescribed knowledge of or about the objects of the experiences associated with the number. An association between a number and experiences of some kind may be held to be necessary for knowledge or held to be a mere psychological or social accident, and the association may go in either, or both, directions: triples might lead to our knowledge of 3 or knowledge of 3 might be a precondition of our experience of triples. There are also intermediate possibilities. "Experience" may turn out to be direct presentation to the conscious thinking subject or it may turn out to be sentences that count as observation sentences according to our best theory of the world. It may be, for instance, experience of moments, physical objects, grammatical objects, sets—triples—of one or another sort of item, or possibilities of performing various operations. With those *caveats* firmly in mind, let me discuss our experiences associated with numbers.

The number 100 may be associated with simple experiences of centuples of objects—say in 10-by-10 array—but the number 43 is more likely to be associated with collections enumerated by a process of counting or with that process of counting itself. Most of us have counted up to a few hundred, and we know—in some sense spelled out differently by different epistemologists—what it would be like to count a good bit higher. In some sense—though there is no agreement about the details—each number is canonically associated with the experience of counting up to it. The artist Jonathan Borossky actually counted to 3,265,772, at which point he stopped. He documented the process carefully. Roman Opalka began painting consecutive "details" of "1–∞" in 1965. At the time I am writing, the last detail he has completed is a canvas with the numbers from 4,776,969 to 4,795,472, but he has no intention of stopping. He speaks each number aloud into a tape recorder, in Polish, as he paints it. The numbers are so minute that each canvas gives one a sense of texture not of figures, until it is investigated closely. With collective effort, much larger numbers are possible: McDonald's began serving hamburgers on 15 April 1955, and served its 90,000,000,000th burger on 7 July 1992.

Some distinctions: A *phenomenological strict finitist* is someone who is skeptical about the coherence of the idea of experiencing particular, sufficiently large numbers. Whether $10^{10^{10^{10}}}$ is large enough to be subject to such skepticism depends on the individual phenomenological strict finitist, but let us suppose that I have chosen a sufficiently large example.

A *phenomenological finitist* is someone who, like the phenomenological strict finitist, is skeptical about the coherence of the idea of experiencing sufficiently large numbers. What counts as sufficiently large to be subject to skeptical doubts may be different for different phenomenological finitists. The phenomenological finitist need not, however, endorse a fixed maximum. A phenomenological finitist might argue against a phenomenological strict finitist as follows: the particular bound $10^{10^{10^{10}}}$ depends on considerations motivated by extramathematical—typically physical—theories. It therefore should not be built into our theorizing about numbers.[6] The question whether

---

6. If the bound were motivated by mathematical considerations, the situation would be quite different. It is not incoherent to allow for the possibility that mathematical considerations could bound the possibilities of numerical experience—a proof of the inconsistency of Peano arithmetic might conceivably produce such a bound. That remote possibility would cause phenomenological finitism, and indeed any other position, to collapse into phenomenological strict finitism.

we have a coherent conception of experience of some number or other should not be hostage to physical theory. The phenomenological finitist endorses that there is an upper bound to the possibilities of numerical experience but can coherently refrain from imposing any particular bound. The *phenomenological liberal finitist* endorses that there is a bound but opposes any attempt to specify it on the basis of extramathematical considerations. Since there seem to be no purely mathematical considerations that issue in a bound to the possibilities of numerical experience, the liberal finitist in fact allows that any given number can be coherently held to be associated with an idea of what it is to count up to it. Phenomenological liberal finitism does not involve any commitment to the possibility of infinitely many numbers or experiences thereof. It does involve not rejecting any actually presented possibility—like $10^{10^{10^{10}}}$—on nonmathematical grounds.

Those who are not phenomenological finitists have—in virtue of the fact that they are not phenomenological finitists—agreed that we have an idea of what it would be like to just keep on counting without stopping.[7] It is hard to see how one might accept that we understand what it would be like to count up to any finite number without allowing the extension involved in denying that there is a bound. Nevertheless, some further idealization is obviously involved, since we all know that no one could actually count without stopping—one would eventually stop because of boredom, sickness, or death. There is serious dispute about what follows from the fact—if it is a fact—that we have the idea of counting endlessly.

If the idealization we made that eliminated the restriction to a finite bound also eliminated the progression of time altogether, then the idea of continuing to count yields an idealized possibility of experience that includes experiences of every number within it—since time has been eliminated from the picture, all of the numbers must be seen as copresent. Moving beyond phenomenological finitism by eliminating the progression of time altogether

---

7. The idea of going on without stopping is not just in the province of recondite philosophy of mathematics. Consider the following children's song [Mar91]:

> This is the song that doesn't end
> Yes, it goes on and on, my friends
> Some people started singing it
> Not knowing what it was
> And they'll continue singing it forever
> Just because [*now repeat* . . .]

leads to accepting an idealized possibility of experience associated with an *actual* denumerable infinite in the sense of classical mathematics.

If we idealize in a different way, keeping the temporal component of the picture intact, then the idea of counting without stopping yields only the potentially infinite—at any particular time there are only finitely many experiences of numbers, but there is no upper bound to the number of them.

The limited experience of the potentially infinite I have just described is the basis of the potentially infinite progression of numbers of intuitionistic mathematics. It is pretty much what Brouwer, the founder of intuitionism, had in mind. According to a modern intuitionist, Michael Dummett [Dum77, p. 56], "it is quite literally true that we can arrive at the notion of infinity in no other way than by considering a process of generation or construction which will never be completed." He continued, "It is, however, integral to classical mathematics to treat infinite structures as if they could be completed and then surveyed in their totality," and he viewed that as a kind of *reductio* of classical mathematics [Dum77, p. 57]:

> the platonist destroys the whole essence of infinity, which lies in the conception of a structure which is always in growth, precisely because the process of construction is never completed. The platonistic conception of an infinite structure as something which may be regarded both extensionally, that is, as the outcome of a process, and as a whole, that is, as if the process were completed, thus rests on a straightforward contradiction: an infinite process is spoken of as if it were merely a particularly long finite one.

Dummett is most certainly right that we do not arrive at the notion of infinity by completing an endless process of generation. But Dummett's criticism of classical mathematics imports a temporal component that is not really present in classical mathematics. As Dummett himself admitted [Dum77, p. 336]: "for [a platonist], no mathematical statement can involve any temporal reference, explicitly or implicitly." In contrast [Dum77, p. 336], "for [an intuitionist] a mathematical statement is rendered true or false by a proof or disproof, that is, by a construction, and constructions are effected in time." One might summarize by saying that the idealization of experience that yields the actual infinite of classical mathematics is that of the Eternal Mathematician, while the one that yields the potential infinite of intuitionistic mathematics is that of the Immortal Mathematician.[8]

---

8. Compare [Kit83, p. 144], and see [VB89] for an analysis of various types of

So far, all I have discussed are the various reasons for admitting or deny-ing the possibility of having one or another sort of experience of mathemat-ical objects. If one makes the fairly plausible assumption that one does not have knowledge that there is a mathematical object with some property unless in principle one has experience associated with an appropriate object show-ing that it has the property, then the phenomenological positions above have epistemological consequences. I shall call the assumption the *epistemic con-straint*. There are associated epistemological positions corresponding to each of the phenomenological ones above: epistemological strict finitism, and so forth. Each of the epistemological positions carries with it all the correspond-ing phenomenological assumptions plus the epistemic constraint. One might also adopt the *ontological constraint* that whatever we in principle cannot ex-perience does not exist. That yields ontological strict finitism, and so forth.[9]

The debate for and against the epistemic constraint is a fascinating one, but it is at right angles to our purposes, since it affects finite mathematical objects as much as infinite ones. For that matter, it can be extended to apply to phys-ical objects, events, and processes as well. I shall therefore confine myself to the briefest of remarks about the constraint. Brouwer accepted the epis-temic constraint because he took mathematical objects and their properties to be nothing other than experiences of a certain sort—"mental constructions." The constraint is a trivial corollary of that ontological assumption. Dummett argued for the constraint on roughly the following grounds: To understand a sentence, we must know what it would be like, in principle, to discover that it is true. Therefore each sentence separately must be associated with appropri-ate canonical "verification conditions." But canonical verification conditions for sentences about the existence of mathematical objects with a certain prop-erty must be conditions in which there are experiences verifying that suitable objects have the property.

The epistemic constraint in and of itself tells us little or nothing about the question of knowledge of the infinite—it just makes the answer depend on what kind of experience is in principle possible. That is my reason for having done something that may have struck you as odd: beginning the discussion of

mathematics (finitism, constructivism, . . .) in terms of various associated idealized mathematicians.

9. It is possible to hold that there are at most $10^{10^{10^{10}}}$ mathematical objects for rea-sons independent of the ontological constraint. That would be a different sort of ontolog-ical strict finitism. There are parallel possibilities for other positions discussed in the text. We shall not be discussing them.

intuitionism without immediately introducing what is usually thought of as its most characteristic feature—the nonclassical logic it entails. As I show below, that logic flows out of the epistemic constraint, and it is required for the intuitionistic characterization of what is in principle possible. Nevertheless, one can endorse the logic without being an intuitionist, and, in the other direction, the intuitionistic employment of the logic can be seen as a consequence of the intuitionistic picture of the natural numbers, which satisfies the epistemic constraint, rather than as a separate, primary assumption of intuitionism.

If we grant the epistemic constraint, then knowledge of any mathematical object of a given type must be associated with an experience that guarantees the knowledge of the object. Since the objects just go along for the ride, with the experiences doing all the epistemological work, we may as well for present purposes do what many of those who accept the epistemic constraint do anyway: identify each object with an associated experience, thereby, in effect, adopting the ontological constraint.[10] With that identification, the experience actually brings the object into existence. It is a useful metaphor to think of the experiences as constructions that build the objects, and so I shall begin writing *construction* instead of *experience*. Mathematical objects are constructions.

Suppose we prove that it is contradictory to assume that every mathematical object of some type under consideration does not have a certain property ϕ, that is, suppose we prove $\neg(\forall x)\neg\phi(x)$. It does not follow that we know that some object of the right type actually has the property ϕ: obtaining such knowledge, according to the epistemic constraint, requires constructing an object that has the property ϕ, which need not even in principle be possible, let alone actually accomplished, especially if it is difficult to verify that an object has ϕ by means of a construction. Thus, it follows from the epistemic constraint alone that $\neg(\forall x)\neg\phi(x)$ does not entail $(\exists x)\phi(x)$. Thus, classical logic turns out to be incorrect for those who grant the epistemic constraint. We shall call the appropriate logic *constructive,* and those who use it *constructivists.*

There is one special circumstance in which $\neg(\forall x)\neg\phi(x)$ entails $(\exists x)\phi(x)$ and in which all of the formal principles of classical logic still apply for the

---

10. I shall not discuss how mathematical objects that are experiences could be intersubjective, whether they came into existence at a time, whether they cease to exist when the experience ceases, and so forth. There are no widely agreed upon satisfactory answers to such questions. The views being outlined here require such answers.

constructivist. If there is in principle a single construction that includes constructing all of the objects of the type under consideration and constructively deciding for each of them whether or not it has the property ϕ, then once we have shown $\neg(\forall x)\neg\phi(x)$, it will inevitably follow that $(\exists x)\phi(x)$: The construction described, once completed, will either yield an object of the type that has ϕ and hence show that $(\exists x)\phi(x)$ or it will show that every object of the type does not have ϕ, that is, that $(\forall x)\neg\phi(x)$. But the second possibility is not a genuine possibility if we have already shown that $\neg(\forall x)\neg\phi(x)$, and so the construction has yielded a proof of $(\exists x)\phi(x)$, as required. In particular, even given the epistemic constraint, classical logic will apply on a finite domain with effectively verifiable properties. It is a subtle question whether the resulting logic, formally identical to classical logic, really is classical logic, or whether it is different because the logical operators have been interpreted in a nonclassical way. That is a topic that need not concern us here. We shall not need further details of constructive logic, which is the correct logic given the epistemic constraint. See [Dum77], [Bee85], or [TD88].

Now I shall argue that an intuitionist must be a constructivist, that is, must endorse only constructive logic.[11] As I have described the position so far, an intuitionist is one who holds that the numbers are potentially infinite in the sense that there is no upper bound that limits the construction of numbers. In addition, the intuitionist rejects any commitments to any time after the present moment as, in Brouwer's term, "metaphysical." For the intuitionist, the only numbers there are the ones that have been constructed, and they are therefore the only ones that can be relevant to the truth of mathematical theorems. The intuitionist will reject talk of the possibility of constructing numbers in the future as speculation that has nothing to do with mathematics and will similarly have nothing to do with the possibility of future changes in truth values. If one were to allow use of some notion of possibility within mathematics, then one *could* hold with the intuitionist that the numbers are potentially infinite and still endorse classical logic.[12]

Consider the following example: A *perfect number* is a number n such that the sum of its divisors, including 1 but excluding n, is n. For example, $6 = 1 + 2 + 3$, and so 6 is a perfect number. It is not known whether there are any odd perfect numbers, but there are none less than 10^{50} [IK80,

---

11. The following characterization of intuitionism owes a lot to discussions with Vann McGee and Sarah Stebbins.

12. We shall be discussing mathematical possibility in §VII.3.

p. 927]. Since there are simple procedures for determining whether a number is odd and perfect, the intuitionist will grant that $(\forall x)(P(x) \vee \neg P(x))$, where '$P(x)$' stands for '$x$ is odd and perfect' and the variable ranges over the natural numbers. Thus, the intuitionist grants that there is a well-defined, computable function that takes each number to, say, 1 or 0 depending on whether or not that number is odd and perfect. The intuitionist need not, however, hold that $(\exists x)P(x)$ is determinately true or false, given our present state of knowledge: No number has been shown odd and perfect, and it has not been shown that the assumption that there is an odd perfect number leads to a contradiction.

Can one hold that $(\exists x)P(x)$ is true? The intuitionist cannot, since that would require the actual existence, that is the actual construction, of an odd perfect number, and no such construction has been performed. In contrast, the classical mathematician, who accepts that all the numbers actually exist, could reasonably suspect that $(\exists x)P(x)$ might be true, on the basis that some number that has not yet been constructed is odd and perfect. In a parallel way a suitable notion of mathematical possibility would make it reasonable to suspect that $(\exists x)P(x)$ is possibly true, on the basis that it is possible that some odd perfect number may be constructed.

On the other side, an intuitionist might coherently hold that $(\exists x)P(x)$ is untrue, for the reason that there is no (known) odd perfect number. As one might put it given suitable "metaphysical" commitments, none has as yet been constructed. But—still speaking outside the range of what the intuitionist allows—the claim that $(\exists x)P(x)$ is untrue held for that reason would be subject to change, should an odd perfect number be constructed, and it therefore cannot be a nontemporal mathematical truth. The intuitionist reacts by giving a nonstandard definition of negation:[13] to show in the intuitionistic sense that $\neg(\exists x)P(x)$ is true requires the construction of a proof that the assumption that an odd perfect number exists—that is, has been constructed—leads to a contradiction. Once again speaking outside the intuitionistic framework, in that way one can be sure on the basis of present information that none will be constructed.

Can one hold that $\neg(\exists x)P(x)$ is true? An acceptance of all the numbers as actually existing would make it reasonable to suspect that $\neg(\exists x)P(x)$ is true on the basis that no number, whether constructed or not, is odd and perfect.

---

13. Brouwer often used expressions of the form "ϕ is absurd" instead of "not ϕ."

Such a route is not available to the intuitionist, since it requires a commitment to unconstructed numbers. In a parallel way a suitable notion of mathematical possibility would make it reasonable to suspect that $\neg(\exists x)P(x)$ is necessarily true, on the basis that no odd perfect number could be constructed. Again, such a notion of possibility is rejected by the intuitionist.

It follows from the above considerations that the assumption that $(\exists x)$ $P(x) \vee \neg(\exists x)P(x)$ is true—an instance of the law of the excluded middle— is incompatible with intuitionism.[14] Since constructive logic does not entail the law of the excluded middle, that will cause the intuitionist who is also a constructivist no trouble. Moreover, an intuitionist cannot endorse the law of the excluded middle and hence classical logic.[15] That leads the intuitionist to a commitment to constructive logic: Without the law of the excluded middle, one cannot give a compositional semantics—a theory of how we understand mathematical language in which the way in which we understand complex expressions is determined by the way in which we understand their parts—in terms of truth conditions. The only conceivable alternative is to give a compositional semantics in terms of verification conditions [Dum75, p. 110], and in

---

14. For simplicity's sake, I have slid by a possibility related to a point emphasized in §1: One might reject existential quantification over the numbers. Then the formula $(\exists x)P(x)$, which played the central role in the argument in the text, is not available. There is a natural way of rejecting existential quantification over the natural numbers once one rejects that they form a completed totality, namely, on the grounds that such quantification is impredicative. If one rejects properties of numbers that are defined in terms of quantification over numbers that have not yet been constructed, one is left with a formalism in which only bounded quantifiers over numbers are useful. (Bounded quantifiers are quantifiers of the forms $(\exists x \leq n)$ and $(\forall x \leq n)$, read "there is an x less than or equal to n" and "for all x less than or equal to n.") In particular, the induction principle will apply only for sentences in which all quantifiers are bounded. That is a limitation both intuitionists and classical mathematicians reject, and so I passed over it in the text. Nonetheless, it is, so far as present considerations show, perfectly coherent. The position is eloquently advocated by Edward Nelson [Nel86], who has shown that a surprising amount of mathematics, including analysis, can be reconstructed along such lines. Since our main task is understanding impredicative set theory, the predicative position outlined in this note is, for us, a side issue.

15. My argument owes a lot to Dummett, who said [Dum75, p. 128], "If someone holds that the only acceptable sense in which a mathematical statement, even one that is effectively decidable, can be said to be true is that in which this means that we presently possess an actual proof or demonstration of it, then a classical interpretation of unbounded quantification over the natural numbers is simply unavailable."

the context of mathematics, verification must here mean constructive proof. Thus, the intuitionist who can provide a nontrivial compositional semantics will also be a constructivist.[16]

I have not shown that it would lead to contradictions for the intuitionist to add on the law of the excluded middle (though Brouwer did argue to that effect). It is in some formal sense possible to add the law to a constructive logic. But on the intuitionistic interpretation mathematical statements often fail to have truth values, and I know of no interpretation of the law of the excluded middle, classical or intuitionistic, on which it is true in the presence of a pervasive failure of statements to have truth values.

It is fairly clear that a generalized version of the above argument shows that to endorse unmodalized unbounded quantification over any mathematical totality while holding that it is potential, not actual, one must be a constructivist. I shall not attempt to formulate that in any precise way, since it is not clear to me what "potential" and "actual" should mean in the more general setting. Nonetheless, the argument illustrates that constructive logic has an ontological motivation as well as an epistemological one: it makes it possible to endorse a potential totality without endorsing the corresponding actual one or the possibility of each of its members.

The two motivations for constructive logic—epistemic and ontological— are independent: One can endorse the epistemic constraint and hence constructivism while accepting the existence of an actual totality for the range of the quantifiers. One such position concerning the natural numbers grants that they form an actual totality and grants that every statement about them is true or false, but denies that we know that either there is an odd perfect number or else there isn't, on the grounds that it is perfectly conceivable that there is no humanly possible way to settle the question, that is, that there is no proof that settles it. According to such a position, any claim that there is a truth of the matter independent of proof must be based on things that go so far beyond humanly conceivable processes or capacities as to be meaningless, incoherent, in principle incomprehensible, or the like. (Compare [Dum75, p. 127].) On

---

16. The position we have now arrived at—intuitionism plus constructivism plus the epistemic constraint in my terminology—is what is usually simply called intuitionism. In *Elements of Intuitionism* Dummett [Dum77] provides a nice detailed introduction to the subject and cites the classic texts. Charles McCarty has written a nice brief chapter giving a quick overview of much of what is philosophically interesting about intuitionism [McC83].

the other side, one can accept that some totality or other is potential but not actual and therefore be a constructivist, independently of any epistemological concerns at all, and hence independently of the epistemic constraint.

I have presented a series of positions that are linearly ordered with respect to their ontological commitments: ontological strict finitism, ontological liberal finitism, potential infinity without actual infinity, and classical mathematics. Despite the fact that it fits into the linear order, the commitment to potential infinity without actual infinity cannot be compared with the others in a straightforward manner. The reason is that to endorse potential infinity without actual infinity requires either intuitionism, which is a constructivist position, or a strong notion of mathematical possibility, while the other members of the list, at least the more or less familiar versions of interest here, employ classical nonmodal logic.

The (ontological) liberal finitist claims that there is an upper bound to the numbers while denying that there is any way to specify the bound. That cannot be acceptable to an intuitionist. After all, an intuitionist will not accept any claim that a number with a certain property, like the upper bound, exists without constructing a specific such number. But that construction is precisely what the liberal finitist rejects.[17]

On the other side, the classical mathematician cannot take over the mathematics the intuitionist allows, precisely because that mathematics relies on the notion of a progression that is not fully determined. Both intuitionists and classical mathematicians can agree on Cantor's definition of the real numbers as Cauchy sequences—convergent sequences of rational numbers. But the intuitionist can prove on the basis of that definition that every function from the real numbers to the real numbers is uniformly continuous (a strong form of continuity) [Dum77, p. 119], while the classical mathematician shows— on the same basis—that there are functions from the real numbers to the real numbers that are not continuous, let alone uniformly continuous, at even a single point. (Dirichlet's function that is 0 on the rationals and 1 on the irrationals is an example of such a function.) Classical mathematics, far from extending intuitionistic mathematics, contradicts it if the two are taken to have a common subject matter.[18] Intuitionistic and classical mathematics are

---

17. I am indebted to Sarah Stebbins for that point.

18. There are constructive mathematicians who accept only a part of intuitionistic mathematics. In particular, there are those who reject arbitrary sequences of rational numbers, accepting only those sequences given by a law that fixes all values "in advance." Such

incomparable—neither is included within the other. So as not to mislead, let me point out immediately that the fact that intuitionistic and classical mathematics are, strictly speaking, incomparable does not mean that they cannot be compared in a more general sense. Much work has been done on giving intuitionistic reconstructions of parts of classical mathematics and *vice versa*. That is discussed in all of the works on intuitionism that have been cited.

My outline of constructive logic was neutral about a crucial point—what counts as a legitimate construction. That is because constructive logic is independent of the details of what counts as a construction. To be an intuitionist, one must be a constructivist, but the converse is far from true: a constructivist who adopts the intuitionistic idealization of experience will be an intuitionist, but there are many other ways to idealize experience, not all of which involve time or modality. The primary mathematical consequence of an idealization is the axioms it allows one to endorse. The axioms, in part, may be taken to specify what counts as a construction. One can, for example, be a constructive ontological liberal finitist. The constructive ontological liberal finitist goes beyond the intuitionist in one key respect—allowing that the finite bound on mathematical objects has been constructed. That narrows the domain to things below the bound. I shall discuss the position in Chapter VIII. One can also be a constructivist while allowing the actual infinite. Let me give two examples of that.

The intuitionist endorses the actual finite, but only the potentially infinite, and as a result allows classical logic only for suitably definite claims about suitably definite finite systems. The constructivist of my first example endorses the actual denumerable, but only the potentially nondenumerable. It is not hard to provide an idealization of experience that yields a notion of construction along appropriate lines: one allows as constructions not only processes that have finitely many steps but also ones that have as many steps as there are natural numbers. One might tell the following story, which connects the relevant sort of potentiality with time: The idealized mathematician not only has no upper bound on the number of performable steps, but also no bound on the speed with which those steps are performed. It is therefore possible to count 0 in one second, 1 in one-half second, 2 in one-quarter second, and so forth. Up to that point, the intuitionist may well agree, but the constructivist of my example now concludes that it is possible to have counted

---

constructivists reject the intuitionistic theorems that conflict with classical mathematics. The topic is too technical to pursue further here. See [Bee85] and [BB85].

every natural number before two seconds have elapsed. The constructivist assumes that the decision by the idealized mathematician to carry out such a
construction, having proved that it takes a bounded amount of time, counts
as a construction of the moment after the construction.[19] Thus, the idealized
mathematician will be able—in principle—to complete the process of counting off any denumerable ordinal number but no nondenumerable ones. That
is a result of Cantor's theorem (constructively interpreted) that a linearly ordered set (and hence a well-ordered one) can be embedded in the rationals
with an upper bound if and only if it is countable. Naturally, "denumerable"—
like everything else—must be taken in a constructive sense: one has constructed a one-to-one correspondence with the set of all natural numbers.

Given the idealization under discussion, classical logic will apply for suitably definite claims about suitably definite *denumerable* systems, and classical elementary number theory will be an example of such constructively justified classical reasoning about a denumerable domain. The theory of the real
numbers will, however, require constructive reasoning. The position outlined
is advocated in [Vel93], though it is argued for in a different way. The position is in some respects like that of Weyl's *Das Kontinuum* [Wey18], which
endorsed full classical arithmetic but put predicative (though not constructive)
restrictions on the theory of the real numbers.

The constructivist of my second example endorses the actual transfinite
but, in Cantorian terminology, only the potentially absolutely infinite. Time
is abstracted from altogether. Even though the collection of natural numbers
is not, on this picture, constructed after each natural number—since "after"
is disallowed—it is still true that the construction of the collection of natural
numbers requires the availability of each natural number. A set is not possible
unless all of its members are actual. As Parsons, who has supported a modal
position related to the constructive one I am now sketching, put it,[20]

---

19. It is the need for the assumption made at this point in the text that prevents the
intuitionist's position from collapsing into the one being described. It is crucial to either of
the two positions that the time series is itself under construction, not given in advance. If
the time series were given in advance, that would already constitute a justification of the
actual infinite within the strictures of the epistemic constraint. Both pictures require that
a construction be a well-ordered sequence of operations. The intuitionist can coherently
reject the sort of "construction of a moment" here considered by imposing the structural
requirement that a construction, and hence every initial segment of a construction, must
have a last step. Compare [Geo91].

20. For Parsons's modal set theory, see [Par83c]. Various constructivist set theories

The idea that any available objects can be formed into a set is, I believe, correct, provided that it is expressed abstractly enough, so that 'availability' has neither the force of existence at a particular *time* nor of givenness to the human mind, and formation is not thought of as an action or Husserlian *Akt*. What we need to do is to replace the language of time and activity by the more bloodless language of potentiality and actuality.   [Par77, p. 526]

If we restrict consideration to well-founded sets—which is, so far as I can see, unavoidable for present purposes—the idea is simply described thus: the availability of all the members of any V_α requires the actuality of all the members of all V_β, for $\beta < \alpha$. (See §V.4 for the definition of V_α.) Constructions do not take place within time, but, in Parsons's phrase, they must respect the intrinsic ordering of relative possibility. We must take the operations of forming unions, power sets, the ranges of functions on sets, and choice sets to be legitimate constructions. (Portions of that may or may not follow from the idealization we have made—the details are related to the justification of the axioms of set theory on the basis of the iterative conception. That was discussed in §V.5. In any event, one can just specify what counts as a construction axiomatically, in the present case by adopting the axioms of ZFC, constructively construed.)[21] In that way, it becomes possible to construct every set, but no proper classes.

Set theory is not being motivated here; it is pretty much just postulated. What *is* being motivated is a certain way of drawing the distinction between sets and classes, with (infinite) sets as actual infinities while classes are potential infinities. Classical logic will apply for suitably definite claims about suitably definite sets, and that will include, for example, full classical analysis right up through Borel determinacy. I refrain from giving further or more precise details here. This is, after all, only intended to be an example.

The position may be attractive to those who would like to view set theory as the theory of all completed collections, since it gives a way of allowing

---

have been discussed in the literature with motivations not unrelated to those described here, including especially [Tai90]. See [Par83c, p. 326] for references to others, and see also [Poz71] and [Lea77].

21. It is not so clear as I may have made it seem what "the axioms of ZFC, constructively construed," should be. There are many classically equivalent but constructively distinct axiomatizations of set theory. For relevant information see Chapters 8, 9, and 12 of [Bee85] and references therein.

all sets and explaining quantification over them while denying in a principled way the coherence of the idea of any completed proper class like, for example, the class of all sets or even the class of all possible sets. The reason for adopting such a position would be ontological, not epistemological, and the position therefore provides an example of the possibility mentioned earlier of a constructivism that is independent of epistemological concerns. The analogy between the position and intuitionism is this: the intuitionist adopts constructive logic to allow the potential infinite without admitting the actual infinite; the advocate of the present position adopts constructive logic to allow the transfinite without admitting the actual absolutely infinite.

As should be clear from the analysis of the combinatorial conception, limitation of size, limitation of comprehensiveness, and the iterative conception in Chapter V, any attempt to provide an explanation of classical set theory compatible with the epistemic constraint is doomed to failure, since the idealized notion of construction involved would have to be so far removed from anything humanly possible that it cannot amount to more than postulation—rhetoric surrounding the simple assumption of set-theoretic principles. The view that set theory concerns idealized possibilities of construction may have its uses,[22] but it just does not provide an adequate basis for understanding the conceptual sources of our set-theoretic principles. The infinite entities of set theory are too remote for such an approach. We are up against the problem of the infinite.

One possible reaction to the apparent problem is to conclude that we have made a mistake. We do not know what we once thought we did. Brouwer reached that conclusion, and he developed intuitionism to correct the mistake:

> From the intuitionist standpoint the dogma of the universal validity of the principle of the excluded third in mathematics can only be considered as a phenomenon of history of civilization, of the same order as the former belief in the rationality of π or in the rotation of the firmament about the earth.   [Bro52, pp. 510–511]

---

22. Kitcher, for example, took it to characterize the epistemic source of our commitment to set theory [Kit83, Chapter 6], and I did something that is somewhat related in the previous section. I am not arguing that such uses are wrong, but only that the problem of explaining how we come to know what the relevant possibilities of construction are amounts to that of explaining the conceptual sources of our set-theoretic principles. The introduction of the notion of possibilities of construction does not solve the particular problem we are facing here.

History has not gone Brouwer's way, at least not yet. There are good reasons: Much of classical mathematics cannot be reconstructed along intuitionistic lines without a great increase in the difficulty both of stating the theorems and of giving the proofs, while some of classical mathematics apparently cannot be reconstructed at all, including most of topology and set theory. None of that poses any logical problem for the intuitionist, who can coherently reject what cannot be reconstructed. But it does represent a huge pragmatic problem: very few have been willing to give up so much mathematics. In addition, much of analysis evolved along with intended physical applications, and the intended applications seem to license the law of the excluded middle: In Fourier's theory of heat, the canonical example of a function from the real numbers to the real numbers is the temperature of a metal bar as a function of the distance from one end. In the intended physical setting, temperature is defined at each point and the bar is continuous, though its temperature need not be. The bar is found in nature, not constructed by us, and so there need be, so far as one can tell, no rule giving the values of the function at each point on the bar. Nonetheless, given any two real numbers a and b, either the temperature a units from the end of the bar is equal to b, or else it isn't. That claim is motivated by realism about the bar, not by mathematical or logical considerations. Thus, applications of mathematics motivate the assumption of many instances of the law of the excluded middle even if it isn't a logical truth. That lends considerable support to classical mathematics, as opposed to intuitionism. One can argue against the idealization of temperature in a metal bar that is used here on *physical* grounds, but considerations like those of the intuitionist, which are not connected with the physics, should not be allowed to block it. The intuitionist might argue that such mathematics is not knowledge but cannot argue that it is not worth knowing. Whatever the ultimate verdict on those issues, our task in this book is to make what sense we can of the prevailing set-theoretic theory of the infinite, and intuitionism can be of no use, since the prevailing set theory cannot be reconstructed intuitionistically.[23]

---

23. A parallel remark applies to Weyl's proposed arithmetical predicative foundation for mathematics. It should go without saying that intuitionistic and predicative mathematics are of interest both for their own sakes and because intuitionistic and predicative reconstructions of parts of set-theoretic mathematics help to clarify what commitments are indispensable for using various parts of mathematics, and may help to give sharper formulations of the set-theoretical principles that are in further need of clarification.

If we do not follow Brouwer, the other natural reaction to the problem of the infinite is to conclude that we do indeed know what we think we know, but that we know it in some way that violates the epistemic constraint.

§3. Getting from Here to There

The theme of this section is Introducing the Infinite: we consider various attempts to show how we obtain knowledge of the properties of infinite mathematical objects in a way that is based on our finite experience. These are attempts to meet the problem of the infinite head on.

There is one approach to the problem we can dispose of quickly. The simplest version is: "You understand 'finite.' You understand 'not.' So you understand 'not finite.' That is all there is to the infinite. What's the problem?" That is, we have experience of the finite, and our knowledge of the infinite is inferentially based on that.

Russell's position is a more sophisticated version. Others have held similar views, but Russell provides an example that is particularly convenient for our purposes. Russell defined what it is for two classes to be of the same cardinal number using one-to-one correspondences, and therefore without reference to the distinction between finite and infinite classes; he defined the finite numbers more or less by induction; and he then took an infinite class to be any whose cardinal number is not finite. He concluded by declaring that to be the (sole) theory of infinity [Rus03, pp. 356–357]. That is: "You understand 'class.' You understand 'finite class.' You understand 'not.' So you understand 'class that is not finite.'" Here, we are taken to have experience of classes (that is the main use for "intensions," compare [Rus03, pp. 66–69] or [Rus19, pp. 12–14]), and it is taken to be the sole source of our knowledge of the properties of the infinite. The basic idea persisted as Russell's views about classes changed [WR57, p. vi]: "cardinal arithmetic is usually conceived in connection with *finite* numbers, but its general laws hold equally for infinite numbers, and are most easily proved without any mention of the distinction between finite and infinite."

I have just described what I shall call the negative account of the infinite. It states that all knowledge about the properties of the infinite consists of general knowledge of both the finite and infinite plus whatever follows from the fact that the infinite is not finite. To be sure, a fair amount of knowledge about the infinite can be obtained in that way—as Whitehead and Russell's *Principia Mathematica* showed. Nonetheless, we have knowledge concerning the properties of the infinite that is *not* obtainable in that way, most notably the Axioms of Choice and Replacement. We should therefore expect them not to be a part of a negative account of the infinite. The fact that neither axiom was derivable in the system of the *Principia* corroborates that expectation.

The next approach to the problem of the infinite, or rather the next group of approaches, are known as *formalist* philosophies of mathematics. The basic idea is that our supposed knowledge of the properties of the infinite is actually knowledge of something much closer to our experience: it is really knowledge of our *descriptions* of the infinite, which turn out to be only apparently descriptions of the *infinite*.

Poincaré held a position related to formalism. He held that our knowledge of the infinite is mediated by finite descriptions, which are what introduce infinite entities. There is no fixed domain of infinite entities; new ones can be introduced in terms of old ones. It follows that impredicative statements about the infinite are incoherent.

There are a number of versions of formalism.[24] *Game formalism* takes mathematics to not be about anything, any more than a game of chess is directly about anything. A bit of mathematics, whether a computation or a proof, is just a series of manipulations of "game pieces" according to arbitrarily stipulated rules. The "pieces" may be numerals, sentences of a formal language, or anything else. Thus, all set theorists do according to a game formalist is see what sentences of first-order logic can be obtained from the axioms of first-order logic plus those of, say, ZFC when playing by the "inference rules" of first-order logic.

There is a technical core to game formalism that is correct: Most mathematicians and philosophers of mathematics grant that the axioms and rules of formalized first-order logic are a correct codification of at least part of informal reasoning and that ZFC, or something like it, is an axiomatization of set theory. There is therefore wide agreement that if a sentence is a theorem of set theory, then it can be obtained by playing according to the rules of the game formalist.

The chief problem with game formalism is that according to it the theo-

---

24. Chapters 2 and 3 of Michael D. Resnik's *Frege and the Philosophy of Mathematics* [Res80] provide a useful discussion of formalism and a history of some of the criticisms of it rehearsed here, many of which originated with Frege. My terminology for different versions of formalism is based on that of Resnik, though I have not followed him exactly.

rems of mathematics are not truths—or even propositions—any more than the positions in a chess game are. But there is no question that we do take the sentences used in mathematics to be true or false.

The game formalist might reply that within mathematics we just "play the game": whatever meaning mathematical theorems take on arises from their applications. But that won't do for many reasons. We apply theorems from one part of mathematics within other parts of mathematics. If neither part is meaningful, we cannot explain the application. We apply the same theorems in many different contexts. For example, we apply group theory to solving polynomials, to characterizing the interactions of elementary particles, to devising and cracking codes, and to working Rubik's cube. But the theorems of group theory do not express different truths in the different contexts. Group theory must therefore be granted to be about its own subject matter, if it has any subject matter at all. Finally—and this is the criticism of game formalism most related to the subject of infinity—for the game formalist, the rules of the mathematician's games are arbitrary, or at least chosen in some extramathematical way. But that leaves the choice of axioms out of the game formalist's account of mathematics and, in particular, out of the game formalist's account of set theory. But we have seen that the choice of axioms and subject matter is very important to the set theorist (and, though it goes beyond our topic here, to mathematicians generally). Thus, game formalism cannot provide any help in understanding some of the most important parts of the activity of mathematicians. At the very least, it is profoundly incomplete.

The next version of formalism might be called "meta–game formalism." It is the view that mathematics, though not itself a formal game in the sense of the game formalist, is the theory of such games. The view is usually simply called *Curry's formalism* after its most ardent advocate, Haskell B. Curry. (See [Cur51].) According to Curry, set theorists, after playing the game formalist's set-theory game to yield a theorem ϕ, make the declaration "ϕ is a theorem of set theory." That immediately solves the problem of giving content and truth to the work of mathematicians.

The position has more substantial advantages. One can, for example, devise a formal game involving strokes, concatenation of strokes, and replacing strokes by strings of strokes. One might then allow as theorems, for example, all sentences of the form $a + b = c$ such that a, b, and c are strings of strokes with c the result of placing a next to b. It seems absolutely reasonable to explain elementary number theory as the theory of a game devised along the indicated lines. It will not matter what "strokes" are, and so one obtains elementary number theory without any mandatory ontological commitments beyond claims like "there are some things suitable to serve as strokes."

Since, in some sense, Curry wanted his theorems to be about actual plays of a game, he required the theory of the games to be constructive. He did not exclude nonconstructive theorizing altogether, but he did not admit it into core mathematics. Curry's restriction to constructive reasoning is not so onerous as it might seem: Since, for Curry, the subject of set theory, for example, is not sets but some theory, say ZFC, the requirement is not that we reason about sets constructively but only that we reason about proofs in ZFC constructively. That is a mild constraint. After all, we do usually actually construct proofs of our theorems.

The position is a considerable improvement over game formalism. Nonetheless, it cannot be made to work. For one thing, the talk of ontological neutrality is less helpful than it might at first seem, for there must be something suitable to serve as "strokes." The need to uniquely characterize the numbers is avoided, but one still must show that there is something suitable to serve as numbers, and even that is not so easy. For example, whatever strings of strokes are, they cannot be actual physical objects, or even types of actual physical objects, if, as seems likely, the universe has a minimum size for objects, a minimum duration for events, and finite spatiotemporal size. If the strings are not physical, we are left with the problem of how anyone could play the games. The familiar problems about the numbers have not been solved or much simplified. In contrast, the analogous problems concerning set theory are indeed simplified, since one does not need an ontology of sets but only one of sentences, which are strings of symbols and are not, it would seem, different in fundamental respects from the strings of strokes needed for number theory.

Curry's formalism supplements game formalism with theorizing about games, but set theory—and for that matter analysis—is still treated as a game, in fact the very same game the game formalist used. Curry's formalism is therefore in exactly the same untenable position as game formalism when it comes to extramathematical applications of set theory or analysis, or indeed any other mathematical theory that is treated as a game instead of as the theory of a game.

In the case of applications of mathematics to mathematics, Curry's formalism does better: The fact that something or other arises in the study of a game might be taken to lend a particular point to the study of a different game related to that thing. As a result, Curry's formalism is better off than game formalism when it comes to the question of choosing which games to study: One might, for example, explain the fact that games in which the rules codify correct inference are particularly interesting to mathematicians by noting that they bear a close relation to our reasoning about games. The set-theory game could be seen as worth studying because so many of the other games can be interpreted in it. The usual claim that set theory is a universal theory is turned into the claim that it is a universal game.

There is a universal game that is simpler than set theory: the game of pure logic, which has only rules that codify logical inference. The game associated with any theory T is embedded in the game of pure logic as the set of theorems of the form $A \rightarrow \phi$ (if A, then ϕ), where A is any conjunction of axioms of T. Set theory is universal with respect to mathematical structures, not just mathematical theories, and it is in that respect superior to pure logic for mathematical purposes. But that advantage cannot be expressed within the strictures of Curry's formalism. One can see in a formalist way that if ZFC is consistent, then so are many other mathematical theories, since those theories can be proved to have models on the basis of ZFC. That is related to the universality of set theory we have been discussing. But it is hard to see how it is an advantage from the formalist perspective: all it does is cast doubt on the consistency of ZFC, since ZFC will be inconsistent if any of the other theories is.

Curry's formalism does have some advantages over game formalism in discriminating which games might be worth playing from the mathematical point of view. But it cannot explain all of the ways in which mathematicians in fact decide about axiom systems. It would be hard to justify, for example, the preference we actually find for set theory over Russellian type theory on formalist grounds. It would, in fact, be hard even to find any sense in which they might be seen as rival theories, since the formalist cannot construe them as theories of collections.

The final formalist position we shall consider, *deductivism,* or if–thenism as it is sometimes called, is suggested by earlier remarks on the game of pure logic. The main change is that one takes it to be more than a game. The deductivist holds that what mathematicians do is deduce the consequences of axioms. A theorem of a theory T is always taken to be an assertion of the form $A \rightarrow \phi$, where A is a conjunction of axioms of T, much as in the game of pure logic. The crucial difference is that here $A \rightarrow \phi$ is taken to be a truth, not just a position in a game. For that reason, deductivism is often not classified as a formalist position. But everyone grants that it is closely related to formalism. The issue is terminological, and for present purposes the similarities are more important than the differences.

The existence of the game of pure logic is a crucial part of the background for deductivism: it provides a standard on the basis of which one can show that a given theorem follows from certain axioms and nothing more. One can show that ϕ follows from T without making any commitment to the truth of T, and hence without any commitment to an underlying ontology. The most one needs in the way of ontology is a formal system with which to play the game of pure logic, and so the ontological commitments of deductivism are more or less those of Curry's formalism.

There is a remaining problem that must be solved before the view can get off the ground. If we accept, say, $(\forall xy)x + y = y + x$ ($x + y = y + x$ for all x and y) as a theorem of some axiomatized arithmetic, what operation does '+' stand for? More generally, if we take certain conditionals to be truths, what do the nonlogical symbols in them stand for? They must, it seems, stand for something, if the conditionals are to be true or false, and that would violate the supposed ontological neutrality of deductivism. There are two possible answers.

According to the first answer, the theorem, correctly construed, is not of the form $A \to (\forall xy)x + y = y + x$, but of the form $(\forall N+)(A^N \to (\forall xy)(N(x) \land N(y) \to x + y = y + x))$. That is, the theorem says that for every domain N and binary function +, if A holds of + on the domain N, then the commutativity of + holds on the domain N.[25] The theorem is not about a particular operation at all. It is a pure logical truth about all possible operations on all possible domains.

According to the second answer, the theorem $A \to (\forall xy)x + y = y + x$, correctly construed, is not actually a theorem but the schematic form of many theorems. To say that it is true is to say that whatever we take the domain of the quantifiers to be, and whatever operation on that domain we take + to be, the resulting sentence is true. A theory T has free nonlogical variables, and we will be able to detach the conditionals to conclude the truth of all of the ordinary theorems of T whenever those free variables are interpreted to stand for things that make all of T true. A theory T provides an implicit definition

---

25. When I say a first-order sentence ψ "holds on the domain N," all that means is that the sentence obtained from ψ by restricting all of the quantifiers in ψ to N holds. I have written out the restricted form of $(\forall xy)x + y = y + x$ in the consequent of the sentence in the text as an example.

of its models, which are just the systems of domain plus interpretations of the nonlogical variables that make T true.

The second version of deductivism assimilates all of mathematics to certain parts of modern algebra, in which one studies the properties of all structures that are models of a certain theory—the groups, rings, fields, and the like. But even modern algebra does not completely fit the deductivist mold: Group theory should be a paradigm of deductivist mathematics on some accounts, but, to take just one striking example, one of the great accomplishments of contemporary group theory is the classification of the finite simple groups. For present purposes, what matters about the example is that the classification came down to discovering and fully analyzing all of the "sporadic" simple groups—the finitely many simple groups that did not fit into a certain quasi-axiomatic classification of simple groups. During the period of discovery, one found summaries of progress that said things like [IK80, p. 539], "Since March 1972, the following [four] simple groups have been discovered: . . . " The work very much concerned the construction of particular groups, not the deduction of consequences from a system of axioms. My point is not that such results cannot be accommodated in a deductivist framework. They can. The point is rather that even the parts of mathematics that are supposed to be paradigms of deductivism do not fit smoothly and automatically into the deductivist mold.

The first version of deductivism says that the formula

$$A \to (\forall xy)x + y = y + x$$

(with '+' and the domain free variables) is satisfied by all domains and interpretations of '+'. The second says that for any domain and interpretation of '+', the sentence $A \to (\forall xy)x + y = y + x$ (with no free variables—domain and + have been supplied) is true. The distinction made here between 'all' and 'any' is much like that suggested by Russell (discussed in §IV.1), though it is used in a quite different way. Either version of deductivism allows the notion of a model of a theory.

Since deductivism takes mathematics to be the activity of deducing consequences from axioms, it can say little about the "extramathematical" activity of selecting axioms. The situation seems to be no better than that for Curry's formalism, and so deductivism cannot help to account for why we employ the theory of the infinite that we do. That, by the way, is the answer to a question that came up in §IV.1: We cannot always account for the mathematical use of questionable assumptions (in §IV.1, Infinity, Choice, and Reducibility; here, most directly, axioms for set theory) by moving them into the hypotheses of theorems. An important part of mathematical activity is the selection of suitable "hypotheses"—that is, suitable theories—for study. It is not the case that just any old theory will do, and to account for that we must resist the characteristic move of the deductivist.

The deductivist will grant that mathematicians should be most interested in consistent sets of axioms—since every sentence is a consequence of an inconsistent set of axioms, inconsistent sets of axioms are not very deductively interesting. Curry will grant that too. But the deductivist has granted the coherence of the notion of a model of a theory, and hence the possibility of showing a theory consistent by showing that it has a model. Curry need not accept that—model theory is, on his point of view, just more formalism.

Our main reason for believing that Peano arithmetic is consistent seems to be that we believe that it has a model. If that is correct, it leaves the deductivist in the awkward position of holding that our main evidence for an important mathematical fact—the consistency of Peano arithmetic—is extramathematical, since the existence of a model would be an extramathematical fact on the deductivist account. That is particularly awkward since whatever that model is supposed to be, it is supposed to be, practically everyone would agree, a *mathematical* entity. But part of the point of deductivism is that it avoids an ontology of mathematical entities.

If one presupposes a background set theory, all of the talk within mathematics about models—in particular, talk of sporadic groups or models of Peano arithmetic—can be explained away as just more drawing of deductive consequences, now from set theory. (What one really needs is a background model theory not a background set theory, but on standard accounts that turns out to amount to much the same thing.) But the use of a background set theory leaves two obligations the deductivist cannot fulfill: First, explaining how one theory, set theory, got to be so important. Weren't mathematicians supposed to be exploring the consequences of *any* mathematical theory? Second, giving reasons to believe the set theory consistent. Once a single theory has been given a primary position, its consistency becomes a pressing issue. For a deductivist using ZFC as a background theory, if ZFC turned out to be inconsistent, then there wouldn't be any mathematics at all.

Whatever is used to explain how set theory got to be so important and why we think it consistent will end up a replacement for deductivism, not a supplement to it. That replacement will have to explain why mathematicians prefer some theories to others, instead of pushing such questions outside mathematics. And it will have to explain why we think that ZFC is consistent. The only options for doing the latter seem to be arguing that ZFC is in some sense true, or proving it consistent. The first option has the consequence that ZFC has a model and therefore gives up on whatever ontological neutrality deductivism had. The second just pushes the problem back to whatever basis is used to prove ZFC consistent. Even though deductivism is not an adequate account of mathematics, any acceptable philosophy of mathematics will have to explain the important part proof plays in it and therefore share some of the more attractive features of deductivism.

Hilbert was, along with Frege, one of the originators of deductivism though he never adopted a fully developed form of deductivism.[26] Hilbert's later philosophy, *finitism,* is more or less a version of one of the options discussed above, namely, proving a formalized set theory consistent on the basis of a prior theory and then adopting deductivism with that set theory as a background theory.

Hilbert's position evolved over many years, and precisely what it was is a subject of dispute. Some of the things Hilbert said are apparently inconsistent with others. What I am about to present is a Hilbertian position extracted largely from his writings in the 1920s. Each of the elements of the position has been influential in part because of Hilbert. Nonetheless, it is controversial whether the position was ever Hilbert's, for reasons that I shall to some extent mention.

Hilbert's philosophy was in part motivated by an attempt to protect classical set-theoretic mathematics from the attacks of Kronecker and later Brouwer. In the 1920s, that was no idle matter. To many, axiomatic set theory seemed threatened by the paradoxes, and some feared it would prove to be inconsistent. There seemed to be a genuine possibility that Brouwer's position would prevail and that classical set-theoretic mathematics would become no more than a curiosity. That was true at a time when mathematics, in large part because of set theory, was enjoying a renaissance. Hilbert therefore wanted both to show that set theory was acceptable on its own terms and to justify set theory to an intuitionist. He had to start on common, neutral ground, and he attempted to do so: he isolated a nontrivial but basic part of mathematics, "finitary" or "contentual" mathematics, that does not involve any commitment to the infinite at any level, either within the theory or on the part of the theorist.

---

26. See [Res80] for a discussion of the history of deductivism.

Since intuitionists and classical mathematicians do not even agree about the principles of logic, except on suitably defined finite domains, and since they do not agree on the sense in which infinite domains like the natural numbers are infinite, a neutral ground must involve no logic and no infinities of any kind. At first sight the only thing left is variable-free equations and inequalities—$1 + 3 = 3 + 1, 3 > 1$, and the like.

There is even disagreement about variable-free equations and inequalities: For the intuitionist, numbers are mental constructions made up of successions of moments. For Hilbert, numbers are sequences of strokes, which he took to be quasi-concrete types of physical objects. Nonetheless, both sides will agree that there is a domain consisting of the first five numbers $(0, 1, 2, 3, 4)$ and a partial operation on it called addition that is such that $1 + 3 = 3 + 1$. Their respective domains will be isomorphic. The two theories of what numbers are represent different solutions to the problem of the finite. Since we are primarily interested in the problem of the infinite, we shall not pay much attention to them. In one sense—up to isomorphism—there is complete agreement about variable-free equations and inequalities. But that seems like too little mathematics to be of any use.

Mathematics cannot even get started without some way of expressing generality. But even a simple general theorem about the natural numbers like $(\forall x)x + 1 = 1 + x$, though intuitionists and classical mathematicians apparently agree on its truth, is not neutral, for the intuitionist will interpret it constructively in terms of proof, which does not require an actual infinity of natural numbers, while the classical mathematician will interpret it in terms of a model, which does. Nonetheless, both sides *will* agree—up to isomorphism—about each instance of the theorem. That is, if c is a natural number, they will agree that $1 + c = c + 1$. Hilbert exploited that using a means of expressing generality that is reminiscent of Russell's "any," as distinguished from the more familiar "every." (Mark Steiner made a closely related observation [Ste75, p. 148]. The analogy to Russell is mine. Hilbert did not mention Russell's distinction.) Russell wrote of an "ambiguous assertion," while Hilbert said [Hil28, p. 470] that in an equation like $1 + c = c + 1$, c is an "unspecified numeral."

It never became clear exactly what Hilbert had in mind [Tai81, p. 543]. Sometimes it seems that he took $1 + c = c + 1$ to be a particular numerical equation, like $1 + 3 = 3 + 1$. It is just that he hasn't told us which. For example, he wrote [Hil28, pp. 469–470] that to infer $1 + 3 = 3 + 1$ from the algebraic equation $1 + x = x + 1$ requires a proof procedure (a rule of substitution) that "already goes considerably beyond contentual number theory." Yet he takes $1 + c = c + 1$ to be contentual. The difference is that in algebra one studies the equations themselves, and so the 'x' in the algebraic equation does not stand in place of a numeral [Hil26, p. 195]. If we take that in the obvious way, then the 'c' in $1 + c = c + 1$ stands for a numeral, though we haven't been told which. Thus, despite appearances, $1 + c = c + 1$ is no more general than $1 + 3 = 3 + 1$.

If 'c' stands for a numeral, then, given a proof that $1 + c = c + 1$, one does not infer $1 + 3 = 3 + 1$ by substituting '3' for 'c.' The letter 'c' is not a variable. What one does instead is specify that c *is* the numeral '3.' That amounts to substituting '3' for 'c' in the entire proof of $1 + c = c + 1$, which results in a proof of $1 + 3 = 3 + 1$ in which 'c' does not occur. Such a view has its temptations. Most strikingly, it enables one to see that $1 + 3 = 3 + 1$ on the basis of a proof that $1 + c = c + 1$ without using a substitution rule.

The reasoning is in effect as follows: The proof that $1 + c = c + 1$ did not depend on whether or not c was 3. Thus, the method we used to prove it also shows that $1 + 3 = 3 + 1$. If one uses the indicated reasoning, anything derivable with a substitution rule is derivable without it. One can therefore take 'c' to be nothing more than a mere symbol, a *placeholder.*[27] The fact that Hilbert never stated a substitution rule during the period under discussion suggests that the use of placeholders is an element of what he had in mind.

The interpretation just suggested cannot be the whole story since, for example, Hilbert wrote [Hil26, p. 194] of "the statement that if c is a numerical symbol, then $c + 1 = 1 + c$ is universally true."[28] Here, 'c' must be allowed to take on more than one value, and so it can be neither a placeholder nor an unspecified numeral. In such a case, Hilbert intended us to interpret the statement "as a hypothetical judgment which asserts something for the case when a numerical symbol is given." It is not completely clear what that is supposed to mean, but at any rate $c + 1 = 1 + c$ is not itself supposed to be a truth, though we are somehow to obtain from it truths of the following sort: $1 + 1 = 1 + 1$, $2 + 1 = 1 + 2$, and $3 + 1 = 1 + 3$. Though Hilbert

---

27. When I say that a letter is a placeholder rather than a variable, all I mean is that there is no associated substitution rule. If 'c' is a placeholder, then $1 + c = c + 1$ is consistent with $1 + 3 \neq 3 + 1$ just as the latter is consistent with $1 + 2 = 2 + 1$. It just turns out that as a matter of fact, for the finitary system we develop, if $1 + c = c + 1$ is derivable, then so is $1 + 3 = 3 + 1$.

28. I have changed Hilbert's notation to conform to that used here.

never specified how we are to accomplish that, it seems clear that he cannot have presupposed that we are to obtain infinitely many such truths, whether actually or potentially. That would build the infinite into the basis of "finitary mathematics," making the introduction of the infinite on the basis of finitary mathematics viciously circular. There has been no general agreement on how to work out Hilbert's idea of a hypothetical judgment.

To spell out my attempt to fulfill Hilbert's intentions, I shall use 'X,' 'Y,' ... as schematic variables for which there is an associated substitution rule permitting them to be replaced by numerals. (When I give formal details below, it will turn out that they can also be replaced by other things, for example, other schematic variables.) The equation $1 + X = X + 1$ is not a truth at all, since it doesn't make any assertion. Nonetheless, when 'X' is replaced by any numeral the result is a truth. The substitution rule is admissible in contentual finitary mathematics because the equation $1 + X = X + 1$ is not a purely symbolic object of study, a particular polynomial like the algebraic equation $1 + x = x + 1$ discussed above, but a way to impart information about something contentual [Hil28, p. 470]. The equation is like Hilbert's "hypothetical judgment." To accept the equation $1 + X = X + 1$ is not to take it to be a truth, most particularly not the truth $(\forall x)1 + x = x + 1$, which is not a part of finitary mathematics no matter whether the quantifier is read intuitionistically, modally, or classically. It is rather to accept any instance of

If ϕ is a variable-free substitution instance of '$1 + X = X + 1$,' then ϕ is true

obtained by substituting an expression for 'ϕ.' (Compare [Res80, p. 83].) The displayed schema exhibits the hypothetical character of accepting $1 + X = X + 1$. The truth obtained by universally quantifying over ϕ cannot be a part of finitary mathematics any more than can the one obtained by universally quantifying over X. Thus, there was little gained by replacing $1 + X = X + 1$ by the displayed expression, which served only to indicate the hypothetical character of the schema. In either case, we have to explain how to understand schemas in such a way that their use is finitarily acceptable.

Here is the explanation I propose: Briefly, by accepting a schema, we become committed to any of its instances that is obtained by replacing each of its schematic variables by a suitable expression. In that explanation, "any" is used in a more or less Russellian sense and must be sharply distinguished from "every." To become committed to an instance of a schema, that instance must actually be obtained. For Hilbert, for whom truths were quasi-concrete sentences (that is, sequences of symbols like the sequences of strokes that served as numbers), it must be written down. Analogously, for the intuitionist, for whom truths are certain mental constructions, the construction would have to actually be performed.

The acceptance of a schema commits us to only finitely many instances of it (since only finitely many instances have actually been obtained) and to the willingness to accept more instances as we are confronted with them. It constrains how we may go on, but it does not commit us to going on to obtain an infinity of instances. It does not even commit us to going on to obtain any more instances than we actually now have. Thus, accepting $1 + X = X + 1$ commits us to $1 + 1 = 1 + 1, 1 + 2 = 2 + 1, 1 + 3 = 3 + 1$, and a considerable number of like equations, but not to infinitely many of them.

Schemas like $1 + X = X + 1$ do not involve any commitment to any kind of infinite. Intuitionists and classical mathematicians accept various schemas and use them to introduce axioms.[29] Each, for example, accepts an induction schema as part of arithmetic. That is, each accepts the familiar induction schema $\phi(0) \wedge (\forall x)(\phi(x) \to \phi(x + 1)) \to (\forall x)\phi(x)$, for suitable ϕ. Each sort of mathematician uses the schema, and introduces it along with a substitution rule. Each will accept the finitely many instances of the schema to which we have become committed so far and more: the intuitionist that there is a potential infinity of instances of the schema and the classical mathematician that there is an actually infinite collection of all the instances. Each sort of mathematician can officially avoid the use of schemas by stipulating that every member of a certain (potentially or actually) infinite totality, namely the collection of all instances of the induction schema, is an axiom. But each will still have to use schemas or an equivalent device unofficially to specify which totality they mean.[30]

Hilbert's finitary mathematics makes use of schemas, but it does not allow any other means of expressing generality (except perhaps those of unspecified

---

29. The intuitionist does not work axiomatically but on the basis of intuition. I should therefore say more carefully that the intuitionist will accept that any instance of a schema is a basic truth justified directly by intuition.

30. One might define a totality in terms of antecedently given totalities, but if one follows the chain of definitions back, there will be a use of a schema or something equivalent somewhere in the chain.

numerals and placeholders, which can be eliminated in favor of schematic variables). Finitary mathematics can be formulated in such a manner as to consist entirely of equations, though Hilbert did not so formulate it. No logical connectives are involved. Thus, no formula of finitary mathematics requires any commitment to more than a small finite number of mathematical entities, and no formula of finitary mathematics involves any logical connectives, like 'not' or 'or,' let alone quantifiers, that may be interpreted differently by constructivists, intuitionists, and classical mathematicians. The device used to express generality, schematic variables, can similarly be accepted by adherents of every ontological position, even ontological liberal finitism and strict finitism, at least the more plausible versions.[31] Moreover, no form of infinity is presupposed by or employed within finitary mathematics. Thus, the formulas of finitary mathematics provide a sort of neutral ground: everyone understands them in more or less the same way.

Finitary mathematics is apparently a rather limited neutral ground. For example, as Hilbert emphasized, though one has a kind of surrogate for universal quantification in finitary mathematics, there is none for existential quantification. Though $1 + X = X + 1$ is in many respects an adequate substitute for $(\forall x)1 + x = x + 1$, what could serve for its negation, $(\exists x)1 + x \neq x + 1$? The obvious candidate, $1 + X \neq X + 1$, won't do. It says the wrong thing. It is the analog of $(\forall x)1 + x \neq x + 1$. As Hilbert put it [Hil26, p. 194], "the statement that if X is a numerical symbol, then $X + 1 = 1 + X$ is universally true, is from our finitary perspective *incapable of negation.*"[32] There are some cases where a surrogate for an existential sentence can be found. For example, one can simply use $2 + 2 = 2 \times 2$ to express that $(\exists x)x + x = x \times x$. The limitations of the approach are obvious.

I have now introduced the full expressive strength of finitary mathematics.

---

31. The (ontological) strict finitist who held that $10^{10^{10^{10}}}$ is a number but that $10^{10^{10^{10}}} + 1$ is not could not accept $X + 1 > X$, since plugging in $10^{10^{10^{10}}}$ for X would not produce a truth. Such a strict finitist could not accept finitary mathematics. The strict finitist who is skeptical about whether $10^{10^{10^{10}}}$ is a number, for example because one cannot produce a string of $10^{10^{10^{10}}}$ strokes, could perfectly well accept finitary mathematics. That sort of strict finitist will be skeptical about whether there is an instance of $X + 1 > X$ of the form $10^{10^{10^{10}}} + 1 > 10^{10^{10^{10}}}$, and so acceptance of the schema need not cause any trouble.

32. I have changed the notation to conform to that used here. Emphasis in the original.

Finitary mathematics is of necessity too weak to express much of anything about functions from the numbers to the numbers, since such functions are defined on all the numbers. That is to say, such functions have infinite domains and so cannot be a part of finitary mathematics. (See [Tai81, pp. 527–528] for a closely related argument.) Certain functions, however, have definitions that take a remarkably simple form. Consider addition. Let S be the function that takes a number to its successor. Then the following two equations define addition recursively:

$$(\forall x)x + 0 = x,$$
$$(\forall xy)x + Sy = S(x + y).$$

They determine, for example, that $2 + 2 = 4$ by the following computation:

$$SS0 + SS0 = S(SS0 + S0) = SS(SS0 + 0) = SSSS0.$$

The recursive definition of addition is simple enough to have a counterpart expressible in a finitary manner:

$$X + 0 = X,$$
$$X + SY = S(X + Y),$$

and the finitary equations suffice to evaluate $2 + 2$, or any other sum of given numbers, in the manner indicated. Those equations do not carry with them any commitment to a function, since they do not carry any commitment to a domain that goes beyond the values actually employed. To paraphrase Hilbert and Paul Bernays ([HB70, vol. 1, p. 287], see also [HB70, vol. 1, pp. 25, 330]), the recursive definition is a brief communication of a procedure that yields a definite number when given definite numbers, and that recursive definition can be imitated within finitary mathematics.[33]

Finitary mathematics allows the use of axioms of the following form, a *definition of the function f by primitive recursion,* when g and h are a part of finitary mathematics:

---

33. The distinction made here between the universally quantified variable x and the schematic variable X is neither in Hilbert's work nor in modern commentaries on it, though Hilbert does distinguish algebraic variables from contentual ones, as discussed above.

$$f(0, X_1, \ldots, X_n) = g(X_1, \ldots, X_n),$$
$$f(SX, X_1, \ldots, X_n) = h(X, f(X, X_1, \ldots, X_n), X_1, \ldots, X_n).$$

Calling this a "definition of a function by primitive recursion" is misleading, since no functions, in the usual sense, are defined within finitary mathematics—they are merely imitated.[34] That is clear when one attends carefully to the interpretation of the schematic variables, since the schematic equations result only in a commitment to finitely many values of the function f. That seems crucial to making tenable the claim that finitary mathematics presupposes nothing infinite, and it is hinted at by Hilbert's distinction between algebraic variables and contentual ones, but the issue has not been adequately discussed in the literature.

Note that the recursion schema *is* a schema—it is given using schematic variables g and h. The use of variables of some kind seems to be essential here—unspecified numerals or placeholders just will not do the job. But that seems to be the only essential use of variables in finitary mathematics—the metalinguistic specification of a system of finitary mathematics. Within the system, unspecified numerals or placeholders are adequate.

Hilbert had high aspirations for finitary mathematics. He argued not that it is neutral but that it is intuitively clear, indubitable, and absolutely secure. That view is not adequately supported by his arguments, in part because he relied on a dubious account of mathematical intuition. Nonetheless, it has been argued recently that finitary mathematics *is* indubitable, on the grounds that it embodies the minimal principles that must be accepted to do mathematics at all [Tai81].[35] Be that as it may, Hilbert took finitary mathematics to be a secure base, and he tried to build the rest of mathematics on top of it. Though Hilbert often emphasized mathematical truth in his writings, I shall focus, somewhat misleadingly so far as historical summary is concerned, almost exclusively on his views concerning mathematical existence.

---

34. Though the function symbols of finite mathematics cannot be seen to stand for *functions* without going beyond the conceptual resources needed for finitary mathematics, they have a perfectly good interpretation that does not go beyond those resources: they can be interpreted as injunctions to perform computations of a certain sort. I believe that is more or less how children in the early school years interpret the signs for addition, subtraction, multiplication, and division.

35. Hallett believes that something like that was actually Hilbert's position [Hal90, pp. 241–242].

There is a familiar procedure for introducing new objects into mathematics on the basis of already given ones. It is familiar in large part because of Hilbert's emphasis on its importance, which was at least initially independent of, and prior to, his finitism. The simplest example to give, though it is not the example Hilbert used, is that of introducing the (nonnegative) rational numbers as pairs of natural numbers. According to the procedure, the fraction $\frac{1}{2}$ is introduced as the pair of numbers 1 and 2, and the rational numbers are introduced as the pairs p and q of natural numbers such that p and q have no common divisors and q is not zero. The operations on the rational numbers are defined in terms of the operations on the natural numbers. The procedure is absolutely familiar, since it is the one actually employed in school to introduce the rational numbers and the arithmetic operations on them.

Hilbert took the process by which the rational numbers are introduced on the basis of the natural numbers to be an example of the following method: First, one produces a new theory. In the example, one produces an axiomatic theory of the rational numbers. Second, one shows that the theory is a reasonable one in the limited sense that it is *consistent,* that is, that it does not lead to any contradictions. Finally, having shown the theory consistent, we may employ it, and the entities it concerns, forthwith. In the example, we have succeeded in introducing the rational numbers as new numbers on the basis of the natural numbers.[36]

How do we show a theory consistent? In the example, we can do that by showing that the theory of the rational numbers holds of the fractions. That suffices because a contradiction in the theory of the rational numbers would lead to a contradiction concerning the fractions, and that in turn, since the theory of the fractions is reducible to the theory of the natural numbers in the way described above, would lead to a contradiction in the theory of the

---

36. A bit more seems to be required to justify taking the rational numbers to be an extension of the natural numbers with the new arithmetical operations extensions of the old ones, instead of a completely different structure that happens to have a substructure isomorphic to, but distinct from, the natural numbers—and that is a move Hilbert certainly did think justified. Let me propose the following criterion on Hilbert's behalf: if it is consistent to take a theory that has an intended model to concern an extension of a given intended model of some other theory, then the intended model of the one theory is an extension of the intended model of the other theory. The criterion is, I think, all that justifies extending the natural numbers to the integers, the integers to the rational numbers, the rational numbers to the real numbers, and many other allied cases.

natural numbers. The consistency of the theory of the rational numbers is thus proved on the basis of the assumed consistency of the theory of the natural numbers.

It is arguable whether Hilbert's analysis of the method for introducing the rational numbers is a reasonable one, but we shall not discuss that here. Hilbert's plan was to introduce all of mathematics, including set theory, starting from finitary mathematics using the method outlined.

Note that it is not clear whether to view the example as introducing the rational numbers or eliminating them. One can take the view that the definition of the rational numbers in terms of the natural numbers shows us what they are, and that they exist in some appropriate sense. On that view the rational numbers are introduced. But one might equally well take the view that the definition of the rational numbers in terms of the natural numbers renders the rational numbers superfluous. Statements about the rational numbers can be reinterpreted as statements about pairs of natural numbers, and so all talk of the rational numbers is eliminable. On that view the rational numbers are eliminated.

It is disputed which interpretation Hilbert intended. Michael Detlefsen has argued [Det86] that Hilbert intended the eliminative interpretation. Nonetheless, much of what Hilbert said suggests the introductive interpretation [Hal90, pp. 226, 235–239]. (See, for example, [Hil26, p. 184].) I shall consider only the introductive interpretation in these pages, since it is more interesting for our purposes and because I think it correct. Nonetheless, much of what I say will not depend on which interpretation is adopted, and even more can be rephrased in simple ways to apply to the eliminative interpretation.

Why introduce the rational numbers? One reason is that it simplifies the theory of division, since every rational number is divisible by every nonzero rational number, while the analogous statement is false for the natural numbers. (Just try dividing 1 by 2.) There is a problem with finitary mathematics that Hilbert saw as analogous to the one concerning division of natural numbers [Hil26, p. 195]: Just as the natural numbers are not closed under division, the finitary statements are not closed under negation. In the case of the natural numbers, we solved the problem by adding new numbers, the rationals. Hilbert proposed that we solve the problem with finitary mathematics analogously, by adding new statements. In detail, he proposed that we supplement the expressive apparatus of finitary mathematics with that of classical logic—in particular quantifiers. The theory of the new statements was to be a standard axiomatization of logical inference of the sort that is familiar today. Such axiomatizations were already well known in Hilbert's day. In order to justify the addition by his own standards, Hilbert had to supply a finitary proof that a suitable axiom system is consistent. Once he managed to do that, it is clear how he would like to have introduced infinitary mathematics: by giving an axiom system and providing a finitary proof that it is consistent. As Hilbert sometimes emphasized (see, for example, [Hil28, p. 474]), such a consistency proof would justify the use of infinitary methods to prove statements of finitary mathematics, since it would make it possible to convert proofs of such statements that use infinitary methods into finitary proofs.[37] It is clear that Hilbert's proposed procedure for introducing infinitary mathematics is subject to many of the same criticisms as formalism and deductivism. After all, many axiom systems are consistent. Why, for example, study set theory with the Axiom of Choice, instead of set theory with its negation? But I am getting ahead of myself.

To prove that the logic that Hilbert wished to introduce, call it $\mathcal{L}$, is consistent on a finitary basis, it is at a minimum necessary that one be able to formulate, or perhaps imitate, the statement "$\mathcal{L}$ is consistent" in a finitary way. Here is Hilbert's ingenious method of doing that. The method was the beginning of the mathematical investigation of proofs—proof theory. The logic $\mathcal{L}$ is usually thought of as having content. Thus, we read $\wedge$ as "and," and $\forall$ as "for all." But, since $\mathcal{L}$ is presented in a purely formal way, we can "deny any meaning to the logical symbols" [Hil26, p. 197] and view the specification of $\mathcal{L}$ as a specification simply of what sequences of symbols are special in that they are called "sentences" and of what sentences may follow other sentences in objects, sequences of sequences of symbols, called "proofs." The idea is familiar today, again largely as a result of the importance Hilbert ascribed to it. We discussed it earlier in connection with game formalism. Now, $\mathcal{L}$ is consistent if and only if no proof ends in the line, say, $c \neq c$. (What sentence we choose will depend on the details of $\mathcal{L}$.) Thus, the question of the consistency

---

37. Here is a sketch of how: Take a finitary statement that has been proved using infinitary methods and assume, for the sake of contradiction, that it is false. Since the statement is finitary, there would in that case be a finitary proof that it is false. Since our finitary methods are to be included among our infinitary ones, that means that there would be proofs within the infinitary system both that the statement is true and that it is false. But that contradicts the finitarily proved fact that the infinitary system is consistent, and so the statement must be true after all. The new proof just offered of the statement using consistency is finitary even if the original one was not.

of $\mathcal{L}$ reduces to one about the combinatorial possibilities of arraying symbols in certain ways. Now the only part of finitary mathematics we have discussed is that concerning numbers, literally involving expressions like $SSS0$. But one can reason in a finitary way about sequences of symbols including $\wedge$, $\forall$, and the rest—as we discuss in some detail in an appendix (§4). There is no necessary reason to restrict oneself to S and 0. The finitary version of "$\mathcal{L}$ is consistent" turns out to take the form, "If X is a proof in $\mathcal{L}$, then the last line of X is not $c \neq c$." I shall not give details about how to express "X is a proof in $\mathcal{L}$" in a finitary way, but it should not seem unreasonable that it is possible to do so, since finitary mathematics encompasses definitions by primitive recursion and standard formalizations of logical systems proceed by the use of just such definitions.

Since the consistency problem can be given a finitary formulation, it was not unreasonable of Hilbert to suspect that it has a finitary solution. It does not, as we discuss two paragraphs below. Let me first note that Hilbert's method of introducing new mathematical entities would not be acceptable to an intuitionist—proving consistency is not at all the same thing as providing a mental construction[38]—and so, even if Hilbert had succeeded in providing consistency proofs, that would not have achieved his goal of justifying classical set-theoretic mathematics to the intuitionist. It would nonetheless have provided a valuable sort of justification that we do not stop to analyze, since the program of providing consistency proofs failed.

A finitary proof of the consistency of a theory would not even convince an intuitionist that the theory is consistent: An inconsistency might arise on the basis of a proof that is not reflected in the formalization of proof. Let me explain. To an intuitionist a proof is a certain sort of mental construction. A proof is intrinsically meaningful, and the pale imitation of some proofs in written symbols that is available within finitary mathematics is just not relevant to the genuineness of proofs, which is a psychological matter for the intuitionist. There is no reason to think that the formalization captures everything we would accept as a proof. Moreover, Hilbert proposed to use a consistency proof to introduce new rules of proof, rules that are formally described. An intuitionist would not accept proofs employing such rules as proofs, unless the rules were antecedently convincing, whatever assurances were provided of their security.

---

38. The completeness theorem, which guarantees that every consistent theory has a model, is of no help here. It is not intuitionistically—let alone finitarily—valid, and even if it were, it supplies a model, not the requisite mental construction.

Why did the program of providing consistency proofs fail? Gödel's second incompleteness theorem showed that no theory that meets some mild conditions can prove its own consistency. Thus, what is generally taken to be the centerpiece of Hilbert's finitary program[39] is hopeless. The program requires proving the consistency of set theory by finitary means. Gödel's second incompleteness theorem showed that that is impossible unless finitary means include some that are not a part of set theory: if finitary means are a part of set theory, then a finitary proof of the consistency of set theory is a proof within set theory of the consistency of set theory, and that is exactly what is ruled out by Gödel's second incompleteness theorem. It seems unlikely that there are finitary means that are not a part of set theory, since set theory includes counterparts of virtually all known mathematics of whatever sort, not just finitary mathematics. Moreover, the existence of finitary means that are not a part of set theory would do as much to cast doubt upon the security of finitary mathematics as they might do to increase our certainty about the security of set theory.[40]

Jacques Herbrand adopted Hilbert's finitary methods and worked on Hilbert's program to the extent of proving the consistency of various theories. His work emphasized something Hilbert had virtually ignored: In familiar consistency proofs, like the one that the theory of the rationals is consistent if the theory of the natural numbers is, one proceeds by producing a model of the theory in question. A theory may have infinitely many axioms. If it has only finitely many, one can conjoin them to yield a single sentence that serves as an axiom for the theory. Herbrand worked out a finitary notion of a model for a sentence (that is, for a finitely axiomatizable theory), and he used it to prove consistency results ([Her30, p. 552], [Her31, p. 275]).

Here is a (slightly modernized) sketch of Herbrand's finitary notion of a model. Suppose a sentence is of the form $(\forall x)(\exists y)\phi$, where ϕ is free of quantifiers. Then we might attempt to build a model by starting with a sin-

---

39. See [Hal90, p. 243] for the minority opinion.

40. The above argument does not show that some parts of mathematics might not be justified on the basis of other, apparently weaker parts, including finitary mathematics. That "modified Hilbert's program" is an active area of current mathematical research. Detlefsen has shown [Det86] that the argument that Gödel's second incompleteness theorem refutes Hilbert's program relies on the assumption that the definitions of proof and consistency take a certain form. He shows that other alternatives that might escape the theorem and be acceptable to a Hilbertian cannot be ruled out. Prospects for an actual development along the suggested lines seem dim.

gle element, say a. Since the sentence is true for any x, it holds for a. That is, there is some b such that $\phi(a, b)$. If $(\forall x)(\exists y)\phi$ has a model, there must be some way of specifying the truth values of the components of ϕ to make $\phi(a, b)$ true. That will determine some of the relations between a and b, which gives us a partial model. Now, much as before, there must be some c such that $\phi(b, c)$ holds, and so, if $(\forall x)(\exists y)\phi$ has a model, there must be some way to make $\phi(a, b) \wedge \phi(b, c)$ true. That yields another, possibly larger, partial model. And so forth. Even in the case of a more complicated sentence, there will be an analogous procedure of starting with finitely many objects and iterating the process of plugging them into the sentence to see that any model must include objects obeying further constraints. If one could satisfy all of the infinitely many constraints simultaneously, one would have a model of the sentence in the familiar infinitary sense. Herbrand said instead that a sentence is *true in an infinite domain* if for every number p there is a partial model (the terminology is not Herbrand's) meeting all the constraints of the pth iteration. That definition has a finitary formulation, since it only involves finite "pieces" of a model in a more usual sense. Herbrand showed that a sentence is consistent if and only if it is true in an infinite domain.[41]

Hilbert claimed it was legitimate to use a theory without reservations once it had been proved consistent. But when we take a theory to be true, we seem to be taking it to be about something. The usual proof of the consistency of the theory of the (nonnegative) rational numbers does more than show it consistent: it shows that we can take rational numbers to be certain pairs of natural numbers. A Hilbertian proof-theoretic proof that some set theory is consistent would have, on his view, entitled us to make use of sets. But it would be no help in seeing what those sets are, since a consistency proof along Hilbertian lines treats a theory as an uninterpreted calculus. Herbrand's methods may not seem like much progress in that respect, since they provide only partial models of a sentence, and the sentence is not even true in the partial models. The methods are nonetheless a step in the right direction. As we shall see in Chapter VIII, modern improvements of Herbrand's work due to Jan Mycielski and Janusz Pawlikowski associate with each sentence a finitary version such that the original sentence has a model in the usual sense

---

41. His proof was not quite correct, though it has been satisfactorily repaired. See the editors' notes to [Her30] for details. The theorem has Gödel's completeness theorem as a simple consequence by an argument that is *not* finitary.

if and only if the finitary version has suitable finite models. Herbrand's partial models are the finite models relevant to the proof.

Though Hilbert's finitary program failed, finitary mathematics retains considerable interest as a neutral ground that involves no commitment to any form of infinity. When we wish to assess whether a certain argument should be acceptable to all parties, one easy way to do it is to show that the argument is finitary. For that reason, we shall make essential use of finitary mathematics in subsequent chapters. I therefore give a detailed specification of a system of finitary mathematics in the following appendix and analyze the system in detail. In fact, I give three systems, PRA, PRW, and PRS. The analysis includes a detailed discussion of schematic variables. The reader who is prepared to take on faith that certain arguments can be carried out in a manner that is acceptable to all parties may skip the appendix.

§4. Appendix

We described and assessed the expressive strength of finitary mathematics in §3. We now turn to its deductive strength. It is necessary to cheat in a standard way. Hilbert never laid down a system of finitary mathematics, or even a codification of a part of finitary mathematics. He only gave examples of finitary reasoning. It is therefore necessary to supply details that go beyond what can be supported by Hilbert's pronouncements. We go along with the tradition (see, for example, [Det86, p. 45], [Par71, p. 50], [Res80, p. 86], [Tai81, p. 544], and [vH67, p. 482]) by taking primitive recursive arithmetic (PRA), a system developed by Skolem [Sko23a], to be a part of finitary mathematics. It has been argued, I think fairly convincingly, that all of finitary mathematics can be formalized within PRA [Tai81], but I shall make no use of that stronger claim. All of Hilbert's examples of finitary reasoning can in fact be formalized within PRA. Hilbert stated that only finitary methods should be used for metamathematical proofs, and he gave his metamathematical proofs within PRA [HB70].

Here is a modern version of PRA essentially due to Reuben Louis Goodstein [Goo57, pp. 19, 21, 104, 105]. I have used schematic variables, since, as I have argued, that is apparently necessary if PRA is to be free of infinitary commitments. There is, however, no reason having to do with the written form of the system why one cannot use ordinary variables and construe them to be implicitly universally quantified. Of course, the ontological commitments might then appear to be very different. When necessary, I shall call the schematic version of PRA presented here sPRA and the parallel universally quantified theory uPRA. As introduced here, PRA is sPRA—the *s* is only for emphasis when the context requires it.

The present version of PRA is considerably simpler than the one employed by Hilbert, since it relies on the observation that the additional apparatus used by Hilbert can be defined within the simpler system. There is a single constant symbol, 0. There are function symbols S, $\mathcal{Z}$, and $\mathcal{P}^i_j$, for $j \leq i$, and variables X_i. I shall use Y_is and Z_is as schematic metavariables that can be replaced by variables. The function symbols have the following associated axioms:

$$\mathcal{Z}(X_1) = 0,$$

$$\mathcal{P}^i_j(X_1, \ldots, X_i) = X_j.$$

The *terms* of PRA are defined recursively in the expected way: 0 and the variables are terms. If f is an n-placed function symbol, then f followed by n terms separated by commas and enclosed in parentheses is a term. (I leave it to the reader to figure out from context how many places each function symbol has.) There is one more sort of function symbol, which must be included in the definition of term: if g is a term with variables among $Y_1, \ldots, Y_n$ and h is a term with variables among Z_1, Z_2, $Y_1, \ldots, Y_n$, then there is a new function symbol associated with g and h, say, for definiteness, the letter $\mathcal{F}$ followed by g and h separated by commas and enclosed in parentheses. Abbreviate $\mathcal{F}(g, h)$ by f. Then f has the following associated axioms:

$$f(0, X_1, \ldots, X_n) = g(X_1, \ldots, X_n),$$
$$f(SX_0, X_1, \ldots, X_n) = h(X_0, f(X_0, X_1, \ldots, X_n), X_1, \ldots, X_n),$$

where $g(X_1, \ldots, X_n)$ is the result of simultaneously replacing Y_i by X_i for $i = 1, \ldots, n$ in g and $h(X_0, f(X_0, X_1, \ldots, X_n), X_1, \ldots, X_n)$ is the result of simultaneously replacing Z_1 by X_0, Z_2 by $f(X_0, X_1, \ldots, X_n)$, and Y_i by X_i for $i = 1, \ldots, n$ in h.[42] The formulas of PRA are equations between terms. There are four rules of inference in PRA:

(1) Infer $f(h) = g(h)$ from $f = g$, where f, g, and h are terms, Y_0 is a variable, and $f(h)$ and $g(h)$ are obtained by replacing every occurrence of Y_0 in f and g by the term h.

---

42. Thus, for example, $\mathcal{F}(Y_1, SZ_2)$ is a symbol for addition, $X_1 + X_0$, as defined by the primitive recursion given earlier.

(2) Infer $f(g) = f(h)$ from $g = h$, where f, g, and h are terms, Y_0 is a variable, $f(g)$ is obtained by replacing every occurrence of Y_0 in f by g, and $f(h)$ is obtained analogously.

(3) Infer $g = h$ from $f = g$ and $f = h$, where f, g, and h are arbitrary terms.

(4) Infer $f(X_0) = g(X_0)$ from

$$f(0) = g(0),$$
$$f(SX_0) = h(X_0, f(X_0)),$$
$$g(SX_0) = h(X_0, g(X_0)),$$

where f, g, and h are arbitrary terms and the substitutions in f and g are for X_0, while those in h are for X_0 and X_1.

The first rule enables us to replace a variable by a term. The second rule allows replacing a term by an equal one (say, $2 + 2$ by 4). The third rule is an elementary property of equality.

The fourth rule, or variants thereof, has excited the most controversy. It is the rule of the uniqueness of primitive recursive definition. That is, it says that if two terms obey the same recursion equations, then they have the same value for any values of the variables. It codifies the idea that primitive recursive definitions enable us to imitate definitions of functions, in that terms introduced via the same definition have the same value for any values of the variables. Let me briefly discuss the expressive power of PRA and then come back to the rule.

One can introduce a function symbol α in PRA with the following properties: One can prove $\alpha(f, g) \leq 1$ in PRA.[43] For any terms f and g one can derive $f = g$ in PRA if and only if one can derive $\alpha(f, g) = 0$ in PRA. Because of those properties, one might think of $\alpha(f, g)$ as a truth value for $f = g$, with 0 being true. Thus, one can define, for example, $f \neq g$ as $\alpha(f, g) = 1$ and $e = f \wedge g = h$ as $\alpha(e, f) + \alpha(g, h) = 0$, where $+$ is defined by the recursion equations written out above. In a similar way one can introduce bounded quantifiers and all the logical connectives. For details see [Goo57].

The connectives and quantifiers are introduced in the manner usual for classical logic, via truth functions. Thus, we will be able to prove many laws of classical logic, for example, $Y = Z \vee \neg Y = Z$, an instance of the law of the excluded middle. That will not faze a constructivist: the schematic law

---

43. I am using 1 as an abbreviation for $S0$, and $\leq$ as an abbreviation for a suitably introduced relation.

commits us only to instances we employ, and in each instance Y and Z are replaced by particular expressions for numbers so that one or the other disjunct can readily be verified by a computation. More is true. Every statement proved in PRA expressed using our logical connectives will have a correct constructive counterpart obtained by reading the connectives constructively and the schematic variables as if they were implicitly universally quantified, with the constructive interpretation of universal quantification. In particular, the constructivist will endorse that $(\forall xy)(x = y \lor \neg x = y)$, since the quantifiers range only over numbers given by definite numerals. In general, since every equation of PRA defines an effectively decidable predicate in a sense suitable to allow the constructivist to conclude that the law of the excluded middle holds, even the constructivist will grant that classical logic applies. (See [Dum77, pp. 20–21] for a discussion of effectively decidable predicates.) Thus, any theorem proved in sPRA will be accepted by an intuitionist or classical mathematician: the theorems are expressed in a neutral language, and all of the rules of inference are accepted by such mathematicians. Moreover, a constructivist and certainly a classical mathematician will even accept the theorems of uPRA, construing the universal quantifiers appropriately, as we have in effect just shown. At least when we construe it classically, uPRA has, on familiar criteria, considerably greater ontological commitments than does sPRA.

Now that we have seen how to extend the expressive power of PRA, we can return to a discussion of the rule of the uniqueness of primitive recursive definition. The rule can be viewed as an inductive principle since it is (modulo the rest of PRA) interderivable with the following induction rule [Goo57, pp. 109, 118]:

$$\text{Infer } f(X_0) = g(X_0) \text{ from } f(0) = g(0)$$
$$\text{and } f(X_0) = g(X_0) \to f(SX_0) = g(SX_0).$$

Hilbert took the induction rule to be part of the basis of finitary mathematics.

The induction rule has been much discussed, for example by Steiner [Ste75, pp. 139–154]. Proving the consistency of a theory of arithmetic finitarily was very much a part of Hilbert's program. Poincaré argued that the chief feature of arithmetic is induction, and that the use of induction in finitary mathematics to prove the consistency of a theory of arithmetic would therefore be circular. Hilbert replied that he justified formal induction on the basis of contentual induction [Hil28, pp. 472–473].

It has been suggested that the difference between the two forms of induction is that contentual induction is only over finitary properties, that is, those that can be defined by equations in PRA, while formal induction allows many more properties. (It is also argued [Tai81, pp. 543–544] that that should have been Hilbert's position.) But, as Steiner noted, that is not a difference in the type of induction, only in what the induction is over.

Steiner argued that the difference between the two forms of induction is in the strength of the conclusion—in my terms, a schematic conclusion versus a universally quantified one. As Steiner put it, echoing Russell's terminology, in contentual induction one concludes that a property holds for any value, not every value. He suggested [Ste75, p. 148] that the difference is that the "any" version cannot be "the minor premise in a *modus ponens* argument," since to conclude Q from $(\forall x) P(x)$ and $(\forall x) P(x) \to Q$ requires that we make use " 'of all the values of the variable at once'; the completed totality of such values is what permits the argument to proceed."

Note that Steiner's argument is principally concerned with the type of generality of the conclusion of an argument by induction. His argument therefore supports the general significance of the distinction between schematic generality and universal generality. Schematic generality is considerably more restricted than universal generality. It is therefore distinct from universal generality.

Steiner's formal criterion for the distinction in terms of modus ponens is questionable, since, as he admits [Ste75, p. 151], it will be sensitive to the deductive system used. Indeed, the formal criterion is useless in the context of PRA, which has no quantifiers. Nonetheless, the motivating description seems right. It suggests the following formal criterion applicable to our quantifier-free system: a variable is used schematically in a proof if every instance of the conclusion of the proof obtained by instantiating the variable can be proved by instantiating the variable appropriately in each place it occurs in the proof.[44] Briefly, the criterion is that one can eliminate general sentences

---

44. More precisely, a variable is used schematically in a proof if there is a simple (primitive recursive) procedure that transforms the original proof and a numeral into a new proof with the numeral substituted for the variable in the conclusion such that every term in the new proof is obtained from a term in the old proof by substituting a numeral for the variable. If the variable occurs in an axiom, one must accept any instance as an axiom.

involving the variable in favor of some of their instances. The criterion has the effect that only finitely many values of the variable can be used, not all the values. We have in effect observed above that the uses of variables of PRA meet the criterion, since they can be taken to be placeholders.

Since a schema commits us to only finitely many instances at a time while a classically construed universally quantified sentence commits us to infinitely many, we can give an example that clearly differentiates between the two: Suppose we want to axiomatize the idea that ω_0 is an indefinitely large natural number. Then we will want ω_0 to be larger than any number we actually use. We can therefore introduce a constant symbol, ω_0, and add the following axiom schema to, say, Peano arithmetic: $X_0 < \omega_0$, where X_0 is a schematic variable, not subject to a rule of universal generalization but only to instantiation by numerals (that is, by expressions of the form 0, $S0$, $SS0$, and the like). We can use the proposed axiom as we intend just if it is satisfiable in the domain of natural numbers. Because we are committed only to finitely many instances of the schema, we can satisfy the axiom by taking ω_0 to be a particular natural number, though which one may have to change if our commitments expand by the addition of more instances of the schema. In contrast, the corresponding universal axiom, $(\forall x_0)x_0 < \omega_0$, is inconsistent with Peano arithmetic. The schematic version can be interpreted as concerning an indefinitely large natural number, the universal version cannot. They are therefore clearly shown to be different. The two sorts of generality are distinct in the commitments that result from them, in the rules of proof that are appropriate to them, and in the consequences of axioms that employ them.

To return to our discussion of induction within PRA, the schematic induction principle is genuinely weaker than the universal one: One can infer each instance of the conclusion of the induction schema without using the schema, the instance for 1 from that for 0, the instance for 2 from that for 1, and so on. Only finitely many values of the induction variable get used. No comparable procedure yields the universal statement. That is all correct, and it seems to me that it is at least closely related to what Hilbert had in mind in his reply to Poincaré. But the axiomatization of PRA given above makes it clear that a different reply is available to the defender of finitary mathematics: The rule of the uniqueness of primitive recursive definition may in some sense be an induction principle, but it is a very weak and special one. All it does is codify that primitive recursive definitions are definitions—that is, it says that two terms with the same primitive recursive definition have the same values. That principle seems to be part of the use of primitive recursive definitions, not a separate principle deserving separate criticism. As we have seen, the induction rule follows from that principle. But Poincaré's criticism applied to the uniqueness rule reduces to the argument that recursive definitions are not justified by a consistency proof, and that is something Hilbert accepted. Hilbert apparently took them to be a fundamental part of mathematical intuition [HB70, p. 286].

There is one more thing that must be checked before accepting PRA as free of infinitary commitments: We are interested not only in the commitments of a theory but in the commitments the theorist must make in accepting the theory. We must make sure that the axiomatization of PRA is itself finitarily acceptable, that is, that no commitment to infinity is required to give the axiomatization. I would like to do that by just showing that PRA can be formalized within PRA, and hence that the commitments of the formalization do not go beyond those within the formalism. Unfortunately, it is not possible to show that for a trivial reason: the formalization of PRA involves many symbols, while PRA works only with numbers in a unary notation.

The problem has a straightforward solution: Suppose that the formalization involves r symbols, $a_1, \ldots, a_r$. (As I did things, there were infinitely many symbols, but that was not essential. For example, the infinitely many variables X_i can perfectly well be replaced by X, X', X'', and so forth, which involves the use of only two symbols, X and $'$.) One can generate all the "words" that can be made up from the r symbols using the following primitive recursively defined "successor function" [Ass60, p. 266]:

$$v(\Lambda) = a_1,$$
$$v(a_1 X_0) = a_2 X_0,$$
$$\vdots$$
$$v(a_{r-1} X_0) = a_r X_0,$$
$$v(a_r X_0) = a_1 v(X_0),$$

where Λ is the empty word and, for example, $a_1 X_0$ is the word that results from prefixing the symbol a_1 to the word X_0. The successive words can be used as a kind of r-ary notation for the numbers, in place of the unary notation of our earlier formalization of PRA. (The notation is closely related to, but not quite the same as, standard base r notation.) We can therefore adapt our axiomatization of PRA into one for the primitive recursive theory of words (PRW) in a straightforward way: We use the same axioms as before, except that we add the schema just given, replace S everywhere it occurs in the axioms by v and 0 by Λ, and add a new uniqueness rule: Infer $f(X_0) = v(X_0)$ from

$$f(\Lambda) = a_1,$$
$$f(a_1 X_0) = a_2 X_0,$$
$$\vdots$$
$$f(a_{r-1} X_0) = a_r X_0,$$
$$f(a_r X_0) = a_1 f(X_0).$$

Our formalization of PRA (or for that matter PRW) can be carried out in PRW, as required.

The axiom system for PRW may seem a bit artificial, since the natural way to implement primitive recursion on words is not to use the v function but to use the same type of primitive recursion on words that we used to define v. But the schema of primitive recursion on words is primitive recursively reducible to the schema of primitive recursive definition in the system we have indicated, and, conversely, there is a primitive recursive word function that takes a definition of a function using our schema to a definition of the same function using primitive recursion on words [Ass60]. Moreover, an appropriate version of the equivalence of the two schemas can be proved in PRW.

The close relation between PRW and PRA is reminiscent of a familiar device—Gödel numbering. Instead of introducing a new set of symbols and the function v, we might as well simply work in PRA and use the number n as a surrogate for the nth word. We will then have a counterpart of a formalization of PRA in PRA that is in fact just a theory about numbers: certain numbers are "terms," certain sequences of numbers are "proofs," and so forth. We therefore do not distinguish carefully between PRA and PRW in what follows.

PRA provides a neutral ground concerning arithmetic, but we shall be primarily interested in showing that certain theorems concerning sets are provable in a neutral way. We therefore need not PRA but a parallel theory for sets—primitive recursive set theory (PRS). We have seen that there is a primitive recursive way to code words as numbers. There is an equally simple way to code hereditarily finite pure sets as numbers [Ack37].[45] A number can be

---

45. Hereditarily finite sets and pure sets were discussed in §V.4. A set is hereditarily finite and pure if it has finitely many members none of which are urelements and if each

written base 2 as a sequence of 0s and 1s. We may think of such a sequence as defining a characteristic function f: $f(n) = 1$ if the $n + 1$st bit of the sequence is a 1, and $f(n) = 0$ otherwise. (We use $n + 1$ instead of n so that $f(0)$ will be defined.) We think of the characteristic function determined by n as determining the "members" of n. That is, we say that m is a "member" of n if the $m + 1$st bit of the binary representation of n is a 1. Thus, $0 = \varnothing$, $1 = \{\varnothing\}$ (since the first bit is a 1, the $1 - 1 = 0$th set is a member), $2 = 10 = \{\{\varnothing\}\}$, $3 = 11 = \{\varnothing, \{\varnothing\}\}$, and so forth.

We shall just use PRA for PRS, taking $\in$ to be the primitive recursively definable relation just described. Since we shall be making strong use of PRS and taking hereditarily finite pure sets to be basic, it would be nice to give a direct axiomatization of PRS. That would show that there is an axiomatization of PRS in which the axioms are natural set-theoretic principles. Unfortunately, no one has, so far as I know, ever worked out an axiomatization of PRS that did not rely on PRA in the way we just did. A certain amount of work has been done on giving a perspicuous definition of the primitive recursive set functions ([Röd67], [Mah67], [JK71]), but all of it has used an unspecified, though weak, background theory.

It is possible to give an axiomatization of PRS using the recursion schema

$$f(\varnothing, X_1, \ldots, X_n) = g(X_1, \ldots, X_n),$$

$$f(X_0, X_1, \ldots, X_n) = h\left(X_0, f\left(\bigcup X_0, X_1, \ldots, X_n\right), X_1, \ldots, X_n\right),$$

$$\text{for } X_0 \neq \varnothing.$$

One can then derive the requisite induction schema from a primitive recursive uniqueness rule in much the same manner as Goodstein did for PRA. Here is the rule: Infer $f(X_0) = g(X_0)$ from

$$f(\varnothing) = g(\varnothing),$$

$$f(X_0) = h\left(X_0, f\left(\bigcup X_0\right)\right), \qquad \text{for } X_0 \neq \varnothing,$$

$$g(X_0) = h\left(X_0, g\left(\bigcup X_0\right)\right), \qquad \text{for } X_0 \neq \varnothing,$$

where f, g, and h are arbitrary terms and the substitutions in f and g are for X_0, while those in h are for X_0 and X_1. Unfortunately, such an axiomatization

---

of its members is hereditarily finite and pure. Thus, a set is hereditarily finite and pure if it and its members, and their members, and so on, are finite and free of urelements.

of PRS seems to require a tediously large number both of primitives and of axioms and rules interrelating them. I therefore omit one. The details seem to be uncontroversial—the primitives can be taken to be things like $\bigcup Y_0$ and $\{Y_0, Y_1\}$ and the axioms and rules are things like $X_0 = \bigcup\{X_0, X_0\}$ and substitution and equality rules. There is no problem mimicking such a system in PRA by defining functions in PRA to serve as the primitives (in such a way that set membership goes over to the counterpart described above) and deriving all the axioms and rules from those of PRA. One can go the other way as well.

Leaps of Faith

The theme of this chapter is What Are We Doing? We consider various attempts to characterize what we actually do know, abandoning all consideration of the epistemic constraint. Instead of using epistemological considerations to put upper bounds on the possibilities of mathematical knowledge, as we did in §VI.2, we use the consideration of our mathematical knowledge to put lower bounds on our epistemic capacities.

§1. Intuition

The most courageous attempt to let the facts of mathematical practice drive epistemology is Gödel's. The clearest statement of his position is found in his 1964 revision of the article discussed in §V.5, the one in which he introduced the iterative conception of set. First, he said [Göd47, p. 474], referring to the skeptical attitudes of Brouwer, Poincaré, and Weyl,

> However, this negative attitude toward Cantor's set theory, and toward classical mathematics, of which it is a natural generalization, is by no means a necessary outcome of a closer examination of their foundations, but only the result of a certain philosophical conception of the nature of mathematics, which admits mathematical objects only to the extent to which they are interpretable as our own constructions or, at least, can be completely given in mathematical intuition.

Thus, Gödel rejected the epistemic constraint because it is in conflict with classical mathematics and its "natural generalization" set theory. In a supplement to the article, Gödel added [Göd47, pp. 483–484],

> But, despite their remoteness from sense experience, we do have something like a perception also of the objects of set theory, as is seen from

the fact that the axioms force themselves upon us as being true . . . The set-theoretical paradoxes are hardly any more troublesome for mathematics than deceptions of the senses are for physics.

It is common to focus on the apparent remarkable consequence of Gödel's argument, namely that there are mathematical objects that we, in some sense, perceive. I should like to emphasize instead the basis for that conclusion: "the axioms force themselves upon us as being true." There is a natural reply to that—the paradoxes show that what is forced upon us as true is at best highly unreliable, and no part of mathematics. Gödel was dismissive of the reply— he characterized the paradoxes as unimportant, and he stopped just short of dismissing the Comprehension Principle (which leads to the paradoxes) as the mathematical counterpart of an optical illusion. An optical illusion loses its power to cause trouble as soon as we understand it. Gödel thought that the iterative conception similarly neutralized the paradoxes—as indeed it does. As we have seen in preceding chapters, no such drastic measure as the iterative conception was in fact needed. The paradoxes had never caused any trouble for the combinatorial conception of set. Our sense of what is true has been a quite reliable guide to how to extend our commitments.

To avoid misunderstanding, it is necessary to digress briefly at this point. Kitcher was surely right when he emphasized [Kit83, pp. 61–62] that we have arrived at the axioms of set theory through a historical, evolutionary process. We have described that process in some detail in previous chapters. No one, not Cantor, not Zermelo, had axiomatic truths forced upon them by revelation or introspection. Indeed, as we have seen, the present-day axioms of set theory were to a large extent selected in a rather *ad hoc* manner by Zermelo to defend his proof of the Well-Ordering Principle. But that does not show that the axioms are not self-evident, or at least based on self-evident principles, and it does not show that that self-evidence is irrelevant to their use as axioms.

In any mathematical theory, there is a network of obvious facts and not-so-obvious facts connected to them. It is most certainly a matter of historical accident, custom, and convenience which are used as axioms and which derived. Von Neumann's axioms for set theory are rather different from those of Zermelo—especially in the original form von Neumann used that took functions to be basic. No one uses the axioms that take functions to be basic, but in principle they are just as good as the axioms we do use, since they are pretty much equivalent. Their lack of popularity is in large part determined by history. Perhaps von Neumann's awkward terminology doomed his innovation. Moreover, von Neumann's Limitation of Size Axiom is not self-evident—it seems too strong, as von Neumann himself noted. Nonetheless, we would be willing to take it as an axiom in part because it has been shown to be derivable from self-evident principles—Replacement and Choice.[1] Axioms need not be self-evident, and not all self-evident principles need be axioms.

In the special kind of axiomatization in which the axioms are intended to be a characterization of an antecedently given subject matter—the natural numbers, the real numbers, the geometric plane, the sets—the axioms must be true to that antecedently given subject matter. That is, they must be suitably related to evident principles about that subject matter. That is what shows that they are a suitable axiomatization of the subject matter. In many interesting cases of axiomatization there is no particular prior subject matter; one simply uses axioms to focus attention on a fruitfully related bunch of structures. That is the case for, for example, the axioms for groups, rings, and fields. In such cases one would expect that considerations of self-evidence will not arise in the justification of the axioms. As Kitcher argued [Kit83, pp. 219–220], the justification rests solely on their use for systematization and generalization. But when an intended application is being axiomatized, that is not the case: self-evidence is then an important factor in the selection of axioms. It cannot be the only factor. Simplicity, strength (that is, unifying power), utility, accidents of history, and other factors all enter in as well.

Choice, which is self-evident for combinatorial collections, was accepted only in part because it is self-evident. The development of mathematics showed that Choice is not only self-evident, but useful. That utility is a part of what led to our adoption of the combinatorial notion of collection and with it the Axiom of Choice, which is self-evident in virtue of the fact that we have adopted the combinatorial notion of collection.

Cantor did not arrive at the set-theoretical principles he employed by a process of historical analysis or systematization or generalization of what came before. He was led to the progression of transfinite symbols by his work on Fourier series, but the further development of his set theory was in large part a working out of the implications of that progression. He developed a picture, something like Kitcher's "local justification" [Kit83, pp. 93–94], building on the core idea of the transfinite progression.

---

1. More carefully, it is derivable from Replacement and Strong Choice in the presence of either Foundation or Anti-Foundation (§V.4).

Cantor's set theory, and ours, is based not on any principles or axioms but on the picture (for want of a better term) to which Cantor was led of infinite orders and sizes and of the uses to which they can be put in understanding functions and series and the real numbers. The axioms we use are self-evident because they are true to a certain picture. That is clearest in the case of Replacement. Replacement received no historical justification as a systematization of prior practice or the like: the fact that its applications are so recondite shows that it had little to do with prior practice. The application of Replacement came only after its acceptance. Both Skolem and Fraenkel saw that Replacement was a self-evident principle concerning combinatorial collections before von Neumann discovered that Replacement was good for something and even before they were convinced of the general utility of combinatorial collections. Even today, the self-evidence of Replacement outstrips its applications. We accept it because it is true to the combinatorial notion of collection, that is, because it is self-evident.

What evolves historically is not only or primarily axioms, as Kitcher sometimes seems to think, but our pictures or notions as well. Claims of self-evidence are not, at least not always, dispensable psychological processes as Kitcher thinks [Kit83, p. 94]: as we saw in §§V.1–V.2, they are claims that certain principles are true to a picture, conception, established practice, or subject matter.

I do not have a general account of what "pictures" are—I wish I did. But to act as if it is just axioms that evolve is to sweep an important problem under the rug. If axioms were the primary locus of mathematical development, then how could it be that so little of the historical development of mathematics has been axiomatic? Instead of trying to give an abstract systematic account of mathematical pictures or whatever, I shall continue to concentrate on the case at hand—the set-theoretic infinite. Gödel confronted what I take to be the same problem.

Gödel started with the fact that the axioms force themselves upon us as being true and tried to develop an epistemological theory to account for that. He identified the fact that the axioms force themselves upon us as being true with the existence of a "kind of perception," namely, "mathematical intuition." After the briefest of indications of what might be the source of such intuition, he continued [Göd47, pp. 484–485],

> However, the question of the objective existence of the objects of mathematical intuition (which, incidentally, is an exact replica of the question of the objective existence of the outer world) is not decisive for the prob-

lem under discussion here. The mere psychological fact of the existence of an intuition which is sufficiently clear to produce the axioms of set theory and an open series of extensions of them suffices to give meaning to the question of the truth or falsity of propositions like Cantor's continuum hypothesis. What, however, perhaps more than anything else, justifies the acceptance of this criterion of truth in set theory is the fact that continued appeals to mathematical intuition are necessary not only for obtaining unambiguous answers to the questions of transfinite set theory, but also for the solution of the problems of finitary number theory (of the type of Goldbach's conjecture), where the meaningfulness and unambiguity of the concepts entering into them can hardly be doubted.

Christian Goldbach's conjecture is the conjecture that every even positive number greater than 2 is the sum of two prime numbers ($4 = 2 + 2, 6 = 3 + 3,$ $8 = 5 + 3, \ldots$). It can be formulated in PRA. The last sentence of the quote refers to the fact that Gödel's incompleteness theorem makes it easy to exhibit a sentence of PRA that can only be shown true using mathematical intuition. The simplest example to describe here is a sentence of PRA that is true if and only if ZFC is consistent. We know it is true only because we know that ZFC is indeed consistent, and we know that only through mathematical intuition. Of course, as Gödel in effect noted, while one might be skeptical about whether the continuum hypothesis is true or false, such skepticism does not seem tenable for sentences of PRA. Gödel recognized that the axioms of set theory *are* self-evident, and that such self-evidence plays a necessary role in mathematics. His challenge is to develop an understanding of the nature of mathematical intuition. Note here that Gödel's mathematical intuition is of things including the axioms of set theory—he is very much concerned with the problem of the infinite, not just that of the abstract.

Gödel's own work toward a theory of intuition is scant and obscure, but he apparently developed a monadology that permitted access to the full set-theoretic hierarchy [Mad90, p. 79]. Even his comparatively uncontroversial remarks concerning the analogy between mathematical intuition and sense perception drew this response [Ben73, p. 415]: "what is missing is *precisely* . . . an account of the link between our cognitive faculties and the objects known. In physical science we have at least a start on such an account."

However one may assess Gödel's theory of intuition, his challenge to develop such a theory remains perhaps the principal problem in the philosophy of mathematics. The lack of a theory is at any rate certainly the chief obstacle to understanding the infinite. Maddy [Mad90] made a very interesting attempt to supply a theory of mathematical intuition, but her theory allows for intuitions only of small finite mathematical objects, failing to confront the crucial problem of intuitions concerning the infinite. My own theory will be presented in the remaining chapters of this book. But we have not yet finished our review of Leaps of Faith.

§2. Physics

The most popular Leap of Faith has been so widely discussed that it requires only brief treatment here: it is the Quine–Putnam indispensability argument. The idea is that our knowledge is not neatly compartmentalized into separate areas, and in particular that there is no sharp dividing line between mathematics and the mathematical parts of the sciences, especially physics. Since even the most well-established parts of physics, not to mention all of the contemporary rival controversial physical theories, make ineliminable use of mathematics, our actual commitment to physics, which in part involves empirical evidence, commits us to some mathematics.[2] So far, Quine. Putnam added that not only does physics use mathematics but the very statements of physical laws and analyses of physical properties make ineliminable use of mathematics. We don't merely use mathematics to draw conclusions about the physical world: our description of that world is ineliminably mathematical. (See [Qui61] and [Put79a], and see, along with many others, [Bur83, Fie89, Mad90, Mad92, Par83b, Ste75] for discussion.)

In the present context the Quine–Putnam view counts as a Leap of Faith, since it is compatible with and frequently accompanied by a denial of anything like the epistemic constraint. But, unlike the other views presented in this chapter, it comes with a perfectly good epistemic theory: we get to know our mathematics in a way that is intimately connected with the way in which we get to know our physics, and so the term Leap of Faith is misleading in this case.

When applied to mathematics as developed from, say, the time of Newton and Leibniz to that of Cauchy and Weierstrass, the Quine–Putnam view is appealing and indeed compelling: Mathematics was as a matter of historical fact developed in intimate association with physics. Mathematicians took physics

---

2. It is controversial just how much mathematics is ineliminably used by physics. See [Fef 88] for a strong argument that not much more than Peano arithmetic is required.

to be directly relevant to mathematics, for example by taking physical arguments to be evidence for mathematical theorems. Most of the mathematics developed during that period has been applied within physics. The epistemological picture that Quine and Putnam suggest looks a lot like the one that was in fact realized in mathematical practice, and their argument is an important one for defusing skepticism about the existence of (at least some) mathematical objects.[3]

Unfortunately, when we turn to the mathematics of the infinite including modern set theory, the Quine–Putnam indispensability argument is much less successful. The most important reason is that parts of set theory have found little or no application in the sciences, and they therefore do not receive the requisite support. Quine has granted that explicitly [Qui86, p. 400]:

> Pure mathematics, in my view, is firmly imbedded as an integral part of our system of the world. Thus my view of pure mathematics is oriented strictly to application in empirical science. Parsons has remarked, against this attitude, that pure mathematics extravagantly exceeds the needs of application. It does indeed, but I see these excesses as a simplistic matter of rounding out. We have a modest example of the process already in the irrational numbers: no measurement could be too accurate to be accommodated by a rational number, but we admit the extras to simplify our computations and generalizations. Higher set theory is more of the same. I recognize indenumerable infinites only because they are forced on me by the simplest known systematizations of more welcome

---

3. Even if Hartry Field [Fie89] is right that the mathematics of physics is largely dispensable for physics, and hence not justified by physics in the way that Quine and Putnam suggest, it will remain true that much of the mathematics of physics was in part suggested by physics and can be pictured in terms of its physical applications, and that that relation to physics provides some sort of epistemic warrant for our interest in and exploration of the mathematics. Field's project of reformulating parts of physics in such a manner as to make them autonomous from mathematics is important for clarifying the physical commitments of physical theories, but so long as the project does not result in changes in physics, the mathematics that is usually used in physics will continue to gain interest and a kind of support (though perhaps not truth) from its close relation to physics. Maddy [Mad92, pp. 280–282, 289] argued that the indispensability argument does not work because (p. 289), even in the case of scientific theories, "indispensability for scientific theorizing does not always imply truth." My attitude is the same: indispensability does guarantee interest and a kind of support, even if not truth.

matters. Magnitudes in excess of such demands, e.g., $\beth_\omega$ or inaccessible numbers, I look upon only as mathematical recreation and without ontological rights.

The reference to $\beth_\omega$ is a pointed one: $\beth_\omega$ is the cardinality of the set

$$\bigcup \{Z_0, \mathcal{P}(Z_0), \mathcal{P}(\mathcal{P}(Z_0)), \ldots\}$$

that Skolem and Fraenkel had observed one cannot prove to exist without using Replacement (§V.2). Quine thus in effect dismissed Replacement as mere "recreation." Whatever the merits of the Quine–Putnam view for less recondite parts of mathematics—and they are substantial—the view is of little or no help for parts of set theory.

Maddy has proposed what she called the modified indispensability argument [Mad92, pp. 279–280], which does better than the Quine–Putnam indispensability argument in that it counts all of set theory as part of our mathematical knowledge. The idea is that although it is physics that leads to our commitment to some mathematics, once we have a commitment to mathematics that commitment can be expanded using techniques internal to the development of mathematics: not every piece of acceptable mathematics needs to be directly tied to physics. Thus, according to the modified indispensability argument [Mad92, p. 280], "the calculus is indispensable in physics; the set-theoretic continuum provides our best account of the calculus; indispensability thus justifies our belief in the set-theoretic continuum, and so, in the set-theoretic methods that generate it; examined and extended in mathematically justifiable ways, this yields Zermelo–Fraenkel set theory."

As Maddy has argued [Mad92], the indispensability view—whichever version—is even in trouble for parts of set theory that might conceivably be accounted for as the result of "rounding out," like Choice or Power Set. According to the view, set theory would have to be anchored in physics, and as a matter of fact it is not. The phenomenon of self-evidence that Gödel emphasized shows that: To paraphrase Parsons [Par80, pp. 151–152] (who was making a similar point about elementary mathematics), "the existence of very general principles that are universally regarded as obvious" that on the indispensability view would "be bold hypotheses, about which a prudent scientist would maintain reserve, keeping in mind that experience might not bear them out" shows that mathematics is in this respect not "continuous with science." The axioms of mathematics do not, contra Quine, "have a status similar to that of high-level theoretical hypotheses." Even though one can arrive at full present-day set theory via the modified indispensability view, that is just not the route mathematicians are actually taking.[4] If we wish to account for actual mathematical practice, we must choose a different way.

§3. Modality

Putnam suggested another view, quite different from the Quine–Putnam view just discussed, in [Put67]. Though he did not advocate the view there, he did advocate it later [Put79b, p. 59]. The view has been developed in detail and advocated by Geoffrey Hellman [Hel89], who called it "modal structuralism." Hellman addressed [Hel89, p. vii] "the idea that pure mathematics is concerned principally with the investigation of structures of various types in complete abstraction from the nature of individual objects making up those structures."

How can such a view, despite its intrinsic plausibility, be coherently maintained? The objects in mathematical structures cannot be required to be physical objects. Among other reasons, there may not be enough of them. If the objects that make up mathematical structures are not concrete, they must be abstract objects, indeed mathematical objects. In that case, individual mathematical objects must be allowed prior to the structures that were supposed to be doing the ontological work. If that is right, then structuralism cannot be the whole answer. Hellman eliminates the need for objects to make up the structures by taking mathematics to be about possible structures, not actual ones. When the modality is taken to be basic, the need for objects prior to the mathematical structures apparently disappears, and structuralism without a background ontology becomes an attractive possibility.

Hellman presupposed a logical modality, second-order logic, and, most important for our purposes here, axiomatic second-order set theory. Let me defer specifying the axioms of second-order set theory until §4. For the purposes of the present section, what matters about axiomatic second-order set theory is that its natural models are precisely Zermelo's normal domains, as Zermelo in effect showed in his work discussed in §V.4. As Hellman said [Hel89, p. 72], "How best to describe and assess our 'evidence' for [the hypothesis that natural models of the axioms of second-order set theory are possible] remains

---

4. Maddy [Mad92, pp. 282–286, 289] made the point in a somewhat different way: If physicists were to stop making any indispensable use of the mathematics of the infinite in any of the fundamental theories, a possibility that she documents cannot be dismissed out of hand, set theorists would continue their investigations.

one of the most difficult challenges confronting mathematical epistemology." That is the problem addressed in this book. It was not addressed by Hellman.

The modal structuralist must suppose that there is no domain of all possible structures or anything similar [Hel89, p. 59], or mathematics could be taken to be the study of that domain and the whole point of the modal-structuralist project would be lost. Since structures can be realized as set-theoretic objects, the modal structuralist must therefore deny that there is a single intended domain of all sets. That denial comes naturally: if there are no combinatorial sets but only various possible models of the second-order axioms of combinatorial set theory, then those possible models will not be a part of an all-encompassing domain of sets, which would constitute a single absolute, not merely structural, possibility.

We want to be able to compare structures mathematically, including models of set theory. That requires that models of set theory be embedded in other structures. The modal structuralist is therefore led to suppose that every model of second-order set theory is a member of some larger structure, which is naturally taken to be a larger model of set theory. Thus, the modal structuralist follows Zermelo ([Zer30], §V.4) by taking each model of set theory to be embedded in a larger one [Hel89, pp. 55-56]. That is the reason for my present interest in Hellman's work.

While Zermelo suggested that "the whole of set theory is only represented by the 'open' totality of all normal domains," he largely confined his mathematical work to individual normal domains and never developed any systematic method of talking about the open totality of them (cf. [Hel89, p. 56]). Hellman, working out a suggestion of Putnam [Put67, p. 310], has provided such a method. Briefly and roughly, the idea is that one interprets "for all sets x" to mean "for all sets in any normal domain" and one interprets "for some set x" to mean "for some set in some extension of the present normal domain." In that way, one takes the possibility of moving to a larger normal domain into account. Thus, we take the assertion that there is some large set to mean not that it is in the model of set theory that we may at present be employing but that we can move to a larger one in which there is such a set.

There is, for example, a set large enough to serve as a model of second-order set theory in every normal domain except the smallest.[5] It is therefore not true to say that there is such a set in an arbitrary model of second-order set theory—the smallest normal domain is a counterexample. But if what we

_____

5. For simplicity, I am considering only pure normal domains.

are interested in is the open progression of models, it seems natural to say that there is such a set, and Hellman's "Putnam semantics" takes that to mean that every model of second-order set theory has an extension in which there is such a set, which is true. More precise technical details follow.

Technical Remark. The proposal is naturally described in modal terms using Kripke semantics. One thinks of each model of second-order set theory as a possible world. As Zermelo showed, of any two such models, one is isomorphic to an end extension of the other. One therefore thinks of one model as being accessible from another if it is an end extension of it. (We neglect the models that are not actually end extensions one of the other, since given any two models, one is always isomorphic to an end extension of the other.)

The Putnam semantics for sentences about the open-ended domain of sets is given by replacing each universal quantifier, $\forall$, first- or second-order, by its necessitation, $\Box\forall$, and, similarly, replacing $\exists$ by $\Diamond\exists$. Thus, for example, $\forall x$ becomes "for any end extension M for any x in M," while $\exists x$ becomes "there is an end extension M in which there is an x."

A model of second-order set theory larger than a given one, M, exists if and only if there is a strongly inaccessible cardinal not in M. The statement that every model is a member of a larger one can therefore be rendered, using Putnam semantics,

$$\forall\alpha\exists\beta(\beta > \alpha \wedge \text{"}\beta \text{ is strongly inaccessible"}),$$

which says, because of the Putnam semantics, that there are unboundedly many strongly inaccessible cardinals in the progression of all possible cardinals, not just in the progression of cardinals in a particular model.

The same sentence may be true in some model, now interpreted in the usual way, but that does not express the right thing: after all, it says nothing about whether that very model is embedded in a larger one. Nonetheless, the truth of the sentence interpreted in terms of Putnam semantics provides a motivation for accepting the sentence with the usual semantics as an axiom of a strengthened set theory. The principle that every model of the strengthened set theory is embedded in a larger one then becomes, "there are unboundedly many hyperinaccessible cardinals," and so modal structuralism provides a kind of argument for the existence of hyperinaccessibles. Hellman has worked out analogous arguments for strong Mahlo cardinals, hyper-Mahlo cardinals, and so forth, though not for weakly compact cardinals [Hel89, pp. 83–91].

The indicated Kripke semantics seems to entail a commitment to a class of all models of second-order set theory, which the modal structuralist cannot countenance. That is avoided by not giving Kripke semantics officially, instead just taking the modalities to be basic, that is, defined by their use [Hel89, p. 15]: "Just how this is to be understood will emerge from the role it plays in the interpretation."

The semantics I have given is not Hellman's. He employed a modality that ranges over all structures, not just models of set theory, and he made them all accessible from each other. His modality is a general one for all mathematical structures, not one special to set theory. That is a key difference between his proposal and that of Parsons [Par83c]. When discussing set theory, Hellman added in the restrictions I have given with explicit second-order sentences. That is, he rendered $(\forall x)(\exists y)\phi$ by [Hel89, p. 74]

$$\Box(\forall M)(\forall x) \,(\text{``}M \text{ is a model of second-order ZFC''} \land x \in M$$
$$\rightarrow \Diamond(\exists M')(\exists y) \,(\text{``}M' \text{ is an end extension of } M\text{''} \land y \in M' \rightarrow \phi)).$$

Since the only aspect of his work I am interested in here is that concerning set theory, I have built the additional restrictions into the modality, which is formally simpler.

§4. Second-Order Logic

In his excellent *Foundations without Foundationalism* [Sha91], Stewart Shapiro did not, as Hellman did, just assume second-order logic. He argued in great detail that ordinary mathematical practice makes essential use of (standard) second-order logic.

While Shapiro accepted Skolem's skepticism as cogent, he argued that mathematicians do as a matter of fact act as if there is an unequivocal understanding (up to isomorphism) of what the natural numbers and real numbers are, and also of the notions of finitude and minimal closure and transitive closure, and even of well-ordered set and cardinalities.[6]

Not one of the notions listed has an adequate definition in first-order logic:

---

6. See [Sha91, §5.1] for definitions. He called the transitive closure the ancestral relation.

the complete first-order theory of the natural numbers has a model that is not isomorphic to the natural numbers, first-order set theory has a denumerable model, and so nondenumerability has no adequate first-order characterization, and so forth for the others.[7] That is the upshot of what Skolem taught us. In contrast, each of the notions has a natural definition in second-order logic. (For example, a set S is finite if *every* one-to-one function from S to S has every member of S in its range.) Thus, Shapiro argued, the natural way to codify mathematical practice is in second-order logic.

According to Shapiro, Skolemite qualms about the coherence of second-order logic are equally qualms about the way we do mathematics. Shapiro endorsed [Sha91, p. 195] Alonzo Church's claim [Chu56, p. 326n] that the assumption involved in employing second-order logic "is not made more acute or more doubtful, but only more conspicuous, by its application" to logic.

Shapiro was unmoved by Quine's argument [Qui70, p. 66] that second-order logic is "set theory in sheep's clothing." He accepted that second-order logic is nothing more than a fancy way of quantifying over all subsets of a domain. He granted Quine's conclusion that second-order logic and set theory are intimately connected, but he rejected the implication that that somehow renders second-order logic unfit to serve as logic. He outdid Quine in his antifoundationalism by denying that there is anything special about logic that requires that it be free of substantial mathematical commitments. If logic were to be in some sense prior to mathematics, that would be necessary. But, for Shapiro, the logical foundations of mathematics are just more mathematics. (See [Sha91, p. 97].)

According to Shapiro, the perspective that just takes first-order set theory to be the "mother tongue" of mathematics, that is, using it while declining to justify it further, is a confirmation of his own view [Sha91, pp. 254–255]. After all, set theory justifies the idea that specifying a domain specifies the relations and operations on that domain, which is what is required to make sense of second-order logic.

Shapiro preferred second-order logic to set theory because second-order logic refers only to the relations and operations on a domain, and not to the other things set theory brings with it—other domains, higher-order relations, and so forth. Thus, for example, second-order arithmetic refers only to relations between and operations on numbers, not to the whole set-theoretic hierarchy. Second-order logic enables one to make clear what the minimum

---

7. See [Sha91, §§5.1, 5.2] for a detailed discussion.

commitments of a given theory are—except, as Shapiro noted, for set theory itself. If the mother tongue is first-order set theory, then one cannot make sense of second-order set theory in that way.

We can agree with Shapiro that the first-order set-theoretic foundation of mathematics is "natural and fruitful" and that second-order theories of arithmetic or geometry are a useful way of isolating just how much set theory is involved in those areas of mathematics [Sha91, pp. 251, 256]. But first-order set theory is not an adequate foundation for set theory. For one thing, first-order set theory has models in which the natural numbers are not well founded,[8] models in which the real numbers are denumerable,[9] and other unintended models with other defects. That is just the modern version of Skolem's paradox. For another thing, first-order set theory includes the different instances of a Replacement schema as separate axioms. There is no general replacement principle to unify those axioms, despite the fact that we inevitably conceive of them as instances of a single unifying principle. But that renders the justification of each of the separate Replacement axioms artificial (compare [Sha91, pp. 117–118]): they do not have separate justifications, but a unified justification shows too much, since it goes beyond what is postulated in first-order set theory.

There is an obvious alternative to first-order set theory that does not have the defects mentioned, namely, second-order set theory.[10] Second-order set theory is just like first-order set theory except that we use the version of Replacement with a universal second-order quantifier over functions from sets to sets, that is, for every function f and set x, the range of f on x is a set. That is, in effect, the theory that Zermelo considered in 1930. (He used second-order versions of Separation and Foundation as well, but the resulting axiom system is equivalent to the one just given, as is well known and straightforward to show.) Second-order set theory does much better than first-order set theory in characterizing what we are talking about: with Foundation, it is quasi- categorical[11] in the sense outlined in §V.4—each model is fully deter-

---

8. That is, there are models that include infinite descending chains of "natural numbers." The chains are not in the model.

9. That is, there are models for which there is a function mapping the real numbers of the model one-to-one onto the natural numbers of the model. Such a function is not in the model.

10. Recall that we are only considering standard or full second-order logic. See §V.3.

11. I have taken the term *quasi-categorical* from [Hel89].

mined, up to isomorphism, by the cardinality of the set of urelements and the least ordinal not in the model.[12] That is not much more than a restatement of Zermelo's results discussed in §V.4. Recall that he also in effect showed that every model of second-order set theory is isomorphic to a normal domain.

To see what is good about quasi-categoricity, assume there are no urelements, for expository simplicity. Then each model is fully determined, up to isomorphism, by the least ordinal not in the normal domain isomorphic to the model and, moreover, given any two models, one is isomorphic to an end extension of the other (that is, up to isomorphism, the larger model extends the smaller one without adding any new elements to the members of the smaller one). Thus, once it is agreed that a certain V_α exists, it is fully determined up to isomorphism what its structure is and how it fits into any successive V_αs. As a result, any two models have isomorphic natural numbers and isomorphic real numbers—the natural numbers and the real numbers are in $V_{\omega+\omega}$ and $V_{\omega+\omega}$ exists in every model. The natural numbers and real numbers are uniquely characterized up to isomorphism, which is all that one can ask for any mathematical object. Moreover, for example, any two models in which there are strongly inaccessible cardinals agree on what is the least such cardinal, and so forth.[13] No room is left for Skolemite relativism.

Second-order set theory is much more suited to the realistic attitude actually adopted by mathematicians than is first-order set theory. But when we take it as a basis for mathematics either we must just dogmatically claim to understand it or we become involved in a regress: to the extent that we understand second-order logic in terms of a background set theory, the second-order logic on which second-order set theory is based is just more set theory.[14] All we have accomplished is to push things back a step by explaining one theory of sets in terms of another behind it.[15] The position now being

---

12. The quasi-categoricity result holds for set theory with Foundation or Anti-Foundation (§V.4). Without either the situation is slightly more complex, but the second-order theory is still a substantial improvement over the first-order one.

13. The results just listed in the text hold even without Foundation or Anti-Foundation.

14. To be sure, the new background set theory need not have levels beyond the subsets of the domain of the first one, and so it need not have a version of Replacement. That might yield a certain amount of simplification.

15. Georg Kreisel argued that the regress is not vicious [Kre71, p. 97]: "it is by no means circular to use a concept in order to *state the facts* about it." He is an advocate of full second-order set theory (see [Kre71] and references therein). He suggested [Kre71,

discussed is Zermelo's of 1930: for him, as we saw in §V.4, the second-order quantifiers over each set-theoretic domain are explained in terms of the sets of a larger domain. The position is in some respects an attractive one.

We would certainly be better off with a theory that had weaker presuppositions than second-order set theory but that was still strong enough to yield quasi-categoricity. If there were no such theory, we might settle for a first-order "mother tongue" for mathematics or for Zermelo's picture, but in fact there is such a theory. We are not obliged to seek further explanation—"all explanations and justifications must eventually end." But neither are we obliged to stop—if we can get answers to our queries, so much the better. (Compare [Sha91, p. 117].)

The key observation toward obtaining a weaker theory is that the only second-order axiom we need—Replacement—has only a universal second-order quantifier at the beginning. It is that that enabled Skolem to give a version of the axiom using a first-order schema in the first place. We may therefore replace the quantified second-order version of Replacement by a free-variable version: we replace $\forall f \phi(f)$ by $\phi(F)$. We use free second-order variables in our language but drop the second-order quantifiers.[16] The only rule involving the second-order variables will then be a substitution rule—any substitutable formula may be substituted for a second-order variable.[17]

Like the schematic variables of PRA, the free second-order variables here have an associated substitution rule, but no other associated rule of inference. (The other rules all involved the quantifiers and so got dropped along with them.) We may therefore take the "free variables" of our free-variable second-order logic to be schematic variables in much the same sense in which we have taken the variables of PRA to be schematic variables. That is why I used a capital letter for the free variable in the free-variable version of Replacement above—we agreed to use capital letters for schematic variables. I shall henceforth usually speak of second-order schematic variables and schemas and of schematic second-order logic, not of free second-order variables.

We can describe universally quantified second-order Replacement as say-

---

p. 96] that we attempt to "formulate the principles involved in the discovery of the existing axioms" and extend them by formulating them "as generally as we can." I take the investigations of the subsequent chapters to be in that spirit.

16. John Corcoran emphasized the usefulness of essentially this logic for proving categoricity theorems [Cor80].

17. One formula is *substitutable* in another if it has no occurrence of a free variable in it that becomes bound in the formula that results from the substitution.

ing that the range of *every* function on a set is a set and describe schematic second-order Replacement as saying that the range of *any* function on a set is a set. Set theory formulated with schematic Replacement differs from set theory formulated with second-order Replacement primarily in that it includes no second-order quantifiers.[18] It differs from first-order set theory in that it includes the Replacement schema in the official formal language, instead of including only the instances of the schema.

Shapiro discussed a deduction system for schematic second-order logic of the sort just described.[19] He showed that the set of logical truths of schematic second-order logic is axiomatizable. In that respect the logic is like the more tractable first-order logic, not like second-order logic. On the other hand, he showed that the consequence relation for the logic is not axiomatizable. In that respect the logic is like the more powerful second-order logic and unlike first-order logic. Schematic second-order logic is therefore seen to be a genuine intermediate possibility.

As you will have assumed, the reason I am rehearsing results concerning schematic second-order logic is that set theory with schematic second-order Replacement is the basis of the theory promised above that yields quasi-categoricity without a commitment to second-order logic. If we restrict the permissible substitutions for a second-order schematic variable to (substitutable) formulas in the language of first-order set theory, then the instances of schematic Replacement are exactly the axioms of the usual first-order Replacement schema, the one suggested by Skolem. With that restriction on substitution, schematic set theory (that is, set theory with schematic Replacement) has the advantage over first-order set theory that it includes a uniform version of Replacement and thus expresses our commitments more accurately, but it has the same deductive strength as first-order set theory.[20] The

---

18. The form of the Replacement Axiom also differs—$\phi(F)$ instead of $\forall f \phi(f)$—but that difference would not be essential if we had second-order quantifiers and a rule of universal generalization to go with them. That would allow inferring $\forall f \phi(f)$ from $\phi(F)$. In allowing that inference I assumed that the F is taken to be an ordinary generalizable variable, not a schematic one. I shall not consider hybrid theories with both ordinary and schematic second-order variables.

19. Naturally, he used ordinary free second-order variables instead of our special schematic ones. He did not discuss the commitments that result from using only free second-order variables in any detail, though he did note the parallel with Hilbert's use of free variables in [Hil26]. See [Sha91, p. 247].

20. The schematic second-order theory has some consequences that the first-order theory does not for a trivial reason: it has consequences in which the schematic letters

same is true even if we allow substituting arbitrary (substitutable) formulas in the language of set theory—including those with schematic letters in them—for schematic variables, as we have been contemplating.[21]

Suppose we insist that the permissible substitutions for a second-order schematic variable include *any* substitutable formula, even one in any expansion of the vocabulary of the present language that we may come to employ, as opposed to the usual substitution rule, which was stated for a fixed vocabulary. Then the schematic second-order version of Replacement acquires substantial additional deductive strength. Set theory with it becomes quasi-categorical and that can be shown in schematic second-order logic [Sha91, p. 248].[22]

We call the new version of the substitution rule just suggested, the one that automatically includes formulas in an expanded vocabulary, *full* substitution (by analogy with full second-order logic), and the fixed-vocabulary version discussed before *weak* substitution.[23] With full substitution, schemas constrain the future commitments that may result from future developments. The point of using full substitution along with schematic Replacement is that when we expand the language of set theory, we automatically obtain new instances of Replacement for functions definable in the expanded language—

---

appear. But the set of consequences of the schematic second-order theory that do not include occurrences of the schematic letters is exactly the set of consequences of the first-order theory. For the straightforward proof, see [Fef91, p. 7].

21. Sketch of proof: Given a proof of a sentence without schematic variables, go through it and replace each schematic variable everywhere in the proof by whatever first-order formula eventually replaces it. If the same schematic variable is substituted for several times, just copy the relevant part of the proof several times, making each substitution in one copy. If a schematic variable never gets substituted away, substitute something arbitrary for it. That yields a proof of the sentence in ordinary first-order logic. Note that the proof sketched shows that the schematic variables meet my proof-theoretic criterion for schematic variables discussed in §VI.4.

22. As before, quasi-categoricity holds only in the presence of Foundation or Anti-Foundation, but most of the nice consequences we use go through without either.

23. Solomon Feferman has investigated theories with full substitution in detail, including, among many other things, the relation between them and the associated first-order and weak schematic theories, as well as some intermediate possibilities not introduced here. He said [Fef91, p. 8] that the extension of schemas beyond an original fixed language "corresponds to a more fundamental view of what it means to accept some axiomatic principles." His motivation was not categoricity but incompleteness, especially of axiomatizations of arithmetic. His and related work is surveyed in [Fef91], which also includes a brief history of the use of schemas.

excluding something or other from the Replacement schema is just not on.[24] Though that consequence of full substitution is not a familiar part of formal mathematics, there is no question that it is something that we had in mind all along. That is closely related to what Shapiro argued [Sha90, pp. 258–259], and, as he observed, it is in line with Benacerraf's observation [Ben85, p. 111] that

> Despite the imagined possible misunderstandings, mathematical practice reflects our intentions and controls our use of mathematical language in ways of which we may not be aware at any given moment, but which transcend what we have explicitly set down in any given *account*—or may ever be able to set down.

The requisites for using the added power of schematic Replacement with full substitution, briefly, full schematic Replacement, do go beyond first-order set theory by assuming that we understand when a given expansion of a language is permissible. But that is a far cry from the commitment required for using second-order universal quantification—the commitment to a determinate but unexplained fact of the matter about what *all* the subcollections of a domain are.[25] Shapiro agreed [Sha91, p. 248]:[26]

> I conclude that one can work with theories formulated in free-variable second-order languages . . . without claiming some sort of absolute grasp on the range of the relation and function variables—or even claiming that there is a fixed range. One only needs the ability to recognize

---

24. The Replacement schema, and, in a parallel way, the induction schema for arithmetic, thus have an air of conventional or analytic truth about them. (Compare [Tap93].) Suppose someone defined a new property of the natural numbers and then proved that it was not definable in first-order Peano arithmetic by showing that it was not inductive in the following sense: zero has the property and, for every n, if n has the property so does $n + 1$, but not every number has the property. That cannot happen for a first-order definable property because of the first-order induction schema, and so a property that is not inductive cannot be first-order definable. A mistake has clearly been made, though it is hard to say exactly what without giving a more detailed example. Part of what it is to define a property of natural numbers is to be willing to extend mathematical induction to it. To fail to do so is to violate our rules for extending and further specifying our arithmetical usage. An analogous remark applies to Replacement.

25. Parsons reached closely related conclusions in a discussion of the natural numbers [Par90].

26. He did not discuss what I have called full substitution. He employed a partly semantic notion instead [Sha91, p. 247] "from $\Phi(X)$, infer $\Phi(S)$, where S is any set."

subsets as they are defined, and in the context of the interpreted formal languages in question this is not problematic. This is a rather weak hold on the range of the second-order variables.

Shapiro was inclined to reject full schematic second-order logic as an adequate basis for mathematical practice because of its limited expressive capabilities [Sha91, pp. 249–250].[27] For example, while "finite" is expressible, "infinite" is not. Shapiro's rejection of full schematic second-order logic as a basis for mathematical practice may well be correct for the purposes of analyzing mathematical practice concerning arithmetic or analysis. But the semantics of the full second-order logic needed for such purposes can be given in a background set theory, and so we may allow that second-order logic even if we do not allow the use of second-order logic for axiomatizing the set theory that underwrites it. That is enough for all ordinary mathematical purposes.

For a set theory to be adequate to underwrite full second-order arithmetic, analysis, and so forth, it is enough that it be quasi-categorical, for it then specifies the required power sets up to isomorphism. It is true that a background full schematic second-order set theory is less expressive than full second-order set theory would be. For example, full second-order set theory can define the satisfaction predicate for first-order set theory,[28] but full schematic second-order set theory cannot. But it is not clear that the comparative lack of expressive power is, on balance, a disadvantage, since full schematic second-order set theory avoids altogether a commitment to collections outside the domain, while second-order set theory does not,[29] at least on standard criteria: The range of the second-order quantifiers just is, after all, the collections

---

27. Shapiro was more kindly disposed toward full schematic second-order logic in [Sha85, p. 734]. In [Sha90], his chief argument for second-order logic is the one given here for full schematic second-order logic.

28. All that is required is an existential second-order sentence, not full second-order logic. Thus, the satisfaction predicate will be available in practically any logic that goes beyond schematic second order.

29. To many readers, it will seem that the natural way to put the point is that full schematic second-order set theory avoids any commitment to classes. But I advocate (see §V.5) a theory with sets and other classes as well. For such a theory the point must be expressed thus: full schematic second-order set theory avoids any commitments to collections of classes. That is the reason for the circumlocution in the text. The theory I provisionally favor is not one with a Replacement schema at all, but the system of §V.5 with Axiom V.5.2 taken to be a full second-order schema.

of members of the domain. Full second-order set theory is in that sense committed to such collections.

Technical Remark. There are certainly logics between full second-order logic and full schematic second-order logic, and so there might be some way to finesse the commitment to all subcollections of the domain. However, practically any intermediate logic will permit one to define the satisfaction predicate for first-order set theory. As Parsons in effect argued [Par74, pp. 212-216], that satisfaction predicate is interchangeable with a predicative theory of definable collections. Such a theory includes a commitment to (some) collections outside the domain, however one construes the second-order quantifiers. Full schematic second-order set theory avoids any such commitment.

Shapiro [Sha91, p. 255] raised an objection to taking full schematic second-order set theory as the background theory for the rest of mathematics, namely, that full schematic second-order set theory is only apparently more general than first-order set theory: "It will do, at least for the present, to 'restrict' the replacement scheme to the language of the background set theory, because this is the most general framework envisaged." Even if we had no commitments beyond the language of a background set theory, it seems to me that there would still be a point to using full schematic second-order Replacement instead of first-order Replacement: it better reflects our commitments, and shows what we would do in any future situation in which we did expand our framework.

That hypothetical argument is not needed. The commitments of present-day physical science go beyond those of the first-order framework that Shapiro argued is *de facto* equivalent to full schematic second-order set theory: It is a part of physics that many physical variables take on real-number values. The appropriate real-number values are the least upper bounds of collections of rational numbers specified in relation to some state of affairs in the physical world. First-order set theory cannot guarantee that the relevant collections of rational numbers are sets—there is no reason why such collections should be first-order definable. It therefore does not guarantee that they have least upper bounds, which is required to ensure that there are enough real numbers for the purposes of physics. Contrary to the usual conclusion, first-order set theory does not suffice to formulate the mathematics used for physics! But full schematic second-order set theory has the requisite power, since it allows us to expand the language of set theory by physically determined predicates that pick out the required collections of rational numbers, and that, given full schematic second-order Replacement, is enough to guarantee that they are sets. Even if we allow adding the same predicates to first-order set theory, they will be of no use without new instances of Replacement.[30]

There is another purpose for which a first-order background set theory will not do, one that is internal to mathematics: One cannot even raise the question of (quasi)categoricity for the background set theory with a first-order background set theory. That is true for two reasons: First, with a first-order background set theory one cannot discuss the background set theory itself. Second, when we discuss the quasi-categoricity of the background set theory we raise the possibility of an expansion of our mathematical commitments beyond the background to include an alternative background, a possibility that cannot be expressed within the language of the background set theory.

The first problem needs to be stated a bit more precisely: one can discuss the syntax and semantics of first-order set theory in a background first-order set theory in precisely the same way one discusses the syntax and semantics of any other theory. The notion of quasi-categoricity, for example, as I have defined it, is a notion introduced in just such a setting. But within such a setting the semantics of a theory is given in terms of the models of the theory, and the universe of any model must be a member of the domain of discourse of the background set theory. That is not what we want when discussing the background set theory itself, since the "model" we have in mind is one with a universe that is not a member of the universe of discourse: it *is* the universe of discourse. The notion of model as it is usually defined in a background first-order set theory is just not up to present purposes. There is, however, an easy way to patch it up—allowing "class models."[31]

Technical Remark. Here is one way to make the idea of a class model precise. We take a class model to be almost exactly like what Joseph R.

---

30. I suppose one could start with a first-order set theory that included all such predicates. Full schematic second-order set theory seems considerably more natural. For one thing, it is not committed in advance to anything about what the physically determined quantities are.

31. Levy takes essentially what I am calling class models to be the basic notion when discussing set theory [Lev79, p. 33].

Shoenfield called an interpretation [Sho67, §4.7]. Let me just consider the case of a model for a language with one binary predicate for simplicity, since that is the case of present interest. The general case does not differ in any important way. A model is usually taken to be an ordered pair with first member a nonempty set, the universe, and second member a set of pairs of elements of the universe, the binary relation. A *class model* is an ordered pair that has a formula of the background theory, $\Phi(x)$, with one free variable, possibly with parameters, as its first component and a formula of the background set theory, $\Psi(x, y)$, with two free variables in a fixed order, possibly with parameters, as its second. We require that the first formula be satisfied by at least one object and that any pair of objects that satisfies the second formula be such that each of the objects satisfies the first formula. (Those requirements parallel the usual ones that the universe be nonempty and that the relation hold only between members of the universe.) One can go on to define satisfaction, truth, isomorphism, categoricity, quasi-categoricity, and so forth for class models by imitating the usual definitions in a straightforward way. Every model in the usual sense determines a class model in a canonical way: Given a model (A, R), the corresponding class model is just $(x \in A, (x, y) \in R)$, where A and R are treated as parameters. But there are additional class models. In particular, the intended interpretation of the background set theory gives rise to the class model $(x = x, x \in y)$, with universe the self-identical things—that is, everything—and with relation the usual membership relation.

From now on we discuss quasi-categoricity in the modified sense that applies to class models. That is the sense appropriate to discussing the quasi-categoricity of the background set theory.

We now have the possibility of discussing the background set theory within the background set theory, but that is only half of what we need for an adequate discussion of the quasi-categoricity of set theory considered as a background theory. What is the challenge to which quasi-categoricity is a response? The challenge is that the background set theory does not adequately characterize its subject matter. The simplest version is this: how do I know, even given that you believe exactly the same set-theoretic axioms that I do as a background theory, that we have the same sets in mind? Given Skolem's worries, couldn't it turn out that the universe of my background (class) model is a mere countable set in your class model? Many will find the picture just sketched to be excessively realist: it presupposes that my use of set theory refers to some particular mathematical domain that can only be fully specified in some way that goes far beyond my expressed theory, and similarly for yours. A rejection of the picture may seem to amount to a rejection of Skolem's challenge as incoherent.

But the challenge seems to me to survive in a more cautious formulation. I use the words, say, *set* and *member*, and you use the words *set* and *member* too. Our usage is similar in many respects, and in particular let us assume that we employ the same axioms. Isn't there a legitimate worry about whether we are using the words in the same way? It is, after all, possible that you could come to believe the continuum hypothesis, while I came to believe its negation. (I am assuming, of course, that the continuum hypothesis is independent of the axioms we agree on, as will almost certainly be the case given the present state of knowledge about sets.) If that were to happen, it would then be clear that we were not using the words *set* and *member* in the same way. But then here is the legitimate worry: is our present usage sufficiently determinate that our present agreement guarantees future agreement?[32] So the question we are raising is whether there is more than one way to use *set* and *member* compatible with present commitments. To raise the question is to envision the possibility of two usages, and to discuss that possibility we must allow ourselves the vocabulary in which to discuss both. A background set theory, with *set* and *member,* has the resources to discuss only one usage, except in the limited case in which a second usage concerns only the very same sets (perhaps not all of them) and a new "membership relation" definable in terms of the original one. We must therefore allow an expansion of the background language beyond a simple background set theory to include, say, *set* and *member* and in addition *set'* and *member'*. We must concomitantly allow that the domain of discourse of the expanded language may include sets' as well as sets, and hence that it may be larger than the domain of either of the separate set theories being compared. When we allow such an expansion, only (two copies of) full schematic second-order set theory—not first-order set theory, not weak schematic second-order set theory—gives the pair of theories that correctly reflects our intentions by adding the entire language into the Replacement schemas for both sets and sets'.

---

32. Actually the worry is whether our present agreement guarantees no future outright disagreement—presumably we would not be seriously disturbed if you came to believe the continuum hypothesis while I remained in suspense.

What is at issue is not the situation with models of some set of axioms within a background set theory but that background theory itself. The universe and membership relations of rival set theories cannot be assumed to be realized, even up to isomorphism, within the domain of one of them without begging the question. Thus, even to raise the question of the quasi-categoricity of the background set theory coherently, one must allow expanding the background language from just $\{\in\}$ to, say, $\{\in, V, \in', V'\}$, where V and V' are predicates for the two domains and $\in$ and $\in'$ are the two membership relations of the two class models $(V(x), x \in y)$ and $(V'(x), x \in' y)$ under comparison. That is exactly the sort of expansion that full schematic second-order logic envisions that is blocked by first-order or weak schematic second-order logic, in each of which a theory has a fixed vocabulary.

Suppose that each of the two theories mentioned—the old background set theory and its rival—includes its respective axioms of full schematic second-order set theory. Then there are instances of full Replacement in either of the theories that involve formulas using the vocabulary of both of them. So far all that we have considered is the minimum apparatus needed even to raise the question of the quasi-categoricity of a background set theory, but that apparatus alone is enough to prove the quasi-categoricity result, that is, enough to prove that one of the two domains is isomorphic to an end extension of the other: no assumption of a fixed interpretation of predicate symbols like that of "standard" second-order logic is required, since the joint Replacement axioms are enough to build the required isomorphism. The apparatus needed to express quasi-categoricity is thus all that is needed to prove quasi-categoricity.

The fact that full schematic second-order logic is the minimum needed to raise the question of quasi-categoricity of the background language shows that the use of that logic in answering the question is not begging the issue. That is what distinguishes the present proposal from one that ensures quasi-categoricity by brute force, for example, by adding the notion of "standard" to first-order logic as a new logical notion with its correct interpretation simply imposed. Full schematic second-order set theory is not only sufficient to prove the quasi-categoricity result, but it is the minimal system in which it is possible to raise the question.

Moving to full second-order logic is not much help.[33] It is no better than first-order logic in that the relevant problem still cannot even be formulated.

---

33. Parsons reached closely related conclusions in a discussion of the natural numbers [Par90].

Exactly as before, when we discuss the quasi-categoricity of the background set theory we raise the possibility of an expansion of our mathematical commitments outside the background. That is a possibility that cannot even be expressed within the language of the background set theory even if that theory is second order, since the second-order quantifiers range only over collections of objects of the background theory. We must therefore allow an expansion of the language. When we do allow such an expansion, the strength provided by the second-order quantifiers is not essentially involved in the proof of quasi-categoricity, and the full substitution rule of schematic second-order logic must still be present to do all the work. Universally quantified second-order Replacement enters into the proof only through the use of finitely many instances—the universal, nonschematic force is irrelevant, as is shown by the fact that the proof can be carried out in full schematic second-order logic. But, more important, the key steps consist in taking certain definitions of functions given in the expanded vocabulary and applying Replacement for those functions. How do we know that those functions are in the range of the universal second-order quantifier (*all* functions)? What licenses the definition of the functions as legitimate? Nothing other than the very same substitution rule used in full schematic second-order logic.[34]

As we have just seen, the question of what expansions of language to allow is as acute for second-order logic as it is for full schematic second-order logic. Thus, the use of full schematic second-order logic has not created any new problems, only focused attention on an old one for second-order logic. Second-order logic has other problems and so the present problem is not a familiar one, but we do not need to saddle ourselves with those other problems to use full schematic second-order logic.

The philosophically satisfying quasi-categorical axiomatization of set theory given here solves, I believe, the chief problem raised by Zermelo's 1908 axiomatization of set theory [Zer08b], that of determining what the definite properties are, the problem that gave Weyl and Skolem and Fraenkel and von Neumann so much trouble. The definite properties are those that are first-order definable in any given expansion of the set-theoretic vocabulary.[35]

---

34. One often uses a comprehension schema in second-order logic instead of a substitution rule, but the two are interderivable in second-order logic. See [Sha91, pp. 66, 68–69].

35. Shapiro has not been so sanguine as I am [Sha90, p. 260]. He said that the Skolemite skeptic would argue that the comprehension axiom (or, equivalently, the substi-

There is a core of truth in Skolemite skepticism. Thoroughgoing relativism is wrong—our commitments do determine what *finite* and *well-founded* and so forth mean, unequivocally. But the way in which those things are determined is rather indirect. If we were today employing nonstandard notions like infinite "integers," or a denumerable "power set of the natural numbers," there would be no way to tell, given our present vocabulary. We do not have, in that sense, a fixed characterization of what is meant by *finite* and the rest. What we do have is a canonical way in which we could be shown that our present notions are, by our own lights, nonstandard. Nonstandard notions are defeasible, but standard ones are not demonstrably standard.

It must instantly be admitted that solving one important problem—that of determining what the definite properties are—does not solve them all. If non-well-founded sets are to be allowed, we must find axioms to characterize them. The Anti-Foundation Axiom (§V.4) is a plausible candidate, but more analysis is required. Even more important, perhaps, the present considerations do not determine how many ordinals there are. Since we want our theory to be as all-encompassing as possible, it seems obvious advice that given two otherwise acceptable domains, we should select the one with more ordinals.[36] But that advice is far from settling the issue: if there is a maximum domain, we should surely use that, but if there is not, we should probably adopt a picture like that of Zermelo in 1930, with each domain a set in a larger one.

To make progress toward settling the question of how many ordinals there are, we need a better understanding of the intuitions that underlie set theory. That project is doubly important because set theory is now seen to be an adequate foundation for (at least most of) the rest of mathematics in a fairly

---

tution rule) changes when the vocabulary expands and that the range of the second-order variables changes with it. Since I, unlike Shapiro, have resisted quantified second-order logic in the present setting, I can reply to the second objection that the second-order variables do not have a range. The first objection, that the substitution rule changes, is more serious. It is a skepticism I cannot unequivocally defeat, but note how much the position has changed: The skeptic must now argue that we do not understand when we are expanding our own language, which is much harder to maintain than the old doubt about whether our claims in the present fixed vocabulary suffice in and of themselves to specify what we mean by "all collections."

36. I am not sure whether the same advice would be appropriate for urelements, though I am inclined to think so. A domain that misidentified some sets as urelements would have extra urelements that we should not allow, and so I have some doubts.

substantial sense. In addition, even though full schematic second-order set theory is quasi-categorical, it is far from having the deductive strength to answer many of the questions it seems to show do have answers. For example, as Zermelo in effect noted [Moo80, p. 132], it follows from quasi-categoricity that the continuum hypothesis is either true or false. But neither its truth nor its falsehood can be deduced from the present-day axioms, as follows from the usual independence results. A better understanding of the infinite is important for both philosophical and mathematical reasons.

From Here to Infinity

(a) . . . [Tarski and I] urged that the only logic to which we could attach any seeming epistemological immediacy would be some sort of finitistic logic. So here we were in epistemology and envisaging a finitistic constitution system.

(b) I argued, supported by Tarski, . . . that . . . we find ourselves on the side of the Platonists insofar as we hold to the full nonfinitistic logic. Such an orientation seems unsatisfactory as an end-point in philosophical analysis, given the hard-headed, antimystical philosophical temper which all of us share . . . So here again we found ourselves envisaging a finitistic constitution system.

How, then, to divorce our essential semantic issues from the questions of epistemology and finitism? As to (b), the answer is immediate: let us *accept,* provisionally, whatever rudimentary Platonism may be embodied in our regular logic and classical mathematics, and so proceed with our semantics, just as we have in the past been proceeding with our regular logic and mathematics. If independent progress should be made sometime in the way of an epistemologically motivated finitistic substructure, so much the better; it would be a case of resolving the Platonic kink without much altering the existing logical, mathematical, and semantical superstructure, perhaps, just as Weierstraß eliminated the nonsense about infinitesimals without wrecking the differential calculus. (Letter from Quine to Carnap, 1943 [QC90, pp. 295–296])

§1. Who Needs Self-Evidence?

It is not necessary that the meaning of intuitive concepts be a function of objects which they represent, and indeed, both Gödel and Kreisel are

241

careful not to insist on this connection. Intuitive analysis is important, but so is the investigation into the *source* of our intuitive concepts, upon which little effort has been expended.   [Kle76, p. 487]

Modern set theory evolved out of two ideas. The first is that of an arbitrary function from the real numbers to the real numbers, not necessarily determined by a rule. The paradigm example is surely Fourier's: the temperature of a bar as a function of distance from a given point on the bar. The values are not computed but determined by factors external to the mathematician, and they are not constrained in any way. That is true in the paradigm case because a temperature curve may be an initial condition, not determined by any given factors. The second idea from which set theory evolved developed out of the attempt to make sense of the first. It is that of Cantor's transfinite progression of indexes

$$0, 1, \ldots, \infty, \infty + 1, \infty + 2, \ldots, \infty \cdot 2, \ldots, \infty \cdot 3, \ldots,$$
$$\infty^2, \ldots, \infty^3, \ldots, \infty^\infty, \ldots, \infty^{\infty^\infty}, \ldots,$$

conceived as numbers counting iterations of an operation, namely, Cantor's derived set operation.

Cantor, Zermelo, and the rest needed to rely on their sense of self-evidence, on their sense of what was implicit in those two ideas, to do their work, work that led to modern axiomatic set theory. But now that we have modern axiomatic set theory, why can't we throw away the ladder? Do *we* need a notion of self-evidence to do set theory? The answer is a resounding yes.

The first and most obvious need for self-evidence is that of the working set theorist. Most set theorists, like most mathematicians, have and rely on a feel for their subject matter to navigate their way through attempts at proofs. That is important because, to give the easiest example, it permits them to postpone the proofs of "obvious" lemmas while working toward a goal. Those proofs get supplied later, if the goal is reached. Every mathematician, and set theorists are no exception, has had the experience of reading a theorem or a proof without really understanding it, which means, without having a feel for what it says. A lot of the daily work of mathematicians consists of attempting to get a better feel for a subject matter or problem.

For most mathematicians most of the time, having a feel for what is evident is important, but it is also enough: There is no further need for a theory of what that feel is. One may arrive at a theorem and technique of proof on the basis of intuition, but an ultimate justification for a theorem or technique is supplied on the basis of rigorous axiomatic proof. Even when one is working with newly developed axiom systems, the justification consists in seeing what familiar objects are models of those axioms and what generalizations and systematizations the axioms permit. Discussions of self-evidence can largely be dispensed with, in favor of proofs or discussions of the utility of axiomatic systems.

Set theorists are not so lucky. There are problems of long-standing interest to set theorists and other mathematicians that cannot be settled on the basis of the present axioms of set theory. The most notable example is the Continuum Hypothesis, though there are many others.

In other branches of mathematics, incompleteness of the axioms usually causes no trouble. To use the most common example, once mathematicians saw that the parallel postulate was independent of the other axioms of geometry, they promptly began to study geometries both with and without the postulate.[1]

No one worries about whether the parallel postulate is really true or false. It is true in some geometries and false in others, and that is that. A crude analogy suggests (as do some other, more refined, arguments) that the situation is analogous for the Continuum Hypothesis: Set theorists should study set theories both with and without the Continuum Hypothesis, and that's all there is to it. There is no further question of absolute truth or falsehood. That attitude is in fact taken by some, and no one would deny that, given our present state of ignorance, it is worthwhile to study set theory both with the Continuum Hypothesis and with its negation. But, for reasons I discuss just below, most of us want to know whether the Continuum Hypothesis is really true or not.[2] That is, we want to know whether or not there is a genuine subset of the genuine real numbers that cannot be placed into *any* one-to-one correspondence with either the genuine natural numbers or the genuine real numbers. Postulation is of no help for that. We need new facts, which we shall presumably incorporate into set theory as new axioms. One might, as a special case, have a direct argument for the Continuum Hypothesis not based on any other prin-

---

1. The story of the parallel postulate is not ours, and so I am taking the liberty of oversimplifying a bit. It was in part the study of geometries without the parallel postulate that led to the recognition that it is independent of the other axioms.

2. The same disclaimers apply here that did in the case of the Axiom of Choice (§V.1, see also §VII.1): truth need not be *a priori* truth, and consideration of the present, historically determined, state of the subject is clearly in order.

ciples suitable to use as axioms. I shall still, for convenience, treat that as a case of accepting the Continuum Hypothesis on the basis of a new axiom—as indeed it is, since the Continuum Hypothesis most certainly follows from itself.

If you grant me that we in fact have a quasi-categorical axiomatization of set theory, as I concluded in §VII.4, and allow a certain amount of realism about mathematical models, the disanalogy between set theory and geometry is easy to explain: The axioms of geometry (not including the parallel postulate) are not categorical. They have nonisomorphic models, some of which obey the parallel postulate and some of which do not. Both kinds of geometric models are worth studying. The axioms of set theory, in contrast, do not have such a variety of models. Because the axioms are quasi-categorical, either the Continuum Hypothesis is true in all the models or it is false in all the models. We would like to know which.

Unfortunately, even though the axioms are quasi-categorical, they are far from complete. To settle the Continuum Hypothesis, we shall need to supplement them. But the only new axioms that can be of any use are ones that are true of the models of set theory that are picked out by the axioms we already have. Utility and fruitfulness are just not good enough for our purpose: we need truth, as determined by an antecedent standard.

As the independence of the Continuum Hypothesis from the first-order axioms of set theory shows, the truth or falsehood of a proposed new axiom that settles the Continuum Hypothesis cannot be determined from the antecedent standard by using the familiar deductive method: we cannot deduce or falsify a proposed new axiom that settles the Continuum Hypothesis on the basis of known axioms and known instances of axiom schemas. That is just what it means to say that the Continuum Hypothesis is independent of those axioms and schemas. We therefore need a nondeductive, nonaxiomatic way of determining whether a proposed new axiom follows from our old standards.

Set theorists have used many heuristics to argue for and against various proposed new axioms. (See [Mad88a, Mad88b].) None of the heuristics has met with much success. Most important, none has provided a persuasive argument either for or against the Continuum Hypothesis. For that purpose, it seems we shall need either a more refined and powerful standard of self-evidence or a more refined and powerful set of heuristics and standards for extrinsic evidence. Hence, I take it, it is important to provide a theory of self-evidence. (See [Mad88b, p. 762] for a forceful expression of a similar sentiment, though Maddy rather inclined toward believing that a solution will lie on the side of heuristics and extrinsic support than on that of self-evidence.)

You need not be convinced that we have a quasi-categorical axiomatization of set theory to be convinced that the Continuum Hypothesis may have a well-defined answer. It is enough to believe that we may have a notion of set that transcends any contemporary axiomatic presentation in some way that settles the Continuum Hypothesis, or even that we have such notions of the real numbers, the sets of real numbers, and the functions from the real numbers to the real numbers. One might believe that we have such notions, for example, on the basis of the intimate involvement of such entities in our physical theories. In such a case, the question of whether a proposed new axiom is really true—or at least really true to a conception or intended physical application—can arise. The only hope we seem to have of answering that question lies in a better theory of the nature of self-evidence.

That was the first reason we need a theory of self-evidence. We need it to help in finding new axioms for set theory in our quest to answer old questions, not just about sets but even about the real numbers and natural numbers. The second reason is that we need a theory of self-evidence to better understand the axioms we already have: The set theory that Cantor developed on a conceptual basis out of his progression of transfinite symbols did not include Power Set, and it therefore, despite his hopes, did not provably include the real numbers. But the theory was developed to classify arbitrary functions from the real numbers to the real numbers (or at least to classify the sets of their exceptional points), and so Power Set was added on anyhow. Set theory is a shotgun marriage between Power Set and Cantor's transfinite symbols. As Zermelo showed, the two parties get along, but they hardly seem made for each other.

The tension is still clear in modern explanations of the axioms. Thus, as we have seen (§V.5), the iterative conception (narrowly characterized) explains Power Set but not Choice or Replacement, which come from the combinatorial ideas of Cantor and Fourier, while von Neumann's version of Limitation of Size can be used to explain Choice and Replacement but not Power Set.

We showed earlier that the problem of how to fit Power Set with Cantor's early combinatorial ideas can be reformulated as the problem of how to see, given the Limitation of Size characterization of sets, that classes that are limited in comprehensiveness are also limited in size. That is easily seen on the domain of hereditarily finite sets, since a class of such sets whose union is finite, that is, a class that is limited in comprehensiveness, must itself be finite, that is, limited in size. That does not, at this stage, seem to be much help, but see Chapter IX.

The first two reasons I have discussed for seeking a theory of self-evidence are reasons internal to the mathematical practice of set theory. The remaining two are external and philosophical. The third is answering the question raised by Gödel—How can the axioms force themselves upon us as being true? The fourth is the problem of the infinite—that of giving an epistemically acceptable account of how we know what we do about the infinite. An acceptable answer to the third question will presumably be at least a good start on an answer to the fourth.

The two questions—Gödel's and Benacerraf's—have an importance that goes beyond the philosophical problems posed by set theory, the mathematical infinite, or even mathematics generally. The apparent difficulties in answering them have been an important part of what has led to a bifurcation in much of philosophy—epistemology, metaphysics, and the philosophy of language. There are those who think that, despite appearances, we must have the knowledge of the infinite that we seem to have. They have embarked upon a far-reaching criticism of epistemology, ontology, and semantics—turning things upside-down by starting with what we know as a given instead of trying to see how we could know it. And there are those who think that, despite appearances, we cannot have the knowledge of the infinite that we seem to have. They have embarked on a far-reaching revisionist criticism of mathematics, science, logic, and semantics.

I do not wish to take sides, or even to discuss whether the polarization is appropriate, but I do wish to say that its origin is closely connected to various attitudes for and against abstract objects. One side wants to see how to live comfortably with abstract objects, while the other side wants to eliminate or tame them. Infinite mathematical objects have often served as the prime example of strange and problematic abstract objects, and so attitudes toward them have had far-reaching effects. Anything that can be done to reduce the apparent strangeness of infinite mathematical objects—or at least to understand the ways in which they are strange—will send subterranean tremors throughout philosophy. A reasonable theory of the self-evidence of our principles concerning such objects will go a long way toward eliminating one source of their apparent strangeness, namely, the apparent difficulties in knowing anything about them. I cannot predict whether any earthquakes will result.

§2. Picturing the Infinite

We have already seen that the infinite is nowhere to be found in reality, no matter what experiences, observations, and knowledge are appealed to. Can thought about things be so much different from things? Can thinking processes be so unlike the actual process of things? In short, can thought be so far removed from reality? Rather is it not clear that, when we think that we have encountered the infinite in some real sense, we have merely been seduced into thinking so by the fact that we often encounter extremely large and extremely small dimensions in reality? (David Hilbert [Hil26, p. 191])

Cantor's development of set theory began with his discovery of the progression

$$0, 1, \ldots, \infty, \infty + 1, \infty + 2, \ldots, \infty \cdot 2, \ldots, \infty \cdot 3, \ldots,$$
$$\infty^2, \ldots, \infty^3, \ldots, \infty^\infty, \ldots, \infty^{\infty^\infty}, \ldots.$$

The progression is immediately appealing and suggestive. Our question in this section is, Why? After all, it is strange that the progression makes any sense to us at all. It is a progression with lots of infinite subparts symbolized by ellipses. The ellipses are no mere abbreviation, since we cannot write out what they stand for.

We certainly do not encounter what the ellipses stand for in daily experience. Indeed, as Hilbert emphasized [Hil26, pp. 185–186], it seems likely that we cannot have encountered what they stand for in any concrete sense at all, since modern physics makes it seem likely that the physical universe is of finite extent and decomposes into parts with a nonzero minimum size.[3] The infinite is ontologically remote from us. So how can we have such an immediate feel for the infinities represented by the ellipses? Why isn't the infinite epistemically remote? It seems miraculous. There is a parallel question on the side of the small: How can we have such an immediate sense of what a point is? That question is related to our feel for the notion of an arbitrary function. I shall neglect it in favor of the question about the ellipsis, since the ellipsis leads to set theory and thus, indirectly, to the notion of an arbitrary function.

We are certainly not born with whatever concept of infinity is implicit in the way we use the ellipsis. Many children in the early school years think there is a biggest number, and to them infinite just means too many to count [Eva83, pp. 136, 169–170]: "some children reported 'counting and counting' in an attempt to find the last number, concluding there was none after much

---

3. It is more usual to think of the universe as being *composed* of parts with a nonzero minimum size, but see [Lav91].

effort." So where does the concept of the infinite come from? Training, to be sure, but how did it begin? My proposal is that it began with an extrapolation from experience of indefinitely large size.[4]

The idea of indefinitely large size is a familiar one that comes out of daily experience. It is the idea of too many to count, Hilbert's "encounter with extremely large dimensions," the idea of a zillion. I know very little about the psychology or history of the idea of indefinitely large size, but it is nonetheless clear that it does not pose the same kind of epistemic problems as the idea of infinitely large size.[5] The problem of explaining our familiarity with indefinitely large size is like the problem of the finite, and very much unlike the problem of the infinite. There are, if anything, too many possibilities to serve as the beginning of an explanation, not too few. There is no skeptical problem comparable to that concerning the infinite since we encounter indefinitely large quantities all the time: zillions of grains of sand on a beach, zillions of stars, zillions of homeless people, zillions of books we wish we had time to read.

The idea of a zillion is in one important respect more complex than that of, say, 3: It is interest or context relative. The number in a zillion dishes that need to be washed might be a small number of grains of sand—they might not even fill a child's pail.

The child is not wrong that an infinite quantity is one that is too large to count. But the child might mean that it is too large to count today, or so large that even Daddy won't be bothered—too large in a context-relative sense. Infinite *is* nothing more than too large to count—too large in a context-independent sense, too large for anyone to count independent of context, abilities, or interest. I said before that our notion of the infinite is obtained by extrapolation from our idea of indefinitely large size. Now I can be more

---

4. The idea is discussed from a slightly different perspective in [Lav95].

5. Others have tried to reconstruct parts of modern mathematics on the basis of a theory of large finite collections, but in an extremely different way that rests on the implausible assumption that there literally are finite collections with no last member [Vop79, p. 33]: "Professor Charles Darwin teaches us that there is a set D of objects and a linear ordering of this set such that the first element in that set is an ape Charlie, each non-first element is a son of the immediately preceding element, and the last element is Darwin himself. The collection A of all apes belonging to D is not a set; otherwise A would have a last element. But, as everybody knows, sons of apes are apes." A. S. Yessenin-Volpin proposed a similar basis for mathematics [YV70]. For an excellent discussion of the paradox suggested by the passage, see [Tap93].

precise about the type of extrapolation that I have in mind—one extrapolates by dropping the dependence on context.

Does the extrapolation from the theory of indefinitely large size provide rational justification for believing our theory of sets? I shall conclude that it does, but let me defer discussion to the end of the section, after I have made it clearer what the nature of the extrapolation is and how intimately connected it is to our actual reasons for accepting set theory in practice.

It is important that extrapolation is intimately connected to actual practice: Suppose we succeeded in giving a justification of set theory that is not connected to actual practice. Then to use it as the justification for set theory would justify *a* practice of set theory—one based on that justification—but it would not justify the *actual* practice of set theory. To meet our goal of a philosophy of mathematics that confronts actual mathematical practice, it will be necessary to establish not only the philosophical thesis that extrapolation can correctly be used to justify set theory but also the psychological and historical thesis that extrapolation was in effect actually used in the process by which we have arrived at present-day set theory.

To sustain my claim that fully developed set theory as a matter of psychological and historical fact rests on Cantor's progression and extrapolation from the idea of indefinitely large size, three things must be shown. The first is that the informal theory I present as a theory of indefinitely large size is in fact nothing more than a more precise version of some of our intuitions that arise from experience of the indefinitely large. Note that I am not claiming that the theory of indefinitely large size presented here is a version of the uniquely determined necessary intuitions we have concerning the indefinitely large. I see no reason to believe that there are such things. Indeed, it seems likely that our intuitions concerning, for example, indefinitely large extent or indefinitely fine divisibility would turn out to yield rather different theories of indefinitely large size than what we are concentrating on here: indefinitely large number, that is, indefinitely large size in the sense of "too large to be counted." The reason for focusing on that notion of indefinitely large size is, in a way that is by now familiar, partly pragmatic and partly historical. We are interested in understanding set theory because of the important part it plays in mathematics, and we are therefore particularly interested in the notion "too large to be counted" because of the origin of set theory as a counting theory.

I shall not show that the informal theory of indefinitely large size is a more precise version of our intuitions. The requisite data are not available. In this work, I shall have to rest content with making that plausible by presenting the theory in such a way that you will see how natural and self-evident it is. As an important part of that, I shall show how natural certain parts of analysis become when they are presented using the theory of indefinitely large size instead of the usual Weierstrassian definitions.

The second thing I must show is that the twin notions of indefinitely large size and extrapolation are sufficiently rich to bear the heavy explanatory burden I am placing on them. That, I shall show in detail. I shall present a mathematical theory of indefinitely large size developed by Jan Mycielski [Myc86][6] and an obvious related notion of extrapolation. The theory yields a formal counterpart of the preformal notion of indefinitely large size, as I shall show by emphasizing the great extent to which its principles are self-evident as principles concerning indefinitely large size. I shall then show that full present-day set theory arises from the theory by extrapolation.

Everyone knows that one can obtain a useful theory of the structural properties of triples, and that that theory is a part of mathematics. What is novel here is that one can find mathematically useful structural properties of indef-

---

6. Though the theory is Mycielski's, he did not interpret it as a theory of indefinitely large size, and he has not commented on its relation to any intuitive picture. He did [Myc81b, p. 632] describe some quantities closely related to those characterized here as indefinitely large but finite ones as "potentially infinite but actually finite." Mycielski's "Poincaréan" philosophy of mathematics is almost entirely different from the position advocated here, and it does not rely essentially on his mathematical results that play a central role here. Mycielski does not ascribe the Ωs on which I rely so heavily (see below) any special intuitive status (though he does conceive of them as "finite but larger and larger universes of mathematics"), and they do not appear in his more recent work. See [Myc89, Myc92]. He does not make use of the Ωs to motivate the axioms of ZFC or any other theory of sets, as I do, and such a motivation would have little relevance to the main thrust of his philosophy. He makes no use of PRA, which plays an important part in my argument—indeed he accepts full Peano arithmetic as finitistic—and it has become central to his position that the sentences of mathematics take their truth values in various boolean algebras, while I retain a classical two-valued semantics. His philosophical remarks in [Myc86]—the article that contains the results used here—are so terse that it is to some extent possible to read my philosophy of mathematics into them, but Mycielski has told me (personal communication, 1991) that the views expressed in a subsequent and considerably more explicit article [Myc89] are fully consonant with those he held when he wrote that article. In particular, though some of the results proved using what I have called the theory of indefinitely large size are relevant to Mycielski's philosophy of mathematics, the concept of indefinitely large size (or potentially infinite but actually finite) is not, except in an incidental way.

initely large collections. I have stressed the analogies between the situation with 3 and the situation with indefinitely large numbers. Let me therefore also emphasize that while I do believe that our theory about 3 is intimately connected to our experience of triples, I am *not* claiming that our theory of infinite collections is as intimately connected with our experience of indefinitely large collections. In the latter case there is an intermediate step: the mathematical theory of the infinite is related to a mathematical theory of the indefinitely large by a process of extrapolation—a kind of simplificatory rounding out. Infinite and indefinitely large are by no means the same.

The third and final thing I must show is that the notion of infinite size that results from extrapolating from the notion of indefinitely large size is the one that actually was operative in the development of modern axiomatic set theory. I am not claiming that anyone ever consciously set out to obtain a theory of the infinite based on experience of indefinitely large size. I *am* claiming that the picture of the infinite—of the use of the ellipsis—with which the founders of set theory started was one that developed out of their experience of indefinitely large size, even though that source was largely unconscious. That claim is not easily verified, or falsified, by a direct examination of textual evidence, though some aspects of Cantor's work are suggestive.

I can only support the claim indirectly, by showing how well the notion of infinite size that results from extrapolating from the notion of indefinitely large size accounts for the actual set theory we have, the analysis from which it developed, and the intuitions we have surrounding it. The major part of the task of supporting the claim, therefore, is showing that the axioms of set theory are extrapolations of evident principles concerning indefinitely large size. But I shall also show how many of our philosophical attitudes about infinite sets can be seen to arise by extrapolation from truths about indefinitely large ones.

To show that a theory of indefinitely large size can serve as the intuitive base on which our understanding of the infinite rests, it is necessary, to avoid circularity, to show that that theory of indefinitely large size does not smuggle in a prior understanding of the infinite. Moreover, whatever aspect of the finite is involved must be rich enough to yield a counterpart of our actual rather substantial and articulated knowledge of the infinite, a counterpart from which to extrapolate to obtain that knowledge. It follows that if some aspect of the finite can underwrite the view that knowledge of the infinite is obtained by extrapolation, it will also be able to underwrite a way of eliminating the need for the infinite in mathematics: Instead of extrapolating, one could rest content with whatever it was that could have been the basis for extrapolation. Thus, the success of a method of extrapolation entails the apparently stronger possibility of an elimination.

Though it is not my motive for developing the finite mathematics of indefinitely large size, the possibility of developing it shows that one could coherently deny the existence of infinite sets without doing violence to mathematical practice, as will become clearer in §§3 and 4. The finite mathematics of indefinitely large size does not, however, commit us to such a denial,[7] and I do not advocate one.

We find ourselves in a position analogous to that of Frege: he developed a formal system to make sure that he hadn't employed unnoticed assumptions about arithmetic, while we shall use one to make sure that we haven't employed unnoticed assumptions about the infinite. To do that we shall, in §§3 and 4, formalize the theory of indefinitely large size in such a way that the resulting axiomatic theory has the following properties: Only finite entities are employed, and all quantifiers have finite ranges. The theory can be described without any covert reliance on infinite sets at the metalevel. As a matter of fact, it will turn out that the theory can be described without using any quantifiers with infinite ranges at all. Nothing infinite, whether actual or potential, is either used or presupposed by the theory.

We have seen a theory, indeed several theories, that are free of infinitary presuppositions, namely PRA, PRW, and PRS,[8] and those theories will, not unexpectedly, provide the starting point for our formal theory of indefinitely large size.

The fact that I show how to introduce set theory as an extrapolation of a formal theory built on PRA makes what I am doing look a lot like what Hilbert proposed. Indeed in many respects it *is* like what Hilbert proposed. The most important similarity is that both permit an interpretation under which what one is doing is introducing the infinite and another interpretation in which what one is doing is eliminating the infinite. I shall favor the interpretation

---

7. Though the existence of a basis for extrapolation does not force us to deny the existence of infinite sets, it does undercut what are usually thought to be the chief positive arguments for their existence. (See, for example, [Mad89] or §VII.2 for a discussion of those arguments.)

8. The theories PRW and PRS are the primitive recursive theory of words and the primitive recursive theory of sets. They have all the desirable features of PRA discussed in §VI.3. See §VI.4.

that introduces the infinite for reasons given below. Those reasons are, however, based on contingent facts about contemporary mathematics and physics, facts that could change as those subjects develop. I shall therefore make occasional remarks germane to the interpretation that eliminates the infinite, primarily in notes.

Beside the similarities, there are also important differences between my philosophy and that of Hilbert. I wish to emphasize some of those differences because, given the tendency to assimilate the new to the familiar, they may be easy to overlook.

The most important difference is that while what I am doing is finitist in the narrow sense that it involves only finite objects, the motivation has nothing to do with showing any portion of mathematics to be secure, certain, free from contradiction, or the like. Mathematical knowledge is like other knowledge in that it simply does not have a foundation on which to rest. In making that claim, I am fully in agreement with Shapiro's *Foundations without Foundationalism* [Sha91]. He argues for the claim in some detail in his preface and chapter 2, and by and large I fully endorse his case. Just as in the case of other forms of knowledge, to admit that mathematics does not have (or need) a foundation is only to acknowledge the uncertainty that is inherent in the human epistemic condition. I am not attempting to secure set theory, and I am not attempting to secure the theory of indefinitely large size. Set theory developed out of Cantor's attempt to work out a theory of a few basic ideas, and I am trying to understand better how we come to have the fairly firm grip that we do on those ideas, despite their apparent remoteness from any experience. But that does not secure the ideas in any final sense. They may yet, for all I know, prove to be bad ideas, even inconsistent ones. Since I do not seek security, my "procedure" for introducing a new theory is just the usual one of seeing whether it looks like a good theory of something interesting. No proof of consistency is required.[9]

Given my interest in how we manage to know as much as we do about the infinite, I am particularly interested in mathematical theories of a special sort: those that are free of infinitary presuppositions and have natural extrapo-

---

9. Like most anyone else, I would be happy to have a proof of consistency should one be available, and I find the project of providing consistency proofs to be of significant mathematical interest. That is particularly true for consistency proofs that take the form of showing that one theory is consistent by interpreting it in another. Such reductions, like that of the rational numbers to the natural numbers, are always useful.

lations to theories that involve the infinite. Let me call the mathematics of all such theories *finite mathematics.* The main examples here will be theories that extrapolate to set theories of the familiar sort. I shall give a more formal and slightly more restrictive characterization of finite mathematics in §3, taking account of Mycielski's work, and I shall show that as a matter of fact *every* theory used by mathematicians has a counterpart in finite mathematics, again relying primarily on Mycielski's work.

Unlike Hilbert's finitist mathematics, finite mathematics is neutral with respect to the nature of mathematical entities (and in particular it involves no nominalist agenda of replacing abstract mathematical entities by quasi-concrete ones). Finite mathematics is just more mathematics, and it has no more need to confront ontological questions than any other mathematics. Finite mathematics in general, like the finite mathematics of set theory discussed above, does not attempt to provide a secure foundation for any part of mathematics (and indeed it seems to me to be neither more nor less likely to be free from contradictions than most of the rest of mathematics). It is motivated by a desire to understand the relation between the combinatorial infinite and the indefinitely large, not by a desire for final justifications. Finally, finite mathematics is in no sense formalist: It will turn out, given the detailed development of finite mathematics in §3, that all of the statements of ordinary mathematics have fully interpreted contentual counterparts in finite mathematics. The statements of ordinary mathematics never play an auxiliary proof-theoretic role, as they do in some interpretations of Hilbert's finitism.

I believe that finite mathematics achieves one of Hilbert's goals, namely, providing an interpretation of modern classical mathematics that should be congenial to many constructivists. The basic idea is that most constructivists would agree that classical logic is obeyed on the finite domains of finite mathematics. Allow me to defer further discussion to §4.

Now that I am done differentiating finite mathematics from Hilbertian finitism, I shall begin to motivate the theory of indefinitely large size with a simple example drawn from arithmetic. We all learned simple addition using some physical model—let's say it was beans. We learned that, say, $2 + 2 = 4$ in part by taking a pile of two beans and another pile of two beans, combining them, and counting the resulting pile to verify that it was composed of four beans. Where did the beans come from? They came from some indefinitely large supply—say, a bucket of beans.

The bucket contained what was for our purposes an indefinitely large supply of beans. It did contain some finite number of beans, but we had no idea how many. All that mattered was that the number was much larger than any with which we were immediately concerned. I shall refer to the indefinitely large number of beans as ω_0 (the reason for the subscript will become apparent in a moment). Now, what would have been the appropriate response if we had run out of beans in the course of a computation? Surely it would have been to go get more beans. We may say that the new, larger, indefinitely large supply contains ω_1 beans. As we shall see, the structure of larger and larger indefinitely large sizes is part of what gives the notion of indefinitely large size the richness that ultimately yields our intuitions of the set-theoretic infinite. In calling attention to it, I am calling attention to a familiar phenomenon, though not one that most of us have paid much mind.

The simplest arithmetical process that lets us use up beans is adding one—applying the successor operation. The fact that our pile of ω_0 beans was exhaustible is expressible by the fact that the following sentence, written using bounded quantifiers, is false:

(1) $$(\forall x < \omega_0)(\exists y < \omega_0)\, y = x + 1.$$

Note, however, that the following sentence, which reflects our actual practice of getting more beans when we run out, is true:

(2) $$(\forall x < \omega_0)(\exists y < \omega_1)\, y = x + 1.$$

We could have started out with ω_1 beans, instead of ω_0, and used up all ω_1. But then we would simply have had to get ω_2 beans:

(3) $$(\forall x < \omega_1)(\exists y < \omega_2)\, y = x + 1.$$

We seem to be forced into an infinite series of statements like (2) and (3), but in fact we are not. Statements like (2) and (3) are more than superficially similar. We know that ω_0 is an indefinitely large number with respect to our present interests, but nothing tells us how large—if we could place an upper bound on the size of ω_0, that would show that ω_0 was not indefinitely large in the relevant sense. Thus, ω_1 could perfectly well have played the role of ω_0. Analogously, ω_2 could perfectly well have played the role of ω_1. The ωs are, in a sense, indiscernibles. (See Lemma 3.1, below.) When we codify that intuition about the indefinitely large, (3) will turn out to be a logical consequence of (2) and our axioms about indefinitely large numbers, not an additional fact. Similarly, given ω_3,

$$(4) \qquad\qquad (\forall x < \omega_2)(\exists y < \omega_3)\, y = x + 1$$

will be a logical consequence of (2) and our axioms about indefinitely large numbers, and so forth for, it seems, as many steps as we like. The layered structure of indiscernible indefinitely large sizes is what will enable finite mathematics to have counterparts for all of ordinary infinitary mathematics.

It is critical to note that in the setting of the argument for indiscernibility we already have use of the four indefinitely large numbers ω_0, ω_1, ω_2, and ω_3, but no more than that. If making sense of finite mathematics required that we postulate some sort of a power to always be able to go on from some already given bunch of indefinitely large numbers to a new, larger one, then we would by that very fact be committed to some kind of potential infinity.

The indiscernibility of indefinitely large sizes will be a crucial part of the theory of indefinitely large size. The indiscernibility is so much a part of one way in which we ordinarily conceptualize indefinitely large sizes that we usually fail to distinguish different indefinitely large sizes in any way at all: Though we can easily see that indefinitely large size is context relative, we often act as if it were not, as if a zillion were a zillion, independent of context. The elimination of context dependence leads to the infinite. In less abstract and somewhat less accurate terms, the thought that the number of grains of sand on the beach might as well be infinite for many practical purposes is what leads to the concept of the infinite—when we ignore the restriction to particular practical purposes.

> **Technical Remark.** The indiscernibility has the result that we can re-label the ωs in any true statement, so long as the relabeling preserves the order of the subscripts. The effect is much like that of systematic ambiguity in Russell's theory of types (§IV.1). In the theory of types, more complicated definitions are of higher order and higher types are required for objects with suitably complicated definitions. Similarly, in finite mathematics more complicated sentences, those with more alternations of quantifiers, will require more ωs. Much as one cannot quantify across types and orders in the theory of types, we shall not quantify over the ωs.
>
> The great usefulness of systematic ambiguity was, given the motivation for the system of Whitehead and Russell, a kind of happy acci-

dent. The different orders and types are genuinely different, and there is no reason why the ways in which they happen to resemble each other should turn out to be more important than the ways in which they differ. One symptom of that is that nothing in the theory of types blocks the addition of laws that are specific to one or another type or order. Such laws would prevent the use of systematic ambiguity. The analogous principle here—the order indiscernibility of the ωs—plays a much more central role. It is one of the principles that is constitutive of the use of successive indefinitely large sizes.

Let us now see how to remove context dependence in a somewhat more precise way. Statements (2), (3), and (4), which are all equivalent, are our counterparts for the statement

$$(5) \qquad\qquad (\forall x)(\exists y)\, y = x + 1$$

of ordinary arithmetic. Statement (5) (together with other parts of elementary arithmetic) guarantees that there are infinitely many numbers, and so it goes beyond finite arithmetic. It is obtained from statement (2) by the process I have been calling extrapolation. Syntactically, the process is trivial—one just drops the bounds ω_0 and ω_1. Moreover, it is just as trivial to go from statement (5) back to (2), (3), or (4): to go from a statement of ordinary arithmetic back to a statement of finite arithmetic, one just supplies arbitrarily chosen increasing ωs to bound the quantifiers.

The process of extrapolation—of dropping the ωs that serve as bounds—is my proposed formal analysis of what we actually do: we fail to attend to the differences between the bounds, the indefinitely large quantities. There is a trivial sense in which dropping the bounds is nothing more than a matter of convenience, since they can be resupplied in a simple, automatic way, but how should the effect of dropping the bounds be understood when we take the theory without them seriously, not as a mere abbreviation? I think the effect of dropping the bounds is best understood as the result of setting all of the bounds equal. When all of the bounds on the quantifiers are equal, the quantifiers all have the same domain, the universe, which then does not need to be explicitly displayed, and so the bounds disappear from the notation. Note that the extrapolated theory is but a special case of the theory within finite mathematics—the special case in which the bounds are assumed equal.

The extrapolated theory is not a departure from the unextrapolated one, a genuinely new theory, but merely a version of the unextrapolated one in which the additional restriction is imposed that a certain dependency—that on the bounds—takes its least complicated form. Nonetheless, setting the bounds equal is making the extrapolation that we can start with so many beans that we can never need more. Of course, that is a heap of beans that is no longer indefinitely large—it is infinitely large. The *theory* is not much changed by the extrapolation, but the class of its *models* is changed in an important way: the class of models of the extrapolated theory excludes finite models.

The example of a theory of the indefinitely large that I have given is too special in two respects. The first is that it is concerned only with numbers. Each ω in the above example was a number. But it did not really function as a number: the point of ω_0 was that it determined which numbers are available at the first stage—the numbers less than ω_0. The other ωs played a similar role. Without the less-than relation or something like it we could not have used a member of our domain—a number—to specify a collection of available numbers. In general the indefinitely large entities will not be members of a domain but indefinitely large subsets of a domain that serve as indefinitely large subdomains of quantification. In most theories a subset of the domain must be treated as a predicate on the domain. From now on we switch almost completely from indefinitely large numbers, ωs, to indefinitely large predicates, Ωs. In our example, the predicate Ω_0 will apply to the numbers less than the ω_0 in the old notation, and so forth. Such predicates for indefinitely large domains may be added to any mathematical theory, whatever its character.

Once we switch from the indefinitely large number ω_0 to the indefinitely large collection of numbers Ω_0, it becomes possible to begin with an indefinitely large collection that is not an initial segment of the numbers. That is the second way in which the example was too special. Once we move beyond addition to multiplication and exponentiation, the "bucket" of numbers with which we begin need not be, and indeed is not most naturally taken to be, an initial segment of the numbers. For example, we readily employ the number $10^{10^{10^{10}}}$, but surely not all of its predecessors. In the finite mathematics of set theory we shall exploit a parallel possibility: there will be indefinitely large collections of sets that contain some set a without containing all of the members of a. That is, the indefinitely large collections will not be transitively closed.

How is it possible to axiomatize a theory of indefinitely large size? There are two apparent problems: the context dependence of the notion of indefinitely large size and the need to avoid infinitary presuppositions. Schemas, interpreted as suggested in §VI.3, are tailor-made for the job. Indeed, we used the axiom that says that ω_0 is indefinitely large,

$$(6) \qquad\qquad\qquad X_0 < \omega_0,$$

where X_0 is a schematic variable that may be replaced by any numeral, as an example of how to take schemas (§VI.4). A schema does not entail that *every* sentence of a specified form is true. Rather, it expresses a commitment that *any* actually given sentence of the specified form be true. When schema (6) is so interpreted, it yields that ω_0 is bigger than any particular number we substitute. That is just a formalization of part of what is meant by saying that ω_0 is indefinitely large—ω_0 is larger than anything that is immediately available to us. Since we cannot actually substitute infinitely many numbers into the schema, we can require that ω_0 be a natural number.[10]

Schemas enable us to formalize the essential aspects of the context dependence of the notion of indefinitely large size precisely because the commitments that a schema yields are themselves context dependent in the relevant way: we use different (finite) sets of instances of various schemas for different purposes.

Nothing about a schema commits us to the existence of a potential infinity of instances of it: no ability to continue is being postulated. If we do in fact continue, the schema tells us something about what that continuation must be like. In short, schemas constrain how we are permitted to go on without imposing any requirement that we do go on. (Compare the remarks in §VI.3 on the acceptance of schemas and those of §VII.4 on the Replacement schema of full second-order schematic set theory.) Schemas in general do not involve covert infinitary universal claims. We can therefore use them freely in axiomatizing finite mathematics. Each axiomatic theory within finite

---

10. Mycielski suggested [Myc86, p. 61] that a finitist theory of the sort he proposed "is interesting only if the theory itself is regarded as a process of formation of axioms and theorems and not as an actually infinite object." He suggested working out that idea by making use of schemas as rules of proof. That suggestion was the origin of my interest in schemas. Despite that, the notions of a "process of formation" and of a "rule of proof" play only an incidental role in my interpretation of schemas, and my focus is more on the objects with which a theory is concerned than on the formal use of axioms to prove theorems. Moreover, in sharp contrast to the position taken here, Mycielski took schemas to involve a commitment to the potentially infinite.

mathematics will be presented using schemas, and so, since schemas do not commit us to the truth of a potential infinity of instances but rather to the truth of the finitely many instances that we actually come to invoke, the axiom systems we shall be interested in will be sets of finitely many instances of the axiom schemas.

Now that I have sketched the basic picture I shall discuss the context relativity of the indefinitely large in greater detail. In some comparatively sophisticated contexts we can place upper bounds on the number of grains of sand on the beach and the number of stars in the sky. That just shows that those numbers are not indefinitely large with respect to those contexts. We can even name the exact number of stars in the sky, not just give an upper bound for it. (I just did name it—"the number of stars in the sky."[11]) The point is that in the unsophisticated contexts in which it counts as indefinitely large we cannot bound the number of stars in the sky in a canonical notation—where what counts as canonical depends on the context.

"Indefinitely large" is a notion that depends on some idea of *availability* of mathematical objects—for example, through description in a canonical notation—that could be taken to be primitive. Availability will not be part of our official version of finite mathematics, but it is worth giving a semiformal account of it nonetheless, both for motivational purposes and to make it clearer what sort of context dependence is being used.

> **Technical Remark.** If one were to permit a suitable notion of availability at the metalevel, finite mathematics could be formalized without the use of schemas: One ordinarily avoids schemas by replacing each with the set of all of its instances. Here, we would instead replace each schema with the set of all of its available instances. I think the use of schemas is considerably more natural. The present remark is intended for those who, like Quine ([Qui61], especially chapter 6), view formalization in conventional style as crucial for assessing the ontological commitments of a theory.

Fix a (finite) set A of available objects and a (finite) set F of availability functions, possibly many valued.[12] A set S is *indefinitely large* (over A and

---

11. Vagueness is not at issue here, and so I shall ignore it.

12. There will be times when we consider that if x is available then at least one y that bears some relation to x is thereby available, without fixing a particular y. Therefore,

F) if $A \subseteq S$, that is, if everything that is available is in the set. Each of the Ωs is indefinitely large in that sense. In addition, each Ω subsequent to Ω_0 is indefinitely large with respect to Ω_0, which means that it includes Ω_0 and the set of things that are available in virtue of the availability of the members of Ω_0. The things available in virtue of the availability of the members of Ω_0 are just those that can be obtained from members of Ω_0 via the availability functions. (In our earlier example, the one availability function was the successor function, which led to the requirement that the successor of every member of Ω_0 be in any subsequent Ω.) A set T is *indefinitely large with respect to a set* S (over A and F) if $A \subseteq T$, $S \subseteq T$, and for every n-ary many-valued function f in F and n-tuple $\overline{x}$ of members of S, some y such that $f(\overline{x}, y)$ holds is in T. That is, a set T is indefinitely large with respect to a set S if everything that is available in virtue of the availability of the members of S is in T.[13] The conditions just discussed are necessary for indefinitely large size, but it is important to realize that they are not sufficient since we did not take account of indiscernibility. I shall give a complete axiomatic treatment of indefinitely large size in §3.

The only indefinite sets we consider are indefinitely *large* sets. Any constraints arising from availability therefore take the form, This set is so large that . . . The more constraints there are, the larger our indefinitely large sets must be. Thus, being liberal about what we allow into A and F will tend to ensure that it is reasonable to think of our "indefinitely large" sets in the technical sense as being really large in an intuitive sense.

The main sort of availability we shall consider for motivational purposes is epistemic availability with respect to a particular purpose, discourse, state of knowledge, or the like. In line with the policy of liberality, we shall not restrict our notion of epistemic availability to be some strong notion of acquaintance, *de re* access, or the like.

The increasing indefinitely large Ωs represent degrees of availability: If Ω_0 is the first indefinitely large set under consideration and Ω_1 the second, then

---

instead of requiring that the members of F be functions, we allow them to be "many-valued functions," where an n-ary many-valued function is just an $n + 1$-ary relation R such that the sentence $(\forall x_1) \cdots (\forall x_n)(\exists x_{n+1}) R(x_1, \ldots, x_n, x_{n+1})$ holds on the domain of interest.

13. The nuisance of having two separate sets A and F can be avoided at some cost of perspicuity by allowing 0-ary many-valued functions into F and defining a set to be indefinitely large if it is indefinitely large with respect to the empty set.

Ω_0 is indefinitely large with respect to what is actually available, while Ω_1 is indefinitely large with respect to what is actually available and also with respect to what would be available in virtue of the availability of what is in Ω_0. Thus, Ωs with higher subscripts include things that are available to a lesser degree. It is a mistake to think of all of the members of Ω_0 as being available. All that follows from our conception of Ω_0 as indefinitely large is that if something is available, then it is in Ω_0, not the converse. Indeed the converse, or for that matter any condition that could serve to exclude members from Ω_0, would contradict our conception of Ω_0 as indefinitely large.

We make the plausible assumption that objects we actually do denote by closed terms in our basic notation are available. In our prime example of availability, epistemic availability, that assumption is clearly reasonable. Thus, for example, schema (6) brings with it the assumption that any number whose notation we employ is available, that is, in A.

To extend our treatment of elementary arithmetic to include addition, for example, we require that (binary) addition be a function in F. In that case in order for Ω_1 to be indefinitely large with respect to Ω_0 every sum of (two) members of Ω_0 must be in Ω_1. If $x + y$ is seen in that way to be in Ω_1, that is, as a result of seeing that x and y are in Ω_0, then $x + y$ is not seen to be in Ω_1 until after x and y are seen to be in Ω_0. (Nothing in the example requires that $x + y$ not be available in a more direct way. If it were, that might show that $x + y$ is in Ω_0. Since Ω_0 is included in Ω_1, the claim that $x + y$ is in Ω_1 would still hold.)

As we have just seen, the order in which things get into the Ωs imposes no necessary constraints on the temporal order in which things can become available. There is no reason why we cannot happen to consider some member of A, and therefore of Ω_0, only after having made use of something in Ω_1. The order of the Ωs is not temporal order.

I now give two examples to make the idea of degrees of availability clearer. First, a nonmathematical example. Let us suppose that the stars that are available to degree 0 (on a given night) are those that a friend points out to you by literally pointing at them. Then those that are available to degree 1 might be those your friend indicates by describing their position relative to stars available to degree 0—"See that star? (*Points at a star.*) Now find the equilateral triangle of stars just below it. Look at the one at the apex." The star at the apex is available to degree 1.

Stars with the next degree of availability, degree 2, would include stars indicated with reference to degree 1 stars—"Remember that triangle? Now look at the very faint star located at about the midpoint of the base." The star at the midpoint is available to degree 2. As before, degree 2 includes degree 1, which, in turn, includes degree 0.

The star at the apex, which was available to degree 1, might also be a star that could have been pointed out directly, which would have made it available to degree 0. Nonetheless, the description given only shows that it is available to degree 1. In the example, the availability set A includes the stars that were pointed out, and the set F of availability functions includes something like a function that takes every star with a visible upright equilateral triangle of stars just below it to the star at the apex of that triangle, similar functions to the other two vertices of the triangle, and a function that takes every pair of stars to a faint star near the midpoint of a visual line between them when such a star exists.[14]

Now, a mathematical example. We can fix a descriptive vocabulary for a mathematical theory in any of various ways. The set of constant and function symbols of the language of the theory would do as a descriptive vocabulary, for example, and so would a set of formulas that define individuals and functions.

Given a finite descriptive vocabulary, we can take each degree of availability to be the set of objects denoted by expressions of length less than some numerical bound. Larger bounds, which allow expressions of greater length, are used for higher degrees. One might also take each degree to be the set of objects that can be generated or represented, in any of various senses, by suitable computer programs of less than some given length, again with increasing lengths used for higher degrees.

We assumed earlier that an object is available if we have actually denoted it by a closed term in our basic notation. That assumption is easy to cash out in the terms of our semiformal presentation: When considering a finite set of axioms of any mathematical theory T, we shall, first of all, always require that all of the objects denoted by constant symbols that appear in the axioms are in A and that all of the functions denoted by function symbols that appear in the axioms are in F. In addition, we shall require that the denotation of every closed term that appears in the axioms is in A and that

---

14. It doesn't matter what the functions do when the relevant descriptions do not apply. For definiteness, let them evaluate to Polaris.

the many-valued functions asserted to exist by the axioms are in F.[15] Each mathematical theory will lead to the requirement that certain objects and many-valued functions be in A and F. As a result, it turns out that no explicit reference to the notion of availability or to A and F will be required within mathematics.

It is crucial to the finite mathematics of the indefinitely large that we are never forbidden to add suitable additional members to A and F, in part because different contexts may give rise to a need for different members of A and F. Suppose A,F is one constraint and A',F' is another. Anything that is indefinitely large or indefinitely large with respect to some set under the combined constraint $A \cup A'$, $F \cup F'$ is indefinitely large or indefinitely large with respect to the set under each of the constraints A,F and A',F' separately. That fact will allow us to smoothly assemble results of finite mathematics that come from different contexts.

Finite mathematics is a supplement to ordinary mathematics not a substitute for it, despite the possibility of an eliminative interpretation. The point of the supplement is to clarify the preformal intuitions that have led to ordinary mathematics, or at least those that have led to ordinary set theory. (Other parts of mathematics, though they have counterparts within finite mathematics, may have other sources, intuitive or systematic, and therefore may require separate analysis.) I hope that a better understanding of the intuitions behind the notion of combinatorial collection will be of help in settling at least some of the open questions that plague set theory: There may be fruitful extensions of the finite mathematics of set theory, extensions that give answers that, by extrapolation, answer important questions of ordinary set theory, that are well motivated, and that therefore, in extrapolated form, can be generally accepted as new axioms of set theory.

It is also possible that there may be fruitful extensions of the finite mathematics of set theory, extensions that give answers that, by extrapolation, answer important questions of ordinary set theory, that do not extrapolate to yield corresponding axioms of ordinary set theory—not because the extrapolated versions are implausible (I claim that could never be the case) but

---

15. For example, if an axiom is of the form $(\forall x)(\exists y)\phi(x, y)$, then there will be a many-valued function f in F such that $(\forall xy)(f(x, y) \to \phi(x, y))$ holds. As a matter of fact, it will always be simpler for mathematical purposes to replace each many-valued function with a single-valued one, a Skolem function. We shall do that after this section, but here I am analyzing what it is in principle plausible to allow.

because there just are no extrapolated versions, for example because the subscripts on the Ωs in the axioms are not in increasing order. Such a state of affairs would be an argument for favoring the eliminative interpretation, that is, for substituting the finite mathematics of set theory for ordinary set theory, at least for some purposes, since there would *ex hypothesi* be important axioms and results of finite mathematics that had no infinitary counterparts. (Compare [Bur83].) But it is premature to take the possibility too seriously.

Why should we take extrapolation to be legitimate? That is, why should we regard the move from finite set theory to ordinary infinitary set theory as a justifiable one? After all, in the absence of a finite set theory that incorporates the indefinitely large, the creators of set theory had ordinary set theory as the only apparent option for making sense of their intuitions, but we have the finite set theory too. Why shouldn't we take the attitude that, while it was once justified to endorse ordinary set theory, it is no longer justified to take it to be anything more than a convenient abbreviation in which the Ωs are suppressed, now that we see what was lurking behind the reasons we apparently had for accepting it?

One important reason for taking the extrapolation seriously is implicit in the discussion above of the possibility of new principles of finite set theory that do not extrapolate. We do not know of any such principles. Everything we know about finite set theory is of a form that extrapolates (or a consequence of things that do). That is to say, in our present state of knowledge, the Ωs do not play any mathematical role.[16] It would therefore, for methodological reasons, be a mistake to require them: one shouldn't burden one's theories with needless technicalities. That is a methodological maxim we have seen operate before in the history of mathematics: it is, I believe, the reason behind the fact that increases in rigor occur only when they are required for the advancement of the subject. At other times such increases are—to the extent that anyone bothers to propose them at all—regarded as unwarranted nuisances that serve "only to conceal the truth and to weary the reader."[17] The extrapolation not

---

16. As will become clear in the first part of §3, any theorem we can prove that involves Ωs may be derived by first proving an extrapolated counterpart in ordinary mathematics and then adding the Ωs on to the result in the appropriate order. The Ωs go along for the ride, doing no proof-theoretic work.

17. A similar methodological maxim is applied in the sciences. We reject operationalized versions of physical theories, accepting the existence of, say, atoms, because the operationalized theories are full of needless technicalities. In precise analogy to our

only results in a simpler—and therefore better—theory, but it also helps to explain why the kinds of sentences of finite set theory that do not extrapolate have played no role in our theorizing. After all, if the extrapolated theory were not of primary interest, it would be difficult to see why only sentences that in fact extrapolate play such a special role. The fact that they do is itself an argument for taking the extrapolated theory seriously. There is a second important reason for taking the extrapolation seriously: we often regard the apparent indefiniteness of some quantity or other as the epistemic reflection of an objective fact. As we spell out in some detail in the second part of §3, within each Ω there are only finitely many real numbers, numbers that we may think of as indefinitely close together. Larger Ωs include more real numbers, and so they permit finer approximations to be made. Hao Wang said [Wan86, p. 147] that it is a part of the "abstract history" of mathematics that "The desire to have an absolutely accurate, or rather indefinitely improvable, measurement leads to the general concept of 'real number'." There is certainly something very right about that. Naturally, I would want to emphasize Wang's idea of an indefinitely improvable measurement—that is the sort of thing the counterparts of the real numbers within finite mathematics provide. But we think of the values of the quantities we measure as being objective, determined independently of our measurements.[18]

It is a fascinating and difficult question in the philosophy of science how we arrived at the belief in objectivity. An adequate discussion would require a book in itself. It is at any rate clear from the very description of objective values of physical quantities as being independent of our measurements that it could not have arisen in any direct way from experience. The measurement of a value has an epistemic component, but the value itself does not.

The counterpart of the real numbers in finite mathematics reflects the idea of indefinite improvability quite well, but—since the effect of extrapolation is precisely to remove dependence on the epistemic situation—it is the extrapolated real numbers that reflect the idea that there are objective values in

---

situation, if anyone found a positive use for those technicalities, that would be a reason for taking the operationalized theories seriously, and it would likely cast doubt on the existence of atoms.

18. I am ignoring those quantum-mechanical measurements for which the claim that the values are independent of the measurements may be controversial. All I need for present purposes is that there are at least some measurements of the sort I am describing. And that is not controversial.

the physical world, independent of our measurements and their accuracy. The point is most easily made concerning the real numbers, but it is equally true of functions from the real numbers to the real numbers, and so forth all the way up. Thus, extrapolation is legitimated by the fact that it is the counterpart within mathematics of what we take to be the relation between our measurements and their objective values.

Now let me return to discussing the finite mathematics on which extrapolation is based. Nothing in finite mathematics relies on any assumption that any mathematical object exists in any sense beyond the usual one that is internal to ordinary mathematics. As I put it earlier, finite mathematics is just more mathematics. As a result, the many structuralist, nominalist, and Platonist attempts to adjust the ontology of mathematics in various ways can be applied to finite mathematics in the same way that they are applied to any other part of mathematics. In the case of finite mathematics, however, any special devices that concern the infinite become irrelevant. Those who wish to base mathematics on perceptual intuition concerning collections need consider only perceptual intuition concerning finite collections; those who wish to base mathematics on the structure of patterns need consider only finite patterns; those who wish to base mathematics on inscriptions of one sort or another need consider only finitely many inscriptions, and so forth.

The possibility of substituting finite mathematics for ordinary mathematics may prove to be of interest to those nominalists who have wished to reduce the ontology of mathematics to various kinds of finite quasi-concrete objects. Analogously, those Platonists who wish to base our knowledge of and intuitions about infinite domains on our experience of concretely instantiated finite patterns may find the process of extrapolation proposed here of interest: it may provide the beginning of a story of how one can get from the relevant finite objects to ordinary mathematics involving the infinite.

§3. The Finite Mathematics of Indefinitely Large Size

§3.1. The General Theory. In the first part of this section I shall present the finite mathematics of arbitrary theories. (Actually, I shall consider only first-order theories, postponing a discussion of second-order logic to §IX.5.) I shall postpone discussing what I find to be the most interesting particular case, the finite mathematics of set theory, to §4, since the general theory is a prerequisite for that discussion.

The finite mathematics of set theory is closely associated with, and illuminating with respect to, the combinatorial conception that resulted in set theory as we know it. The counterparts in finite mathematics of other theories need not bear any illuminating relation to our motives for developing those theories. At the present time, for example, I see no reason to think that the finite mathematics of group theory has anything much helpful to say about our ways of thinking about groups.

Set theory, however, is not the only case in which finite mathematics is illuminating: we shall consider the finite mathematics of analysis in the second part of this section, and the finite mathematics of geometry has some suggestive features. But set theory will be the main case of interest.

Mycielski [Myc86] showed that to any set T of sentences there corresponds another set of sentences Fin(T) with certain finitist characteristics. My claim that every theory used by mathematicians has a counterpart within finite mathematics relies on that—the counterpart of a theory T is Mycielski's Fin(T). The theory Fin(T) will be of potential interest for finite mathematics only if T is axiomatized by finitely many schemas, but that restriction is not so onerous as it may seem: by a theorem of Robert L. Vaught [Vau67], in a finite language every sufficiently rich theory that is axiomatizable (with a recursively enumerable set of axioms) is axiomatizable by a single schema.[19] What is new is not Fin(T) but the argument that Fin(T) can be described in a metalanguage that does not involve infinite sets. Mycielski simply introduced Fin(T) for arbitrary T in a standard background of ordinary mathematics.

Fix an arbitrary first-order theory T. We shall discuss Fin(T) for that T. Every quantifier in Fin(T) will be bounded by an indefinitely large Ω, and hence every quantifier will have a finite range. Thus, the theory of indefinitely large size enters into the finite mathematics of any theory whatsoever—indefinitely large domains serve as the counterpart of infinite domains. The vocabulary of Fin(T) is that of T plus unary predicate symbols Ω_p for the indefinitely large domains, where p is any rational number,[20] plus the vo-

---

19. If T is not recursively axiomatizable, then neither is Fin(T). In that case, Fin(T) is not in a direct sense a part of finite mathematics. All of the theories mathematicians work in (as opposed to those they work on) are, of necessity, recursively axiomatizable. Mathematicians do study theories that are not recursively axiomatizable, but they do so within some background theory, typically set theory. Given a background set theory, the theories that are not recursively axiomatizable are just taken to be certain sets. Those sets *do* have counterparts within finite set theory, and so no part of the practice of mathematics is lost as a result of present considerations.

20. Since none of our work will ever involve more than finitely many indefinitely large sets, the business of the rational numbers is only a convenience. We could work

cabulary of first-order logic. We do not include equality, which it is simpler to present separately for expository reasons. We write $(\forall x \in \Omega_p)\phi$ for $(\forall x)(\Omega_p(x) \to \phi)$ and $(\exists x \in \Omega_p)\phi$ for $(\exists x)(\Omega_p(x) \wedge \phi)$. Those notations are slightly misleading since Ω_p is a unary predicate symbol, not a constant symbol, and since the membership relation need not be a part of the vocabulary of Fin(T). We write, for example, $(\forall xy \in \Omega_0)\phi$ for $(\forall x \in \Omega_0)(\forall y \in \Omega_0)\phi$. I shall make a few definitions and state some axioms and then discuss them. The reader who is not concerned with complete mathematical detail may wish to skip to the discussion.

A formula ϕ of the language of Fin(T) is *regular* if it is bounded (that is, all of the quantifiers in it are bounded by Ωs) and whenever Ω_q occurs within the scope of a quantifier bounded by Ω_p in ϕ, then $p < q$. In particular, when a quantifier bounded by Ω_q occurs within the scope of a quantifier bounded by Ω_p in a regular formula, then $p < q$.

A formula ϕ of the language of Fin(T) is a *relativization* of a formula ψ of the language of T if ψ can be obtained from ϕ by deleting all the bounds—that is, by changing $(\exists x \in \Omega_p)$ and $(\forall x \in \Omega_p)$ to $(\exists x)$ and $(\forall x)$ for all variables x and subscripts p, wherever they occur in ϕ. Thus a formula is a relativization of some formula if the only Ωs that occur in it are bounds of quantifiers. *Regular* and *relativization* have straightforward primitive recursive definitions, and so they can be defined in PRW.

Any instance of each of the following schemas is an axiom of Fin(T):[21]

---

with Ω, Ω', Ω'', …. The rational-number subscripts merely enable us to insert a new indefinitely large set between two members of our sequence without having to stop to renumber.

The careful reader may object that I have given up on my restriction to finite objects by allowing Ωs with rational subscripts—I seem to be allowing the set of rational numbers in the metalanguage. That is correct, but it is easily fixed: we can use the reduction of the rational numbers to the natural numbers familiar from the discussion of Hilbert in §VI.3 to replace each rational subscript with a pair of natural-number subscripts, and each natural number is a finite object.

One might still object that I have eliminated the rational numbers in favor of all of the natural numbers, which is still an infinitely large set. That is not correct. What I have said is that if a given symbol is of a certain type (an Ω with a rational-number subscript), then it is a symbol of our language. I have not said anything characterizing all symbols of the language; I gave a test for whether an antecedently given symbol is a symbol of the language. That is, the specification of the predicate symbols is itself schematic. It can be carried out in PRW, and it is therefore free of any commitment to the infinite set of natural numbers, as argued in §VI.4.

21. More precisely, the predicate Fin(T) holds of any instance of each of the follow-

(7) $\Omega_p(c)$,

for any constant symbol in the vocabulary of T.

(8) $(\forall x_1, \ldots, x_n \in \Omega_p)\Omega_q(f(x_1, \ldots, x_n))$,

for $p < q$ and for any function symbol f in the vocabulary of T, where n is the number of places of f.

(9) $(\forall x \in \Omega_p)\Omega_q(x)$,   for $p < q$.

(10) $(\forall x_1, \ldots, x_n \in \Omega_p)((\forall x \in \Omega_q)\phi \leftrightarrow (\forall x \in \Omega_r)\phi)$,

where $p < q$, $p < r$, q and r are less than s for all s such that Ω_s appears in ϕ, the formula ϕ is regular, and the free variables of ϕ are among $x, x_1, \ldots, x_n$.

The axioms just given are minor and inessential variants of those of My-cielski. I changed Mycielski's axioms slightly to make it clearer that all quantifiers are bounded and to avoid reference to the notion of relativization—since that notion is not internal to finite mathematics. The four axioms stated so far give necessary and sufficient conditions for the Ωs to be indefinitely large and for Ω_q to be indefinitely large with respect to Ω_p whenever $p < q$. They may be referred to collectively as the axioms of indefinitely large size.

Axiom (7) says that Ω_p is large enough that it includes the objects denoted by the constant symbols. In effect, it declares that those objects are in the set A of available objects.

Axiom (8) says that if $a_1, \ldots, a_n$ are in Ω_p, then the result of applying the function denoted by f to $a_1, \ldots, a_n$ is in Ω_q for $q > p$. It has the effect of declaring that each f is in the set F of availability functions, or in other words that if $a_1, \ldots, a_n$ are in Ω_p then $f(a_1, \ldots, a_n)$ is in any set indefinitely large with respect to Ω_p.

Axiom (9) says that Ω_q includes Ω_p when $q > p$.

Axiom (10) says that the Ωs are in a certain sense indiscernibles (see Lemma 1 below): Suppose without loss of generality that $q < r$. The axiom

---

ing schemas. Fin(T) is to be thought of as a predicate of sentences, not a set of sentences, for the obvious reason that the set would be infinite, and we therefore cannot make use of it. The same trick of introducing predicates independently of set-theoretic extensions needs to be exploited at other points in our development as well, but I shall not continue to mention it explicitly.

formalizes the intuitive idea that all that matters about Ω_q is that it is indefinitely large. We can replace Ω_q with the larger Ω_r without changing the situation in any way that matters, that is, without changing the situation with respect to any property ϕ and objects $x_1, \ldots, x_n$ under consideration. If we could not replace Ω_q with any given larger Ω_r, that would mean that there was some upper bound to the possibilities for Ω_q, contrary to the idea that it is indefinitely large.

A semiregular formula will be one that allows blocks of like quantifiers with the same bound, for example, $(\forall xy \in \Omega_0)$, while requiring the bound to increase when the quantifiers are not alike: $(\forall xy \in \Omega_0)(\exists z \in \Omega_0)$ is not allowed. In detail: An occurrence of a quantifier in a formula is *general* if the quantifier that occurs is universal and it occurs in the formula positively[22] or if the quantifier that occurs is existential and it occurs in the formula negatively. An occurrence of a quantifier in a formula is *restricted* if the quantifier that occurs is existential and it occurs in the formula positively or if the quantifier that occurs is universal and it occurs in the formula negatively. (The terminology is from Herbrand's dissertation [Her30].) A formula is *semiregular* if every quantifier in it is bounded by an Ω and whenever Ω_q occurs in the scope of a quantifier bounded by Ω_p, then $p \leq q$ and $p < q$ if Ω_q is also the bound of a quantifier and the occurrences of the quantifiers bounded by Ω_p and Ω_q are of distinct types (that is, one is general and the other restricted).

We can now define Fin(T). It consists of Axioms (7)–(10), plus semiregular relativizations of the sentences in T.

The following two lemmas establish that the Ωs are indiscernible in the sense indicated in §2.

LEMMA 3.1 (MYCIELSKI [Myc86]). *The above axioms yield*

$$(\forall x_1, \ldots, x_n \in \Omega_p)(\phi \leftrightarrow \phi'),$$

where ϕ and ϕ' are both regular, differ only in the Ωs that occur in them, and have no occurrences of any Ω_q such that $q \leq p$.

---

22. A *positive* occurrence of a quantifier is one in the scope of an even number of negation signs. A *negative* occurrence of a quantifier is one in the scope of an odd number of negation signs. For those definitions of positive and negative to work, one must consider only formulas without conditionals or biconditionals in them, since conditionals have a hidden negation buried in the antecedent and biconditionals have conditionals hidden in them.

The proof is essentially a straightforward induction on the rank of ϕ. It can be carried out in PRW, and so it is acceptable as a part of finite mathematics (see §VI.4).[23]

Lemma 3.2. *The above axioms yield*

$$(\forall x_1, \ldots, x_n \in \Omega_p)(\phi \leftrightarrow \phi')$$

for any semiregular ϕ and ϕ' that differ only in the Ωs that occur in them and that have occurrences only of Ω_qs with $q > p$.

Lemma 2 can be proved in PRW.

Sketch of Proof of Lemma 2. A straightforward induction on the rank of ϕ using Lemma 1 shows that there is a regular formula ψ such that

$$(\forall x_1, \ldots, x_n \in \Omega_p)(\phi \leftrightarrow \psi)$$

is a theorem, where ϕ is as in the statement of Lemma 2 and ψ differs from ϕ only in the choice of Ωs and has occurrences only of Ω_qs with $q > p$. Lemma 2 is then immediate from Lemma 1. $\square$

If the theory T uses equality, then $\mathrm{Fin}(T)$ will include regular relativizations of the axioms of equality:

(11) $(\forall x \in \Omega_0)\, x = x;$

---

23. In more detail, we can define terms σ, τ, and ρ in the vocabulary of PRW such that (i) $\sigma(p, \phi, \phi')$ is equal to 0 if p, ϕ, and ϕ' are as required in the statement of the lemma and equal to 1 otherwise; (ii) if p, ϕ, and ϕ' are as required, then $\tau(p, \phi, \phi')$ is a proof as required; and (iii) we can prove (ii) in PRW, that is, ρ is a term such that $\rho(x, p, \phi, \phi')$ is equal to 0 if x is a proof of a statement of the form required by the conclusion of the lemma for p, ϕ, and ϕ' and equal to 1 otherwise, and it is a theorem of PRW that if $\sigma(w, x, y) = 0$, then $\rho(\tau(w, x, y), w, x, y) = 0$.

The careful formulation of the lemma in PRW is awkward to state. Nonetheless it can, after a certain amount of practice, be read off the original statement in the text that involves apparent and inessential use of quantification over an infinite domain. In what follows, I shall just note that various theorems and lemmas can be proved in PRA, PRW, and PRS. That is an abuse of language—what I mean is that there are suitable reformulations that can be proved in those theories.

(12) $(\forall xy \in \Omega_0)(x = y \to y = x);$

(13) $(\forall xyz \in \Omega_0)(x = y \wedge y = z \to x = z);$

(14) $(\forall x_1 y_1, \ldots, x_n y_n \in \Omega_0)(x_1 = y_1 \wedge \cdots \wedge x_n = y_n$
$$\to f(x_1, \ldots, x_n) = f(y_1, \ldots, y_n)),$$

for any function symbol f in the vocabulary of T, where n is the number of places of f; and

(15) $(\forall x_1 y_1, \ldots, x_n y_n \in \Omega_0)(x_1 = y_1 \wedge \cdots \wedge x_n = y_n \wedge P(x_1, \ldots, x_n)$
$$\to P(y_1, \ldots, y_n)),$$

for any relation symbol P in the vocabulary of T, where n is the number of places of P. The motivation for Axioms (11)–(15) is clear, and so I shall not discuss them in any detail.

I can now make good on my claim that every theorem of T has a natural counterpart in Fin(T):

THEOREM 3.3 (MYCIELSKI [Myc86]). *If ϕ is a sentence in the language of T and ϕ' is a regular relativization of ϕ, then ϕ is a theorem of T if and only if ϕ' is a theorem of Fin(T).*

Thus, the counterpart of a sentence of the language of T is any regular relativization of it. The situation discussed in §2 for a fragment of elementary arithmetic is perfectly typical. Theorem 3 can be proved in PRW. The proof is elementary, given the right axiomatization of logic. Roughly, a proof of ϕ' from Fin(T) becomes a proof of ϕ from T when the Ωs are deleted, and the other direction is an easy induction on the lengths of proofs.[24] It follows

---

24. There is a quadratic function f such that given a proof of a theorem of T that contains n symbols, there is a proof of a regular relativization of that theorem from Fin(T) that contains at most $f(n)$ symbols. Thus, if it is feasible to think of the proof in T, it is also feasible to think of the corresponding proof in Fin(T). (See, for example, [GJ79, pp. 6–11] for an introductory discussion of the relevant notion of feasibility.) In the other direction, there is a linear function g such that given a proof that contains n symbols of a regular relativization of a theorem of T from Fin(T), there is a proof of the theorem from T that contains at most $g(n)$ symbols. Those bounds can be extracted by a straightforward analysis of Mycielski's proof of Theorem 3. I am indebted to [Myc89] for the point that the analysis given in this note is called for, in order to show that T and Fin(T) are so closely connected that one can extrapolate not only theorems but their proofs.

from Theorem 3 that T is consistent if and only if Fin(T) is consistent. Thus, the statement in §2 that finite mathematics is neither more nor less likely to be free from contradictions than most of the rest of mathematics is true in a precise sense.

Theorem 3 by itself is useless: all it shows is that Fin(T) is an ordinary mathematical theory with the same sort of status as T. Fin(T) and T are equiconsistent and their proof theories are entirely comparable because of Theorem 3. So far, there is no reason to prefer one of them over the other. The way in which they differ significantly is with respect to their model-theoretic properties: Fin(T) has models in which the Ωs demarcate domains that are indefinitely large but finite in the precise sense we used to motivate Axioms (7)–(10).

In fact, Fin(T) has such models if and only if T (or, equivalently, Fin(T)) is consistent, which is the best one could ask. The following theorem is the key step toward establishing that.

THEOREM 3.4 (MYCIELSKI AND PAWLIKOWSKI). *T is consistent if and only if every finite subset of Fin(T) has a finite model.*

Mycielski proved Theorem 4 in Peano arithmetic [Myc86], and Janusz Pawlikowski showed how to modify Mycielski's proof so that it can be carried out in PRW [Paw89].[25] Their proof shows (still in PRW) that there is a function * (not dependent on T) such that for every finite set S of axioms of Fin(T), S^* is a finite set of axioms of Fin(T) such that $S \subseteq S^*$ and such that S^* is consistent if and only if S^* has a finite model. The model of S^* is *a fortiori* a model of S. That version of the theorem is of interest because it refers only to Fin(T), not to T.

One could use Theorem 4 to show that every finite subset of Fin(ZFC) has a finite model on the basis of our conviction that ZFC is consistent. What I am doing is reversing that procedure. I claim that we have reason to be convinced of the consistency of ZFC because we have an autonomous reason to believe that finite parts of Fin(ZFC) have finite models: We see that finite parts of Fin(ZFC) have finite models based on our experience of the indefinitely large.

---

25. The primitive recursive version exhibits a function that takes each string that lists a set of axioms of Fin(T) to either a derivation of a contradiction from the corresponding axioms of T or a (code for) a model of the set of axioms of Fin(T). The stipulations that the set and the model are finite are superfluous, since everything in PRW is finite.

Technical Remark. Here is a rough indication of the proof of Theorem 4. The direction from right to left is easy using Theorem 3, and so we discuss only the forward direction. Assume that T is consistent, and let S be a finite subset of $\text{Fin}(T)$. Let A be the set of constant symbols that appear in S. (Add one if necessary to ensure that A is not empty). Let F be the set of function symbols that appear in S plus the Skolem function symbols required to Skolemize the sentences obtained from those in S by dropping the Ωs that bound the quantifiers. Let $\Omega_{q_0}, \ldots, \Omega_{q_n}$ be the list of all the Ωs that appear in S, in ascending order of subscript. To build the required model, let the interpretation of Ω_{q_0} be the set A. Let the interpretation of $\Omega_{q_{i+1}}$ be the union of the interpretation of Ω_{q_i} and the set of all terms of the form $f(\tau_1, \ldots, \tau_k)$ with f in F and the τs in the interpretation of Ω_{q_i}. Construct interpretations for as many Ωs as appear in S plus one more. The construction of the interpretations of the Ωs is just as required to ensure that each is indefinitely large with respect to its predecessors in the informal sense of §2. The Skolem functions that arise from instances of Axiom 10 ensure the required indiscernibility. (That observation is Pawlikowski's contribution [Paw89]—Mycielski's original technique to ensure indiscernibility was not primitive recursive [Myc86].) The terms in the interpretations just constructed for the Ωs form the domain of the required model. Interpret the constant and function symbols in the obvious way. (The obvious way breaks down for terms in the interpretation of the last Ω, but, since that Ω doesn't appear in S, it doesn't matter. Supply the required values for the functions in any way at all.) Since every element of the model under construction is a Skolem term for the consistent theory T, there is some interpretation of the relation symbols that results in a model of S, as required.

The interpretations of the Ωs constructed are pretty much the partial models that Herbrand used. But Herbrand's partial models were not models of the theory of interest, T. That defect is remedied here by supplying an intermediate theory, $\text{Fin}(T)$, obeying Theorem 3. (Compare [Myc86, p. 61].) Axiom 10 is the key to the improvement.

If T has infinite models, then so will $\text{Fin}(T)$. More precisely, what I mean is that $\text{Fin}(T)$ will have models in which the indefinitely large sets are infinitely large. That is trivial to prove: any model of T, with each Ω interpreted as the universe of the model, is as required. The point of finite mathematics is not that its theories have no infinite models. The point is rather that its theories do have finite models. What is required to explain how we extrapolate from the finite to the infinite is that there are suitable finite models of the theories of finite mathematics, not that there are no infinite ones.

I have argued at length that theories of finite mathematics do not involve a commitment to potential infinity. We can now make that more precise. Our commitments within a theory Fin(T) of finite mathematics are always consistent with a statement of the form "there exist at most n objects" for some natural number n: Let S be the finite set of axioms of Fin(T) to which we are actually committed. Then (if S is consistent) S has a finite model $\mathfrak{M}$ with, say, m elements in its domain. The model $\mathfrak{M}$ is a model of "there exist at most m objects," which is therefore consistent with S.

To make the point as strongly as possible that theories of finite mathematics do not involve a commitment to potential infinity, suppose that Peano arithmetic is inconsistent and that there really are only finitely many numbers. (I do not take the possibility very seriously and can't imagine what the situation would be like, but bear with me.) Then nothing would be left of ordinary classical or constructive mathematics. One might be able to reconstruct some parts, but nothing would survive in its original form. In sharp contrast, it would be perfectly possible for all of finite mathematics to survive: since axiom schemas have only finitely many instances and the hierarchy of Ωs comes to a stop, the finitely many axioms with their finitely many finite Ωs might still have a perfectly good finite model.[26] Thus, finite mathematics is in no way incompatible with the assumption that there are only finitely many numbers, and hence it surely does not presuppose any form of potential infinity. The necessary presuppositions of finite mathematics are those of the liberal finitist, in the terminology of §VI.2. The Ωs play the role of the indefinite bound discussed in that section, but there is an unexpected twist: the practitioner of finite mathematics grants that the constraints on the bound may change in a context-dependent way.

Even Theorem 4 is not enough to show that the model theory of Fin(T) is of interest for our purposes: for that, Fin(T) must have finite *reasonable models,* not just crazy ones.[27] Some infinitary theories have only one reasonable

---

26. The finitely many axioms of finite set theory could even have a natural model in the sense of §4, as a consequence of Theorem 4.1, below.

27. Theorem 5 is messy, and the situation is simpler in the main case of interest,

model, the intended model, while some infinitary theories have no unreasonable models, since what is in fact of interest is arbitrary models of the theory. It is possible to prove the desired result even without an abstract characterization of a reasonable model:[28]

THEOREM 3.5. *Let $\mathfrak{M}$ be a model of T. Let S be a finite set of axioms of Fin(T), and let p be such that Ω_p occurs in S but no Ω_q with $q > p$ occurs in S. Then S has a finite model with universe a finite subset of the universe of $\mathfrak{M}$ such that the interpretations of the relation, constant, and function symbols of the language of T in the finite model restricted to members of Ω_p agree with those of $\mathfrak{M}$.*[29]

Thus, for example, any finite set of axioms of Fin(PA) (PA is Peano arithmetic) has a model that has a finite set of numbers for its universe and that is such that the interpretation of + on members of the indefinitely large sets really is addition, and so forth.

Theorem 5 cannot be proved in PRW. Since the theorem refers to a model of T, it cannot even be formulated in PRW. In the absence of a general definition of *reasonable model*, I do not see how to avoid the reference to a model of T. However, given a particular theory T one can focus on a particular model of interest. In that kind of setting, the results are often more satisfactory. It follows from Theorem 5 that Fin(ZFC) has reasonable models. But, by using a proof adapted especially to the case of set theory, one can prove in PRS that Fin(ZFC) has reasonable models. See §4.

---

set theory, as we shall see in §4. You might therefore wish to skip to the end of the next technical remark.

28. Here is a proposal for such a characterization: As Richard Montague emphasized [Mon65, p. 136], in the familiar cases (Peano arithmetic and Zermelo–Fraenkel set theory) a reasonable model, which in those cases means a standard model in the usual sense, can be characterized as a model of the strengthened theory obtained by replacing the schematic axioms by their second-order universal closures. One can take that to be the general procedure.

29. The conclusion is essentially that the finite model of S is a substructure of $\mathfrak{M}$, but that is not quite correct: the finite model has additional relations, the Ωs, and, more substantially, the function symbols may not behave correctly on members of their domain that are not in Ω_p. The axioms of S have all their quantifiers restricted to (subsets of) Ω_p, and so they don't say anything about the rest of the model. Thus, the values of the functions on objects outside of Ω_p are of no concern.

Technical Remark. Theorem 5 is immediate from the proof of Theorem 4. Instead of building a model out of Skolem terms, as we did in the proof of Theorem 4, choose Skolem functions on $\mathfrak{M}$ and let the elements of the model be the denotations of the Skolem terms.

There is one detail remaining. The Ωs are supposed to be indefinitely large, and so we should make sure of the following: Let S be a finite set of axioms of $\mathrm{Fin}(T)$ involving $\Omega_{q_0}, \ldots, \Omega_{q_n}$, and let $A_0, \ldots, A_n$ be finite sets of variable-free terms in the language of T. Then there should be a finite model of S in which $\Omega_{q_i}(\tau)$ is true for every term τ in A_i, $i = 0, \ldots, n$. That will show that there is no restriction on what is in the Ωs and hence that they can be thought of as indefinitely large. But the required result follows immediately from Theorem 4 and the fact that $\Omega_{q_i}(\tau)$ is a logical consequence of an easily specified finite set of axioms of $\mathrm{Fin}(T)$ for any q_i and closed term τ.

That concludes my presentation of general features of finite mathematics. It is now time to consider some examples, which will, I hope, make it clearer what is going on.

§3.2. Some Examples. The first example is really not so much an example as a suggestion of a place in which to look for examples. In physics we frequently model indefinitely large things as infinite ones: We model long rods as infinitely long to avoid having to consider the ends; we model a cloud of particles as if the particles were continuously distributed, taking there to be infinitely many particles instead of indefinitely many; and so forth. The justification for applying the physics of an infinitely long rod to a concrete case is that in the concrete case the rod is supposed to be sufficiently long that the effects of the ends are negligible. There are ends, and they do have an effect, unlike the official model. What we really have in mind in using the ideal rod of infinite length as a model is that however long the actual rod needs to be to reduce the effect of the ends below some level, the rod is longer. If in doing an experiment the effects of the ends turned out to be too large, we wouldn't give up—we would get a longer rod. The analogy to the bucket of beans in §2 is clear. That suggests that it might be fruitful to consider physical models that are more accurate in that they use the indefinitely large instead of the infinite, but I do not know of any case in which there is a clearcut advantage.

I now develop some of the finite mathematics of analysis. That is worth doing because the finite-mathematics counterparts of many of the basic ideas of analysis admit of simplifications that make them illuminating examples of finite mathematics.

I shall not make any attempt to provide a fully general treatment that can mimic every part of set-theoretically based analysis. Indeed, I shall not even attempt to make it clear how all of my examples fit together. That is not necessary since it is already clear that it is possible to give a fully general treatment: Present-day analysis can be done in set theory, say ZFC, and so, by Theorem 3, there is a finite-mathematics version of analysis in Fin(ZFC) that is a precise counterpart of the usual treatment. Of course, the point is more general—virtually all of mathematics can be developed within ZFC, and so virtually all of mathematics has a counterpart within Fin(ZFC). Moreover, those few parts of contemporary mathematical practice that go beyond ZFC or that contradict the Axiom of Choice can be handled in the finite-mathematics counterparts of the theories in which they are usually handled.

Let us start with the notion of a limit—the notion Cauchy made central to analysis. To keep life simple, let us consider only functions of one variable from the real numbers to the real numbers. The modern Weierstrassian definition of a limit is as follows:

> The limit of the function f as x approaches a is l if for every ϵ there is a δ such that for all x if $|x - a| < \delta$ and $x \neq a$, then $|f(x) - l| < \epsilon$.

If that definition isn't terribly clear, bear with me for a while—the finite-mathematics version will end up a lot more perspicuous. The finite-mathematics version of the definition of a limit is a regular relativization of the definition just given:

> The limit of the function f as x approaches a is l if for every $\epsilon \in \Omega_0$ there is a $\delta \in \Omega_1$ such that for all $x \in \Omega_2$ if $|x - a| < \delta$ and $x \neq a$, then $|f(x) - l| < \epsilon$.

That definition isn't any better looking. If anything it is worse. But Ω_0 is a finite set. It therefore has a smallest member, call it ζ_0, which we think of as an indefinitely small quantity. Since there is supposed to be a δ for every ϵ in Ω_0, there will have to be a δ for ζ_0. But the δ that works for ζ_0 will work for every ϵ in Ω_0, since ζ_0 is the smallest. We can therefore simplify the definition as follows:

The limit of the function f as x approaches a is l if there is a $\delta \in \Omega_1$ such that for every x in Ω_2 if $|x - a| < \delta$ and $x \neq a$, then $|f(x) - l| < \zeta_0$.

We are about halfway home: if there is any δ in Ω_1 as required, then the smallest member of Ω_1, call it ζ_1, must be as required. Now the definition looks like this:

The limit of the function f as x approaches a is l if for all $x \in \Omega_2$ when $|x - a| < \zeta_1$ and $x \neq a$, then $|f(x) - l| < \zeta_0$.

The restriction that x must be in Ω_2 is vacuous—since Ω_2 is arbitrarily large, any real number at all could be in it, and if we add one, there are no subsequent Ωs that might require adjustment. Thus, our definition is equivalent to this:

The limit of the function f as x approaches a is l if for every x when $|x - a| < \zeta_1$ and $x \neq a$, then $|f(x) - l| < \zeta_0$.

Now we are getting somewhere.

Let us introduce a notation for approximate equalities, or *adequalities*. We will write $x =_p y$ for $|x - y| < \zeta_p$. With that definition, $a =_0 b$ means that a is approximately equal to b at level of approximation 0 (they differ by less than ζ_0), and $a =_1 b$ means that a is approximately equal to b at finer level of approximation 1. We can now just say,

The limit of the function f as x approaches a is l if for every x when $x =_1 a$ and $x \neq a$, then $f(x) =_0 l$.

The new version is much easier to understand than the Weierstrassian one with which we started, though it is exactly as general. It says that the limit of the function f as x approaches a is l if the value of $f(x)$ is near l when x is very near, but distinct from, a.

The relation between the ζs and the Ωs is a little indirect. We can simplify it by starting with a variant of the Weierstrassian definition of a limit given above:

The limit of the function f as x approaches a is l if for every natural number e there is a natural number d such that for all x if $|x - a| < 1/d$ and $x \neq a$, then $|f(x) - l| < 1/e$.

We are in effect using $1/e$ in place of ϵ and $1/d$ in place of δ. If we let ω_p be the largest number in Ω_p (in place of letting ζ_p be the smallest), then we obtain, in much the same way as before, that

The limit of the function f as x approaches a is l if for all x if $|x - a| < 1/\omega_1$ and $x \neq a$, then $|f(x) - l| < 1/\omega_0$.

The new definition is not yet an improvement, since the relation between ω_p and Ω_p is just as indirect as was the relation between ζ_p and Ω_p. But we know how to define indefinitely large numbers without reference to indefinitely large sets. They are what our exposition of indefinitely large size started with. So we now just take the ωs to be indefinitely large numbers on their own, without reference to any indefinitely large sets. That suggests redefining ζ_p to be $1/\omega_p$. We can then give the same definitions of adequalities and of limits as before, reinterpreted using the new definition of the ζs.[30]

Cauchy defined the derivative as a certain limit, and we can now apply our new definition of a limit to obtain the counterpart of Cauchy's definition in our present setting. The following auxiliary definition proves useful: for any function f and any x, $d_p f(x)$ is $f(x + \zeta_p) - f(x)$. Note that it is a consequence of the definition that $d_p x = \zeta_p$. The definition of the derivative of f with respect to x is now just this:[31]

---

30. The present definition of the limit and the definitions of derivative, of irrationality, and of $\mathbb{R}_p$ to follow are suggested by Mycielski's articles [Myc81a, Myc81b], which also contain a development of enough analysis to make it clear that one can do most of analysis quite directly essentially along the lines I am describing. The motivation given here in terms of a relation to Weierstrass's definitions is not in Mycielski's articles. I do not know how he would view it. The analysis described in Mycielski's articles is presented without any sort of "derivation" from the more familiar Weierstrassian analysis. The fact that Mycielski's articles on analysis appeared before the one on Fin(T) [Myc86] suggests that it was the analysis that led to the theory of indefinitely large size, not the other way around.

31. The definition I am using is parallel to the one usually given in beginning calculus courses. That definition of the derivative is the most familiar one. It is actually the definition of the one-sided derivative from the right. A more careful definition is avoided in

The derivative of f at a is l if $\dfrac{d_1 f}{d_1 x} =_0 l$.

Let us compute the derivative of x^2 at $x = 1$ as an example:

$$\frac{d_1 x^2}{d_1 x}(1) = \frac{(1+\zeta_1)^2 - 1^2}{\zeta_1} = 2 + \zeta_1 =_0 2.$$

The first equality holds because of the definition of d_1, the second as the result of algebraic simplification, and the third because $\zeta_1 < \zeta_0$ and because of the definition of adequalities. Unlike Newton's derivation, ours is perfectly rigorous.

Since we are dealing only with finite quantities, it may seem that only rational numbers can be discussed. But consider the following more or less standard definition of irrationality:

The real number r is irrational if for every pair of integers m and n there is a nonzero real number ϵ such that $\left| r - \frac{m}{n} \right| \geq \epsilon$.

It leads, in the same way indicated above for the definition of a limit, to the following definition of irrationality in finite mathematics:

The real number r is irrational if for every pair of integers m and n,

$$\frac{m}{n} =_1 r \Rightarrow |n| > \omega_0.$$

One can prove, for example, that the square root of 2 is irrational by a familiar method [Myc81b, p. 632].

It is immediate from the above definition of irrational number that ζ_0 is rational, while ζ_1 is irrational.[32] That suggests, in particular, that we can take

---

beginning classes because such a definition looks a bit more complex. A counterpart in finite mathematics, likewise, looks a bit more complex, but there is no difficulty whatever in giving it. The nice appearance of the derivative as a ratio of indefinitely small differentials is, however, lost.

32. I have been using just two or three indefinitely large quantities at a time, and so I have just subscripted them with 0, 1, and 2. In a fully general treatment, the number

the rational numbers to be the numbers

$$-\omega_0, -(\omega_0 - \zeta_0), -(\omega_0 - 2\zeta_0), \ldots, -2\zeta_0, -\zeta_0, 0, \zeta_0, 2\zeta_0, \ldots,$$
$$\omega_0 - 2\zeta_0, \omega_0 - \zeta_0, \omega_0.$$

Briefly put, it suggests that we can take the rational numbers to be the numbers between $\pm\omega_0$ with the smallest interval equal to ζ_0. Similarly, it suggests that we can take the real numbers to be the numbers between $\pm\omega_1$ with the smallest interval equal to ζ_1. Those definitions of the rationals and reals are not very good ones for a technical reason: the number ω_0 might not divide the number ω_1, and so with the definition contemplated the rational numbers might not be real numbers. We therefore use the numbers $\omega_0!$ and $\omega_1!$ instead of ω_0 and ω_1, where ! is the factorial—that is, $n!$ is $1 \cdot 2 \cdots (n-1) \cdot n$. The use of factorials ensures that $\omega_0!$ divides $\omega_1!$, as desired. We correspondingly redefine ζ_p to be $1/\omega_p!$, which does not necessitate any further changes in any of the definitions or derivations above. The ζ_p in the definition of adequality must, of course, be reinterpreted in the new sense.

We can now make the above considerations official. We define $\mathbb{R}_p$ to be

$$\left\{ \frac{m}{\omega_p!} : |m| \leq \omega_p!^2 \right\}.$$

Then $\mathbb{R}_0$ is (our counterpart of) the set of rational numbers, while $\mathbb{R}_1$ is the set of real numbers.

We can now take on Dirichlet's challenge to compute the area underneath the function f with value 0 on rational numbers and value 1 on irrational numbers in the interval from 0 to 1: There are $\omega_1!$ real numbers between 0 and

---

of indefinitely large quantities should not be fixed in advance and the various indefinitely large quantities should be indiscernible in the sense described in detail in §2. The theorems should therefore be true not just of ω_0 and ω_1 but of ω_p and ω_q, for $p < q$. That requirement has just been violated in the text. In a fully precise treatment, the definition of irrationality I just gave would actually be the definition of 0,1-irrationality, and an irrational number would be defined to be a number that is p,q-irrational for any p and q. Then ζ_1, though it is 0,1-irrational, is not 1,2-irrational. The square root of 2, of course, is p,q-irrational for any p and q, as one shows by proving it irrational for schematic letters p and q. The additional niceties I am describing in this note are not needed for the examples in the text, and so I just ignore them.

1.[33] The area under the curve at each rational point—there are $\omega_0!$ of those—is 0. The area under the curve at each irrational point—there are $\omega_1! - \omega_0!$ of those—is $\zeta_1 \cdot 1$. Thus,

$$\int_0^1 f(x)dx = \sum_{i=0}^{\omega_1!-\omega_0!} \zeta_1 = \sum_{i=0}^{\omega_1!-\omega_0!} \frac{1}{\omega_1!} = \frac{\omega_1! - \omega_0!}{\omega_1!} = 1 - \frac{\omega_0!}{\omega_1!} =_0 1,$$

where the final equality holds because $\omega_1 > \omega_0^2$. Even Lebesgue measure—the refined notion of area required to meet Dirichlet's challenge—is reduced to a finite sum using indefinitely large and small quantities.

As a final example, let us see what the usual proof that not every set of real numbers is Lebesgue measurable gets us in our present setting. The theorem is worth presenting here since it says that not every set of real numbers has a well-defined length, and it may look inevitable that every subset of $\mathbb{R}_1$ does indeed have a well-defined length. In addition, the proof uses the Axiom of Choice.

I begin by reviewing the usual proof before introducing any finite mathematics. Say that two nonnegative real numbers that are less than 1 (that is, real numbers in the interval ≥ 0 and < 1) are *equivalent* if their difference is rational. Using the Axiom of Choice, pick a set B consisting of one member from each equivalence class. Then B is the required bad set: Suppose, for the sake of contradiction, that B has a well-defined length (that is, Lebesgue measure). Let B_q, for any nonnegative rational number q that is less than 1, be the set of real numbers in our interval that differ from a member of B by q. Every real number in our interval is in exactly one B_q. Thus, since there are only denumerably many B_qs (because there are only denumerably many rational numbers q), the length of the set of all real numbers between 0 and 1 (that length is 1) is the sum of the lengths of the B_qs. But every B_q has the same length, namely, the length of $B = B_0$, since the B_qs are all obtained essentially by sliding B_0 over to the right q units.[34] The length of B cannot be 0, because the sum of denumerably many zeros is not 1 (it is 0), while the union of the denumerably many B_qs is the interval, which has length 1. Similarly, the length of B cannot be greater than 0, because the sum of a positive num-

---

33. I am taking some liberties about the endpoints. More careful analyses get the same result by the same method.

34. And wrapping around the part that sticks out over the end of the interval.

ber with itself denumerably many times is not 1 (it is infinite), while the union of the denumerably many B_qs is the interval, which has length 1. Thus, there is no possible value for the length of B. That is the required contradiction.

We now repeat the same argument in the finite setting we have been developing. Say that two nonnegative real numbers less than 1, that is, nonnegative members of $\mathbb{R}_1$ that are less than 1, are *equivalent* if they differ by a rational, that is, a member of $\mathbb{R}_0$. We let B, once again, consist of one member from each equivalence class, but now, since everything is finite, the Axiom of Choice is not needed. In fact, we can give a simple definition of a suitable B: let B be the set of all nonnegative members of our interval that are less than ζ_0. Choice arises by extrapolation from a simple logical principle concerning finite sets—we can pick members of finitely many nonempty finite sets. We can simplify the proof considerably at this point, since it is clear that in whatever sense the length of our interval is 1, the length of B would have to be ζ_0.[35] Once again the result is that the length of B is neither 0, since $\zeta_0 \neq 0$, nor is it a real number greater than 0, since ζ_0 is less than any given real number r exactly because ζ_0 is indefinitely small.

The last step of the proof just presented is not correct, since ζ_0 is in fact a real number, that is, a member of $\mathbb{R}_1$. The conclusion should be more carefully obtained as follows: In an obvious notation, we showed that length$(B_{01}) = \zeta_0$, but the argument in fact showed that if $p < q$, then length$(B_{pq}) = \zeta_p$, where p and q are schematic letters. That concludes our reconstruction of the argument within finite mathematics. It is not complete, since no contradiction has been obtained. A complete reconstruction would be more complex, but not more illuminating. Our reconstruction captures the essential details of the argument: If we were now to extrapolate, dropping the subscripts, we would obtain that there is a set B of length ζ, which is simultaneously nonzero and less than every positive number, a contradiction. We conclude, as above, that there is a set B that does not have a length. That concludes our examples from analysis.

The presentation of finite analysis involved indefinitely small quantities (the ζs) that are strikingly like infinitesimals, both the infinitesimals of analysis before Weierstrass and the infinitesimals of modern nonstandard analysis. Moreover, the treatment of the area under a curve was strikingly like that of Leibniz: we treated a curve as a succession of adjacent points, just as he did.

---

35. The length of B is ζ_0 because the interval consists of $1/\zeta_0$ copies of B.

Our treatment of the derivative also resembles his, in that both take the derivative to be a ratio of "differentials." There is, however, a big difference. We used indefinitely small quantities, while Leibniz used infinitely small quantities, infinitesimals.

The resemblance between finite analysis and Leibnizian analysis is not accidental. The following is quoted in [Kli72, pp. 386–387] as from an unpublished manuscript of Leibniz from about 1695:

> Yet a state of transition may be imagined or one of evanescence, in which indeed there has not yet arisen exact equality or rest . . . but in which it is passing into such a state that the difference is less than any assignable quantity; also that in this state there will still remain some difference, some velocity, some angle, but in each case one that is infinitely small . . .
>
> For the present, whether such a state of instantaneous transition from inequality to equality . . . can be sustained in a rigorous or metaphysical sense, or whether infinite extensions successively greater and greater, or infinitely small ones successively less and less, are legitimate considerations, is a matter that I own to be possibly open to question . . .
>
> It will be sufficient if, when we speak of infinitely great (or, more strictly, unlimited) or of infinitely small quantities (i.e., the very least of those within our knowledge), it is understood that we mean quantities that are indefinitely great or indefinitely small, i.e., as great as you please, or as small as you please, so that the error that any one may assign may be less than a certain assigned quantity.
>
> On these suppositions, all the rules of our algorithm, as set out in the *Acta Eruditorum* for October 1684, can be proved without much trouble.

The paper of 1684, Leibniz's first publication on his calculus, concerned only differentials, not quadratures (integrals). It gave rules for computing the differentials of sums, products, quotients, powers, and roots, and applied them to problems concerning tangents, maxima and minima, and points of inflection. (See [Boy49, p. 208].) Leibniz said things similar to those in the passage just quoted in a reply to Bernard Nieuwentijdt also written in 1695 [Boy49, pp. 214–215].[36]

In the second paragraph of the quotation, Leibniz frankly admitted that

---

36. As was mentioned in §II.3, Nieuwentijdt criticized the calculus forty years before Berkeley, and his work was widely noticed. But the chief response Nieuwentijdt provoked,

he was not sure whether his infinitesimals are tenable, but in the third paragraph he said that indefinitely large and small quantities suffice to develop the differential calculus. Just as some of Newton's pronouncements suggest the theory of limits, some of Leibniz's suggest the theory of indefinitely small quantities. If Weierstrass's foundation of analysis using limits has Newtonian roots, then the foundation proposed by Mycielski has Leibnizian ones.[37]

The idea of the indefinitely small did not originate with Leibniz. Nieuwentijdt had related ideas [Boy49, p. 214], and so did, for example, Galileo and Descartes. The idea of the indefinitely small can apparently be traced back as far as Nicholas of Cusa in the mid-fifteenth century [Boy49, p. 92]. Some of the early pronouncements about the relation between the indefinite and the infinite seem to provide powerful support for my thesis that the idea of the infinite arose from that of the indefinitely large. I cannot, at this stage, rely on such evidence. A great deal of historical research will be required to give a satisfactory account, since the theological, mathematical, and physical need to be carefully disentangled before one can draw any conclusions concerning the implications of pre-Cantorian pronouncements for the history of the modern, mathematical infinite.

The similarity between the Mycielskian theory of indefinitely small quantities and the Leibnizian one may make the history of what led up to the Leibnizian idea seem more interesting. It does something more substantial for the subsequent development of the Leibnizian idea: It makes a new kind of reconstruction of some of the old arguments involving infinitesimals possible. That may lead to a radical reevaluation of those arguments and hence, for example, to a reevaluation of our idea that Euler did things that have no rigorous reconstruction and of our idea that Cauchy made "mistakes." For a picture of just how radical the changes in our assessment of the history of mathematics might be, see [Lak78].

I am not the first to propose a new look at the use of infinitesimals. Abraham Robinson developed a "nonstandard analysis" that includes infinitesimals, and a fair amount of work has been done reassessing the use of infinitesimals on that basis. But Robinson's nonstandard analysis is based on an application of modern mathematical logic that could not possibly have been

---

so far as I know, is the one I am about to discuss. It would have been premature to discuss it in Chapter II, for reasons that should now be obvious.

37. One might claim that Newton's ideas were a part of what actually led to the theory of limits. I am not making such a claim for Leibniz and the theory of indefinitely small quantities.

what Leibniz and his successors had in mind, in whatever indirect sense. The history done with nonstandard analysis as a touchstone is therefore suspect, as has been argued in detail ([Bos74, pp. 81–84], [Dau88]).

Finite analysis of the sort outlined here can make a much better claim to be the sort of thing Leibniz and his successors had in mind, as we have seen. The relation between Robinson's nonstandard analysis and Mycielski's finite analysis is subtle and not completely direct. There actually are infinitesimals in nonstandard analysis, while there are none in finite analysis. Nonetheless, the actual mathematical development of analysis from the points of view of nonstandard analysis and of finite mathematics are in most respects strikingly parallel. While the historical work done with nonstandard analysis may be a useful guide, there is little reason to think that it will all hold up from the perspective suggested here. Mycielski's work has provided a whole new way of seeing an important part of the history of mathematics.

§4. The Theory of Zillions

The main thesis of this section is that ordinary infinitary set theory arose essentially as the result of extrapolation—in the sense of §2—from natural principles of finite set theory. To establish that thesis we must establish a subsidiary thesis, namely, that the principles of finite set theory can be motivated as self-evident principles concerning finite sets, including indefinitely large ones. Without the subsidiary thesis, the main thesis would be open to the charge of being backward, that is to the charge that infinitary set theory is like finite set theory because finite set theory developed out of infinitary set theory, not the other way around.

The main thesis enables us to make sense of how we can know so much about the combinatorial infinite, despite the obstacles discussed in Chapters VI and VII. We know about the combinatorial infinite by extrapolating from our experience of the indefinitely large. More precisely, what has happened is that we have systematized our experience of the indefinitely large without attending to the context-dependent features of that experience, and that has led to our picture of the combinatorial infinite.

In this section we study a particular finite set theory, Fin(ZFC), in detail in order to argue for the subsidiary thesis. I have chosen to use ZFC instead of some other set theory because ZFC is the most widely used and well-known version of set theory, not because it is the best or most natural. In §IX.4, I shall discuss the finite-mathematics counterpart of the natural axiomatization of set theory proposed in §V.5, but first I want to make it clear that the main thesis of this section is independent of the argument of §V.5 or of that axiomatization.

If the main thesis is correct, a lot of things that were mysteries suddenly begin to make sense, in addition to the main mystery of how we have managed to come into such a clear picture of the infinite. Cantor's "finitism," in Hallett's courageous turn of phrase, no longer seems so odd if his heuristics are based on experience of the indefinitely large. Cantor, recall, thought of infinite sets as ones that can be counted in more than one way, while Frege, quite sensibly, objected that infinite sets cannot be counted at all. Cantor's procedure was forthrightly one of attributing properties of finite collections to some infinite ones, namely the transfinite ones.

Cantor's distinction between the transfinite and the absolutely infinite, which has its heir in modern set theory in the distinction between sets and proper classes, is the extrapolated counterpart of the distinction between finite (including indefinitely large but finite) and infinite. Naturally the ordinals of a model of set theory have to turn out to comprise a proper class: that is just the extrapolated counterpart of the familiar fact that there is no number of numbers! Note also that that is virtually Cantor's reasoning. As he put it [Hal84, p. 42], "just as . . . given any finite number . . . the power of the finite numbers following it is *always* the same, so following each supra-finite number . . . there is a totality of numbers . . . which has lost nothing as regards power with respect to the whole of the absolutely infinite totality of numbers beginning with 1." (See §III.4 for more of the passage.)

I am not claiming that Cantor was self-consciously relying on a picture of the indefinitely large, though, as the passage just quoted makes clear, he was indeed self-consciously relying on the method of extrapolating from the finite. He was working out what the transfinite progression

$$0, 1, \ldots, \infty, \infty + 1, \infty + 2, \ldots, \infty \cdot 2, \ldots, \infty \cdot 3, \ldots,$$
$$\infty^2, \ldots, \infty^3, \ldots, \infty^{\infty}, \ldots, \infty^{\infty^{\infty}}, \ldots$$

suggested. To do that he made sense of the ellipses in the way that I believe we all ultimately do—by assimilating the infinite progressions represented by them to indefinitely long ones.

Cantor thought of sets in terms of what God could do with them, and his Omnipotent Mathematician is still very much with us, even if theology has been taken out of the picture. Virtually every attempt to motivate the principles of combinatorial set theory still relies on some notion of idealized manipulative capacities, capacities of the Omnipotent Mathematician. Those capacities are, inevitably, idealizations of our finite capacities. But appeals to omnipotence are of little use for understanding the nature of the idealization.

There is one thing that is right about faith in the powers of the Omnipotent Mathematician: it is undoubtedly the case that combinatorial sets share many features with finite ones. But to see how they come to share those features, we need to look at extrapolation from the indefinitely large, not at any mysterious powers. The ideal capacities required to manipulate indefinitely large sets require no great leaps of faith, since the sets are, after all, finite. The idealization of those capacities to infinitary ones is part and parcel of the extrapolation from the indefinitely large to the infinite. That is the idealization that leads to set theory and along the way, if desired, to the ideal powers attributed to the Omnipotent Mathematician.

Though we can explain the Omnipotent Mathematician along with set theory, the Omnipotent Mathematician is of no use in explaining set theory. For example, the "explanation" that Choice holds because the Omnipotent Mathematician has the power to pick members out of infinitely many nonempty sets is not so much an explanation of Choice as an exhortation to believe that Choice is true. On the other hand, the explanation that Choice arises by extrapolation from a simple fact about the finite, since we can supply the details of what is meant by extrapolation, does indeed explain the source of the self-evidence of Choice.

Cantor's early theory of sets foundered on the Power Set Axiom. Roughly, he could not see how to well-order the power set of a well-ordered infinite set.[38] Zermelo showed that Power Set got along with a Cantorian picture rather nicely, but to this day it is not completely clear how the two are related. That is the problem of why Limitation of Size and Limitation of Comprehensiveness are principles about the same things, the combinatorial sets, instead of principles of two separate set theories about two different kinds of sets (§V.5).

As I mentioned in §1, the counterpart of the problem in finite set theory is answerable: A hereditarily finite set that is limited in comprehensiveness is clearly limited in size—the size of the union of a set places an upper bound on the size of the set. The two principles go together because they both arise by extrapolation from the hereditarily finite sets. That was far from a precise

---

38. Actually, what he could not see was how to obtain the existence of a well-ordering of the set of functions from a well-ordered infinite set to a two-element set.

description, since I have paid no attention to the detailed requirements of extrapolation. See §IX.4.

The kinds of considerations just presented are the chief evidence for the main thesis that the notion of a combinatorial set is just an extrapolated version of the notion of an indefinitely large set. The fit between the two is just too good to be coincidental. I shall give more evidence of the same kind in Chapter IX. I shall discuss Gödel's single universe of all sets vs. Zermelo's succession of universes (§IX.3), the distinction between sets and classes (§IX.3), and the status of second-order logic (§IX.5) from the perspective of finite set theory. The discussion must be more technical than that here, since in each case one must pay detailed attention to the structure of the sentences in finite mathematics, with their indefinitely large bounds on the quantifiers.

None of the above considerations can cut any ice without the subsidiary thesis of this section, which says that the principles of finite set theory, that is, the axioms of Fin(ZFC), can be motivated as a theory of the indefinitely large independently of their historical source in the theory of the combinatorial infinite. We now turn to the subsidiary thesis.

In broad outline, it is a trivial matter to see that the axioms of Fin(ZFC) are truths of finite set theory. After all, almost all of the axioms are true of finite sets. There is one exception—the Axiom of Infinity. Its counterpart in finite set theory is the Axiom of a Zillion, which says that there are indefinitely large sets.

Theories in finite mathematics all use indefinitely large domains to bound the quantifiers (§3). What is in those indefinitely large domains is determined by the particular theory being used. The indefinitely large Ωs of finite mathematics replace the fixed infinite domains of ordinary mathematics, and they permit us to have many of the advantages of quantifiers that range over an infinite domain. The Ωs appear, for example, even in the finite mathematics version of Peano arithmetic.

The feature of finite set theory that distinguishes it from other theories within finite mathematics is the Axiom of a Zillion, which asserts that there is an indefinitely large collection that is a member of an indefinitely large domain. We have experience of particular indefinitely large collections, and so that distinctive feature of finite set theory is firmly tied to ordinary experience. The notion of an indefinitely large collection is context dependent, and it therefore could not have received an adequate formalization like the Axiom of a Zillion until the mathematical mechanisms needed to handle context dependence became available.

So far we have only shown that finite set theory can be motivated in a rather rough and careless way. We must now take on a more arduous task—checking that the actual formal axioms of Fin(ZFC) reflect that motivation.

It should not be expected that the axioms of Fin(ZFC) will themselves look particularly obvious or natural. For one thing, ZFC is not a particularly natural axiom system, and so there is no particular reason why its counterpart in finite mathematics should be. The principles are clear, but not chosen in any systematic way.

Second, and much more important, even in the case of ZFC—or for that matter Peano arithmetic—the axioms of the final precise system are transcriptions into an awkward formal language of something preformal expressed in natural language. One can understand, say, the following form of Choice, "every collection of pairwise disjoint sets has a choice set that contains exactly one member of each nonempty set in the collection," perfectly well and still have to go to a considerable amount of trouble to see that the following is an appropriate formalization of it:

$$(\forall x)((\forall yzu)(y \in x \wedge z \in x \wedge y \neq z \rightarrow \neg(u \in y \wedge u \in z)) \rightarrow$$
$$(\exists y)(\forall z)(z \in x \wedge z \neq \varnothing \rightarrow$$
$$(\exists u)(u \in z \wedge u \in y) \wedge$$
$$(\forall uv)(u \in z \wedge u \in y \wedge v \in z \wedge v \in y \rightarrow$$
$$u = v))).$$

That is a problem that we shall have to confront in a particularly acute form in the finite mathematics of set theory: not only are the axioms of ZFC— and hence their Fin(ZFC) counterparts—complicated, but the formalization is taking place in the new and unfamiliar setting of finite mathematics.

Let us begin. I shall take any hereditarily finite pure set that I have use for to be an antecedently given unproblematic object, and I shall use PRS, as developed in §VI.4, as a background theory. (As usual, nothing turns on the restriction to pure sets. That is largely a matter of custom and convenience.)

The term *hereditarily finite pure set* is a bit unfortunate for present purposes since it suggests that we have started with some broader notion of set and narrowed it down to those sets with some special properties. The reader may therefore find it helpful to think of *hereditarily finite pure set* as one word.[39] I shall henceforth use *hfpset* as that word. It is to be pronounced with

---

39. I am indebted to Sidney Morgenbesser for that way of putting it.

the *hfp* silent. Our Ωs will themselves be hfpsets, the sort of things suitable to be members of the domain as well as subsets of the domain, since they will be finite collections of hfpsets. Though I shall frequently refer to "indefinitely large hfpsets" without explicit restriction to a kind or degree of availability, that is just for the sake of simplicity of language. The relevant restrictions are never absent, though they are often implicit.

Mycielski [Myc86] discussed Fin(ZF) and various extensions of it. What is original here is not Fin(ZFC) but the argument that Fin(ZFC) can be described in a metalanguage that does not involve infinite sets[40] and the subsidiary thesis that Fin(ZFC) can be motivated within finite mathematics as a theory of hfpsets including indefinitely large hfpsets. Mycielski introduced Fin(*T*) for arbitrary *T* in a standard background of ordinary mathematics. The only motivation he provided for Fin(ZF) is that it is the finitized version of ZF, which is normally motivated on the basis of infinitary considerations. Once having obtained Fin(ZF) he proposed to take it to be a finitist theory.[41]

In [Myc86] Mycielski announced a paper in which he would argue on the basis of finite intuitions that every finite subset of Fin(ZF + AD)[42] has a finite model, but that paper has not appeared. I do not know what finite intuitions Mycielski had in mind, but they must be different from the ones presented here, since AD contradicts Choice.

The vocabulary in which I shall present Fin(ZFC) will consist of a binary relation symbol for membership, $\in$, a constant symbol for the empty set, $\varnothing$, a binary function symbol, ;, that will be used so that $x \,;\, y$ is what would be written $x \cup \{y\}$ in a more familiar notation, and unary predicate symbols, Ω_p, for indefinitely large hfpsets, where *p* is any rational number.

The symbols $\varnothing$, $\in$, and ; are special in a way that is familiar from infinitary set theory—they have an intended interpretation, just specified. The Ωs, on the other hand, are just ordinary predicate symbols of the sort familiar from elementary logic—their interpretation will be constrained solely by the axioms. We take the language of ZFC to be the same as the language for Fin(ZFC), except that we omit the Ωs and hence the quantifiers bounded by them.

The reason for having $\varnothing$ and ; is that it will be convenient to be able to form

---

40. That is true of Fin(ZFC) for the same reason that it is true of any other theory of the form Fin(*T*), and so it does not require separate discussion. See §3.

41. His subsequent work [Myc89] suggests that he holds that we can construe ZF itself to be a finitist theory.

42. AD is the Axiom of Determinacy, a rival to the Axiom of Choice.

a notation for any given hfpset. In particular, if x is one of the Mirimanoff–von Neumann natural numbers, then $x ; x$ is its successor. *Natural number* shall henceforth mean Mirimanoff–von Neumann natural number. That will save us many repetitions of "the set of predecessors of," but it is otherwise inessential.

The axioms of Fin(ZFC) are just Axioms (7)–(15) of §3 (that is, the axioms of indefinitely large size and the axioms of equality) plus particular semiregular relativizations of the axioms of ZFC, including the definitions of $\varnothing$ and ;. We show that the axioms can be motivated independently of the relation to ZFC as we state them.

First, the definitions of $\varnothing$ and ;:

(16) $$(\forall x \in \Omega_0)x \notin \varnothing.$$

(17) $$(\forall xyz \in \Omega_0)(z \in x ; y \leftrightarrow z \in x \lor z = y).$$

No discussion seems to be required.

EXTENSIONALITY:

(18) $$(\forall xy \in \Omega_0)((\forall z \in \Omega_1)(z \in x \leftrightarrow z \in y) \to x = y).$$

As will be required below by Axiom (24), the Axiom of a Zillion, we do *not* assume that the Ωs are transitive. That is, we allow $\Omega_0(a)$, $b \in a$, and $\neg\Omega_0(b)$ to be simultaneously true. Though related kinds of things are often rejected in the literature, I believe that in our setting the failure of transitivity is perfectly natural, as can be seen if we take the notion of availability that underlies the Ωs to be epistemic availability. It seems clear, for example, that $10^{10^{10^{10}}}$ is epistemically available in many actual contexts even though there is no actual context in which all of the natural numbers below it are epistemically available. (It is easy to concoct various related examples.) The example of $10^{10^{10^{10}}}$ is only intended to show that it is reasonable not to require transitivity, not that it is necessary to require intransitivity. Further and rather different considerations will be required below to motivate that.

Since Ω_0 need not be transitive, when x and y are in Ω_0, it is not enough to look at members of x and y that are in Ω_0 to see if x and y have the same members. We can, however, stipulate that Ω_1 is not indefinitely large with respect to Ω_0 unless it is so large that if x and y from Ω_0 are distinct, then there is a set in Ω_1 bearing witness to that. That is exactly what Axiom (18),

Extensionality, says. It reflects the intuition that a set that distinguishes two sets is available in virtue of the fact that the two sets are. In effect, Axiom (18) is the postulate that there is a many-valued function in F that takes any two distinct sets to members of their symmetric difference (that is, to the sets that are in one of them but not in the other).[43]

FOUNDATION:

(19) $(\forall x \in \Omega_0)(x \neq \emptyset \rightarrow (\exists y \in \Omega_1)(y \in x \wedge (\forall z \in \Omega_2)(z \in x \rightarrow z \notin y)))$.

Axiom (19) says that Ω_1 is so large that every nonempty hfpset x in Ω_0 has a member y in Ω_1 such that there is no z in Ω_2 that is a member of both x and y. That is, Ω_1 is so large that every nonempty hfpset x in Ω_0 has a member in Ω_1 that is $\in$-minimal with respect to Ω_2.[44] Since the members of Ω_0 are hfpsets they have $\in$-minimal members. One could therefore ensure the truth of Axiom (19) by adding a many-valued function to F that relates every nonempty set to at least one $\in$-minimal member, thereby making sure that Ω_1 includes an $\in$-minimal member of each member of Ω_0.[45]

WEAK UNION:

(20) $(\forall x \in \Omega_0)(\exists y \in \Omega_1)(\forall zu \in \Omega_2)(z \in u \wedge u \in x \rightarrow z \in y)$.

The Weak Union Axiom should be fairly self-explanatory by now: it says that Ω_1 is so large that for every x in Ω_0 there is a y in Ω_1 that has in it every member of a member of x such that we have both in Ω_2. That could be guaranteed by putting the union function into F, but that is not necessary. Using the variables used in the axiom, what is necessary is not "that every member of a member of x be in y," but something like "that every available member of

---

43. See §3 for A and F, the sets of available objects and of availability functions.

44. y is an $\in$-minimal member of x (with respect to Ω_2) if y is in x and no member of y (that is in Ω_2) is in x.

45. I could have omitted Foundation. It does not seem unnatural to have it in finite mathematics: it says that our epistemic process of getting hold of sets and then of their members and of the members of their members and so forth has an end. That is not necessary, and neither is Foundation. It follows from Theorem 3.3 and the fact that ZFC minus Foundation is equiconsistent with ZFC that Fin(ZFC) minus Foundation is equiconsistent with Fin(ZFC). The fact that ZFC minus Foundation is equiconsistent with ZFC was discussed in §V.4.

an available member of x be in y." That expression is not precise since there is, so far as I can see, no perspicuous way to render the subscripts to the Ωs in English. Moreover, the quantifiers do not range over the available members but over the members of an indefinitely large set that includes the available members. Nonetheless, what I have said does, I hope, serve to convey the idea that a set that includes the available members of available members of a set is available in virtue of the availability of the set.

The explanations of Choice and Power Set are similar to those for Weak Union, and so we just list Choice and Power Set. Choice poses no special problems since we are interested only in finite sets.

CHOICE:

(21)

$$(\forall x \in \Omega_0)((\forall yzu \in \Omega_1)(y \in x \wedge z \in x \wedge y \neq z \rightarrow \neg(u \in y \wedge u \in z)) \rightarrow$$
$$(\exists y \in \Omega_1)(\forall z \in \Omega_2)(z \in x \wedge z \neq \varnothing \rightarrow$$
$$(\exists u \in \Omega_3)(u \in z \wedge u \in y) \wedge$$
$$(\forall uv \in \Omega_2)(u \in z \wedge u \in y \wedge v \in z \wedge v \in y \rightarrow$$
$$u = v))).$$

POWER SET:[46]

(22) $(\forall x \in \Omega_0)(\exists y \in \Omega_1)(\forall z \in \Omega_2)(z \in y \leftrightarrow (\forall u \in \Omega_3)(u \in z \rightarrow u \in x)).$

REPLACEMENT:

(23) $(\forall x_0, \ldots, x_n \in \Omega_0)$
$$((\forall xyz \in \Omega_1)(\phi(x, y, x_0, \ldots, x_n) \wedge \phi(x, z, x_0, \ldots, x_n) \rightarrow y = z) \rightarrow$$
$$(\forall x \in \Omega_1)(\exists y \in \Omega_2)(\forall z \in \Omega_3)$$
$$(z \in y \leftrightarrow (\exists u \in \Omega_4)(u \in x \wedge \phi(u, z, x_0, \ldots, x_n))))),$$

where $\phi(x, y, x_0, \ldots, x_n)$ is a schematic symbol that can be instantiated by

---

46. Power Set requires that for every hfpset x in Ω_0 there is a hfpset y in Ω_1 such that the members of y in Ω_2 are exactly the subsets of x in the sense of Ω_3. The axiom is far from requiring that the actual power set of x be in Ω_1. That requirement cannot be formulated within finite mathematics. It may be of interest to investigate a hybrid theory that goes beyond Fin(ZFC) by requiring that the power set of x in Ω_1 be the real power set, that is, the power set determined by the power-set operation of PRS. We do not consider such a theory in this book.

any regular formula with free variables among x, y, $x_0, \ldots, x_n$ that has no occurrences of Ω_p for $p \leq 4$ and no occurrences of u or z.[47]

Replacement gains its plausibility in much the same way that the preceding three axioms did. An instance of the axiom for a particular formula replacing ϕ has the effect of adding to the availability functions one that takes any set x to the range of the partial function defined by the formula (with parameters) on x, if it does define a partial function. However, there is a new worry for Replacement: The formula that replaces the schematic letter in an instance of the axiom may involve quantification over indefinitely large sets. Since those sets are, it seems, indefinite, it may seem that the formula cannot define a definite partial function on members of x. If that were true, it might cast doubt on whether our usual intuitions about partial functions are adequate to underwrite the axiom. The worry is the counterpart in finite mathematics to standard doubts about Replacement based on its impredicativity (§§V.1 and V.3).

The worry raised by indefiniteness is groundless. The partial function defined by the formula is perfectly definite. In any natural model of a set of axioms of finite set theory, perfectly ordinary hfpsets play the role of the indefinitely large ones. They obey the axioms of indefinitely large size in the model, but there is nothing in the slightest ill-defined or indefinite about them.

The sense in which there is indefinitely large size does not involve any new sort of entity, an indefinite set or indefinite function. The indefiniteness is epistemic, not ontological. The indefiniteness in any theory within finite mathematics is reflected by the fact that the theory has, even up to isomorphism, many models. That is a more general statement of the fact noted in §3 that even though finite analysis uses indefinitely small quantities, there are no new numbers in finite analysis, that is, no indefinitely small numbers. Such numbers would be infinitesimals, after all. The indefiniteness enters into finite analysis through the fact that there are many possible values for the indefinitely small quantities. The worry that indefinitely large sets are indefinite is parallel to the worries engendered by the thought that variables stand for variable numbers.

The remaining axiom, the Axiom of a Zillion, is the main point.

---

47. If we require that the formula substituted for ϕ be a formula of the language already specified, we get the weak Replacement schema. If we allow the formula to be a formula of any extension of the language we come to use, then we get the full second-order Replacement schema.

Zillion:

(24) $(\exists x \in \Omega_0)(\varnothing \in x \wedge (\forall y \in \Omega_1)(y \in x \to y\,;\ y \in x)).$

The other axioms are compatible with the Ωs not being transitive, but the Axiom of a Zillion requires intransitivity.

The axiom requires that there be an x in Ω_0 such that $\varnothing$, or 0, is in x and such that if 0 is also in Ω_1 then 0 ; 0 (that is, the number 1) is in x, and that if 1 is also in Ω_1 then 1 ; 1 (that is, 2) is in x, and so forth. Thus, either x will have to be infinite, a familiar possibility we can afford to ignore here, since Fin(ZFC) is to be a theory of hfpsets, or—to break the chain—there will be some number in Ω_1 and x whose successor is in x but not in Ω_1. To be concrete, say that 9 and 10 are in x but that 10 is not in Ω_1.[48] Then 11 could be in Ω_0 to serve as the required x. More generally, if x is a nonzero natural number in Ω_0 whose predecessor is not in Ω_1, then x can fulfill the condition of the axiom, for the greatest member of both x and Ω_1 can have its successor in x but not in Ω_1, breaking the chain. To return to the example, the greatest member of 11 in Ω_1 would be 9, which has its successor in x but not in Ω_1. The axiom will hold if there is a natural number available whose predecessor is not available to any actual degree.

Zillion declares that there are gaps in the set of available numbers: the first number after a gap is an available number that does not have an available predecessor. That seems obviously correct—self-evident—once we grant that there is a number that is available that is so large that not all of its predecessors are available to any degree we actually employ. In granting that such a number is available we go beyond the usual intuitions that ground the theory of numbers, and so it should not be surprising that the theory that results is rather different from ordinary arithmetic. The number $10^{10^{10^{10}}}$ is my standard candidate for an available number such that not all of its predecessors are available. Since not all of its predecessors are available, there must be a number less than $10^{10^{10^{10}}}$ that is available without its predecessor being available to any actual degree, and that is a number of the sort being postulated by the Axiom of a Zillion. The Axiom of a Zillion says that there is a number available to us that is not available from below (in the precise sense already spelled out that its predecessor is unavailable, and hence we have not counted up to it).[49]

---

48. Larger numbers would be more appropriate, but life is already too complicated.
49. It seems to me that if you have a zillion things and you take one away, then you

If Ω_1 were a temporal stage that came after Ω_0 then the Axiom of a Zillion would not be true: we could just choose Ω_1 to include all the members of members of Ω_0, and then every number in Ω_0 would have all of its predecessors in Ω_1. But the increasing Ωs are not successive temporal stages—they represent the varying degrees of availability in a single fixed context. It seems self-evident that in any fixed context there will be a number whose predecessor is not available to any degree employed in that context, and that is all the Axiom of a Zillion says.

Since the Axiom of a Zillion requires that a large set be in Ω_0, the other axioms might contradict it by requiring that other sets related to that large set be in Ω_1. For example, the following sentence, which is consistent with the other axioms, would clearly contradict the Axiom of a Zillion:

(25) $(\forall x \in \Omega_0)(x \neq \varnothing \rightarrow (\exists y \in \Omega_1)(y \in x \wedge y; \; y \notin x))$.

When x is a natural number, the only y that meets the condition given is the predecessor of x. Thus, sentence (25) has the effect of adding the predecessor function to the availability functions. Sentence (25) is not plausible in a theory that allows indefinitely large sets into Ω_0, since it imposes a restriction on what may be found to be a member of any hfpset in Ω_0: for any x, it asserts that there is a y such that $y; \; y$ is not in x, and that exclusion is contrary to the possibility that x is indefinitely large.

The exclusion that rules out sentence (25) does not come out of general considerations of the nature of availability; there may well be some theory for which it would be natural to take predecessor to be one of the basic availability functions. The exclusion is special to the axiomatic theory of indefinitely large size. In that theory, which is precisely concerned with the existence of gaps, the exclusion is natural, not just an *ad hoc* maneuver to avoid contradictions: It is precisely the nature of an indefinitely large number that we have not counted up to it—otherwise it wouldn't be indefinitely large. But that is just to say that not all the predecessors of an indefinitely large number are available. In particular, the least indefinitely large number will

---

still have a zillion things. But not everyone will agree. Some will insist that if you take one away from a zillion things, then you are left with a zillion minus one things. Thus, the sort of intuition mentioned in this note is not robust enough to provide significant support for the self-evidence of Zillion, and I shall not make use of it in the text. It is, at any rate, intuition concerning the cardinal number a zillion, not the ordinal number, which is the more important one for our theory.

not have a predecessor. Predecessor therefore is not an availability function. None of that should be especially surprising in light of the earlier remarks about $10^{10^{10^{10}}}$.

One can, as a matter of fact, adjoin predecessors of the ordinal numbers to the system of ordinal arithmetic (so that there are new "numbers" like $\omega - 1$, the predecessor of ω) using the very same construction usually used to define the integers from the natural numbers—a construction whose main point, after all, is to add predecessors to zero. But note that here one has predecessors in a new sense: new numbers are added to serve as predecessors, which is not at all like the procedure of identifying which of the numbers we already had is a predecessor of another. One can even produce a real closed field containing the ordinal numbers using a standard construction of the real numbers [Knu74]. In such a field, ω has not only a predecessor but even a cube root! As one might expect, such constructions have found comparatively few applications. The ordinal numbers were invented to count the iterations of an operation (the derived-set operation), and—as Cantor taught us—for counting it is the process of finding the next number after a succession of numbers that is central. The predecessor operation on numbers used for counting is the operation, "take the number counted just before this one."[50] That operation is designed to pick out something that was already supposed to be available, not to create something new, and that is why one should not expect it necessarily to be one of the availability functions.

In a theory of indefinitely large counting numbers—numbers that occur after a succession that is indefinitely long—the predecessor operation cannot even be definable from the availability functions. After all, to say that a number is the one that comes immediately after an indefinitely long succession just is to say that the numbers that immediately precede it are not available. The following discussion of whether the other axioms contradict Zillion is in part designed to show that the predecessor operation is not definable from our availability functions.

To simplify the discussion of whether the other axioms contradict the Axiom of a Zillion, let w be a witness to the axiom that is a natural number. That is, w is a natural number in Ω_0 and

---

50. In PRA, the predecessor of X, $P(X)$, has the following definition by primitive recursion: $P(0) = 0, \quad P(SX) = X$.

(26) $\varnothing \in w \wedge (\forall y \in \Omega_1)(y \in w \to y; \, y \in w).$

(I am thus arguing for the consistency of the stronger version of Zillion that says there is an x in Ω_0 that is a natural number and . . .) Note that every member of w is a natural number.

The following discussion of consistency is not intended to be a proof of the consistency of Fin(ZFC), any more than discussions of the iterative conception of sets are supposed to be proofs of the consistency of ZFC. The present discussion does not need to meet a higher standard of explanatory rigor than those do.[51] In each case, the idea is to provide intuitive, extra-logical considerations in support of the theory in question. In our case, there is the additional goal of showing that the theory Fin(ZFC) makes intuitive sense without reference to ZFC.

Our finite set of axioms might include Ωs with negative and fractional indexes. For expository simplicity, assume that in our finite set of axioms the predicate symbol Ω_0 is the Ω with the least index that appears, that Ω_1 is the next, and so forth.[52]

An axiom cannot come into conflict with sentence (26)—in effect, the Axiom of a Zillion—unless it entails the existence of a member of Ω_1 that is in w. Such a member of Ω_1 must, given our simplifying assumption, be a

---

51. It must, however, succeed on its own terms, which I have argued (§V.5) that the usual discussions of the iterative conception fail to do.

52. By the indiscernibility of the Ωs formulated in Lemma 3.2, that amounts to little more than a reindexing of the Ωs that were actually in the finite set of axioms with which we started. In more detail: Suppose we start with a finite set R of axioms and that the Ωs that appear are $\Omega_{q_0}, \ldots, \Omega_{q_n}$, with $q_0 < \ldots < q_n$. Reindex all the sentences in R by replacing Ω_{q_0} by Ω_0 and so forth everywhere in R to form R'. If we had taken every semiregular relativization of an axiom of ZFC to be an axiom of Fin(ZFC), we could just use R' instead of R. Unfortunately for present purposes we have not taken all the semiregular relativizations as axioms, and so R' need no longer be a set of axioms. More specifically, the reindexed versions of instances of Axioms (11)–(24) need not be axioms. Let S be the set of axioms obtained from R' by replacing each of the nonaxioms in R' with the corresponding axiom and then adding to S the axioms required to prove, according to Lemma 3.2, that the nonaxioms replaced are equivalent to the axioms replacing them. A detailed analysis of the proof of Lemma 3.2 shows that all of the Ωs that will then occur in S are among $\Omega_0, \ldots, \Omega_n$, and hence that S is as required. The "reindexing" referred to in the text consists of replacing R by S. The metatheorem that permits the procedure says that if $\phi(\Omega_{q_0}, \ldots, \Omega_{q_n})$ is derivable from R, then $\phi(\Omega_0, \ldots, \Omega_n)$ is derivable from S. The proof of the metatheorem is routine and can be carried out in PRW.

natural number. Thus, Axioms (7)–(17), which are purely universal and hence do not assert the existence of anything, are compatible with the Axiom of a Zillion.

Axiom (18), Extensionality, requires that if a in Ω_0 is a proper subset of w, then there is a member of $w - a$ in Ω_1. (If a is not a proper subset of w, then Extensionality applied to it does not entail that anything is a member of w.) All that is required to avoid conflict with Zillion is that a is not the predecessor of w: for any other possible value of a, there will be a member of $w - a$ that is not the predecessor of w, and such a member of $w - a$ can be in Ω_1, which will satisfy Extensionality, without violating Zillion. But it follows that there need be no conflict just because the predecessor of w is not in Ω_1, and therefore not in Ω_0, by our choice of w.

Axiom (19), Foundation, is trivially verified for $x = w$—just let y be $\varnothing$, which is in w by sentence (26). Then $\varnothing$ is as required by Axiom (16) (the definition of $\varnothing$) and Lemma 3.1.

Checking Axiom (20), Weak Union, is subtler than checking Axiom (18), Extensionality. There is an apparent problem: the union of a natural number is the predecessor of the number, and the predecessor of the natural number w is precisely what we must avoid. When applied to w, Axiom (20) yields

(27) $(\exists y \in \Omega_1)(\forall zu \in \Omega_2)(z \in u \wedge u \in w \to z \in y).$

We may, by the indiscernibility of Ω_1 and Ω_2, assume that the predecessor of w is not in Ω_2, just as we have assumed before that it is not in Ω_1. The set y that sentence (27) says exists therefore need not have the predecessor of the predecessor of w in it, and it therefore need not be the predecessor of w. Thus, even if y does turn out to be a natural number (whether it will depends on accidental features of Ω_2), it can be one that is sufficiently small by comparison with w, and so there is no problem. A similar analysis checks the application of Weak Union to sets other than w.

Axiom (21), Choice, can be analyzed in a related fashion: Briefly, members of a choice set must be in the Ωs and so the predecessor of w will not appear.

Axiom (23), Replacement, yields to a similar analysis. The worst case appears to be that according to Replacement the range of the predecessor function on w ought, it seems, to be in Ω_2.[53] That range is the predecessor of

---

53. I am referring to the Ωs that are actually used in the axiom. My informal description is intended to refer to the Replacement Axiom in which $n = 0$, ϕ is the formula y; $y = x$, and the consequent is applied to the case $x = w$.

w. But what is actually required is that the range of the predecessor function on that part of w that is in Ω_4 be in Ω_2, and, by the indiscernibility of Ω_1 and Ω_4, that part does not include the predecessor of w.

Axiom (22), Power Set, is no problem: it does not require that Ω_1 contain any large natural numbers and *a fortiori* it does not require that Ω_1 contain any large members of w.

We have now considered all of the axioms, and so I have demonstrated the subsidiary thesis of the section—that Fin(ZFC) is indeed a theory of hfpsets including indefinitely large hfpsets that can be motivated without going beyond finite mathematics.

The motivation for Fin(ZFC) admittedly falls short of a complete explanation of the consistency and utility of Fin(ZFC), but that is even more true of the comparable explanations generally offered for ZFC. My claim is not that Fin(ZFC) is a final foundation for mathematics—whatever that would be—or even a complete explanation of what has led to set theory. What I do claim is that the explanation of Fin(ZFC) and of its relation to ZFC is illuminating about the relationship between the notion of a combinatorial collection and that of a finite, but possibly indefinitely large, collection and that that relationship shows how nontrivial principles concerning the infinite are self-evident: they are extrapolated from principles concerning the indefinitely large. The relationship between the notion of a combinatorial collection and that of a finite, but possibly indefinitely large, collection also explains the relationship between the principles of limitation of comprehensiveness and size, that is between Power Set and Cantor's early theory of transfinite combinatorial sets (§IX.4). It would certainly be desirable to provide a more natural theory of hfpsets that included indefinitely large hfpsets than the one given here, but initially it is most important to provide a theory within finite mathematics that is easily compared with the standard theory of sets.

We have now completed the discussion of both the subsidiary and main theses, but there are two issues we postponed earlier that still need to be discussed: The notion of a reasonable model appropriate to set theory and the question whether constructivists should accept finite set theory.

The least satisfactory part of the general metatheory of finite mathematics provided in §3 was that connected with the notion of a reasonable model. The difficulty lay in giving a general characterization of what a reasonable model is. In the specific case of set theory, it is easy to do much better. A reasonable model of finite set theory in the sense of §3 is a model in which all of the elements are sets and the relation and function symbols have their intended interpretation. But that is not enough to sustain the picture of finite set theory presented in this section: the elements of a model need to be hfpsets, not arbitrary, possibly infinite, sets. The finite model must be a *natural model—* one in which the elements are hfpsets and '$\varnothing$' and '$\in$' are interpreted in the intended manner (that is, as the empty set and the membership relation), and ';' is interpreted in the intended manner on members of any Ω. (The term *natural model* is used without the usual implication that the power set is the "real" power set.) The improved version of Theorem 3.5 we require is

THEOREM 4.1. *ZFC is consistent if and only if every finite subset of Fin(ZFC) has a finite natural model.*

A further, substantial, improvement over the general theory of §3 is possible. Unlike Theorem 3.5, Theorem 1 is provable in PRS.

The intuitive motivation for Fin(ZFC) was not constructive, but the proof of Theorem 1, as a bonus, gives Fin(ZFC) some constructive content: The proof (see below) in effect specifies a procedure that, given a finite set of axioms of Fin(ZFC), yields a finite set of finite natural structures. If ZFC is consistent, one of the finite natural structures will be a finite natural model of the given set of axioms. Note that one need not rely on the consistency of ZFC to check whether any of the structures satisfies the given set of axioms—that can be checked in a completely constructive way within PRS. The consistency of ZFC is only needed to prove the general theorem. Thus, there is a function that can be proved within PRS to have the following property: it takes each finite set of axioms of Fin(ZFC) either to a natural model of those axioms or to a formal derivation of an inconsistency from the axioms of Fin(ZFC) (or, equivalently, of ZFC).

I shall give a brief sketch of an easy proof of Theorem 1 that can be carried out in PRS using the following:

THEOREM 4.2. *Every well-founded relation (A, E) is isomorphic to (the membership relation on) a set. If a is in A and there is no b in A such that bEa holds in (A, E), then (A, E) is isomorphic to some set $(S, \in)$ via an isomorphism that takes a to $\varnothing$.*

If A is finite, then (A, E) is isomorphic to the membership relation on a hfpset. If we consider only hfpsets A, the theorem can be proved in PRS.[54]

---

54. Theorem 2 is a simple consequence of the Mostowski collapsing lemma (for which see, for example, [Mak77, p. 247] or [Lev79, p. 70]). Here is a sketch of a proof. Let

SKETCH OF PROOF OF THEOREM 1. The conditional from right to left follows from Theorem 3.3. Suppose that ZFC is consistent, and let S be a finite set of axioms of Fin(ZFC) all of whose Ωs are among $\Omega_{q_0}, \ldots, \Omega_{q_n}$, with $q_0 < \ldots < q_n$. Let q_{n+1} be a rational number greater than q_n. Let S^+ be the union of S with (a) Axiom (8) for $p = q_n$ and $q = q_{n+1}$, and with (b) the Axioms of Equality (11)–(15) and Axioms (16) and (17)—the definitions of $\varnothing$ and ;—all with Ω_0 replaced everywhere it occurs by $\Omega_{q_{n+1}}$. Those variants of Axioms (11)–(17) are theorems of Fin(ZFC) by Lemma 3.2.

The proof by Mycielski and Pawlikowski of Theorem 3.4 shows that it is possible to construct a model of the theory S^+ with a universe of at most $M = (e + 1)^{d^N + d^{N-1} + \cdots + d + 1}$ elements, where e is 2 plus the number of restricted occurrences of quantifiers in S^+, d is the maximum of 2 and the depth of any of the restricted occurrences of quantifiers in S^+, t is the maximum rank of a term that occurs in S^+, and N is $t + n + 2$.

As Mycielski essentially observed, one can add any universal theorem of ZFC to S^+ and still get a model by the same construction. Let W be the set of all sentences of the form

$$(\forall x_1, \ldots, x_k) \neg (x_1 \in x_2 \wedge \cdots \wedge x_{k-1} \in x_k \wedge x_k \in x_1),$$

where $k \leq M$, and let T be $S^+ \cup W$. The Mycielski–Pawlikowski construction yields a model $\mathfrak{A}$ of T with at most M elements since the values of e, d, t, and n are not changed by the addition of W. The model $\mathfrak{A}$ is well founded since it has at most M elements and is a model of W. Let $\mathfrak{A}'$ be the model obtained from $\mathfrak{A}$ by discarding the elements outside $\Omega^{\mathfrak{A}}_{q_{n+1}}$ and dropping $\Omega_{q_{n+1}}$ from the language. (If a and b are in $\mathfrak{A}$ but the old value of a ; b has been discarded, then patch that up, for example, by defining a ; b to be a.) The model $\mathfrak{A}'$ is well founded. But every finite well-founded relation is isomorphic to the membership relation on a hfpset by Theorem 2, and moreover, because $\mathfrak{A}$ is a model of the ap-

---

$\mathfrak{A} = (A, E)$ be well founded. Let $\kappa = |A|$, $a^E = \{b : bEa\}$, and let $a_0, \ldots, a_\alpha, \ldots, \alpha < \kappa$, be a well-ordering of A. Let $\mathfrak{A}' = (A', E') = \mathfrak{A} \,\dot\cup\, (\kappa^+ + \kappa, \in)$ ($\dot\cup$ is disjoint union), and let $E'' = E' \cup \{(\kappa^+ + \alpha, a_\alpha) : \alpha < \kappa\}$. Then (A', E'') is well founded and extensional. Now we apply the Mostowski collapsing lemma: for $a \in A'$, let $f(a)$ be $\{f(b) : bE''a\}$. The range of f on A is as required.

If $a_\alpha^E = \varnothing$, one may modify the above construction by omitting the pair $(\kappa^+ + \alpha, a_\alpha)$ from E'' to ensure that $f(a_\alpha) = \varnothing$.

propriate version of Axiom (16), one can require that $\varnothing^{\mathfrak{A}'}$ be taken to the empty set by such an isomorphism. Any model $\mathfrak{A}''$ isomorphic to $\mathfrak{A}'$ such that the membership relation is the membership relation on a hfpset and $\varnothing^{\mathfrak{A}''} = \varnothing$ is as required: the interpretation of ; is as required because $\mathfrak{A}$ is a model of the appropriate versions of Axioms (8) and (17). $\quad\square$

There is a more direct proof of Theorem 1 based on Theorem 3.5 which, however, cannot be carried out in PRS: Theorem 3.5 shows that there is a finite model of S with domain a finite set (not necessarily a hfpset) with $\varnothing$, $\in$, and ; as required. Since $\in$ is well founded on the set, the model is isomorphic to one with universe a hfpset, as required, by Theorem 2.

One cannot prove that ZFC has a standard model even assuming that it is consistent. Finite mathematics is in a superior situation. One can prove the following version of Theorem 1 within finite mathematics, indeed within PRS: for every finite set S of axioms of Fin(ZFC) there is a finite set $S^* \supseteq S$ of axioms of Fin(ZFC) such that if S^* is consistent, then it (and hence S) has a finite natural model. In sharp contrast to the situation for ZFC, one can show in an acceptable way, given the appropriate consistency assumption, that Fin(ZFC) has natural models in the appropriate sense.[55]

There is one detail remaining. The Ωs are supposed to be indefinitely large, and so we should make sure of the following: Let S be a finite set of axioms of Fin(ZFC) involving $\Omega_{q_0}, \ldots, \Omega_{q_n}$, and let $A_0, \ldots, A_n$ be finite sets of hfpsets. Then there should be a finite natural model of S in which $A_i \subseteq \Omega_{q_i}$, $i = 0, \ldots, n$—that will show that there is no restriction on what members can be in the Ωs and hence that they can be thought of as indefinitely large. We discussed the analogous problem for the general case of an arbitrary Fin(T) in §3. But the answer there concerned reasonable models, and we need the analogous result for natural models. The proof is analogous to that of the comparable result in §3. The required result follows immediately from Theorem 1 and the fact that $\Omega_{q_i}(\tau)$ follows from Fin(ZFC) for any q_i and closed term τ: Just augment S with enough axioms to prove $\Omega_{q_i}(\tau_a)$ for each $a \in A_i$, $i = 0, \ldots, n$, where τ_a is a term built up from $\varnothing$ and ; that denotes a. Then Theorem 1 does the job.

---

55. One cannot show that all of Fin(ZFC) has a natural model—indeed it does not. But to ask for such a model would not be in the spirit of finite mathematics. On the other hand, asking for a standard model of ZFC is very much in the spirit of ordinary set-theoretic mathematics.

I now return to the question whether Fin(ZFC) provides an interpretation of contemporary set-theoretic mathematics that will prove acceptable to constructivists. The question is a difficult one to answer since there are nearly as many motivations for constructivism as there are constructivists, and motivations will play an important role in deciding whether to accept Fin(ZFC). Presumably constructivists will not accept the process of extrapolation, and so they would at best accept Fin(ZFC) without thereby accepting ZFC.

The objects under discussion in Fin(ZFC)—the hereditarily finite sets—are so closely related to the natural numbers that they are acceptable to all constructivists of whom I am aware, with the possible exception of those who doubt whether exponentiation is a total function. We have been taking any theorem of Fin(ZFC) to be a schema asserted to be true of any finite natural model of the axioms used to prove the theorem. The use of classical logic to prove theorems of Fin(ZFC), when the theorems are construed in that way, seems to be uncontroversial since classical logic is applied only to well-defined relations on particular, fully specified finite models—the fact that a set plays the role of an indefinitely large set in a model does not make that set indefinite, any more than assigning a set to a variable makes the set variable. Any constructivist who is willing to accept the axioms of finite set theory should therefore be willing to accept the use of classical logic in drawing inferences from them. The central issue is the grounds for accepting the axioms of finite set theory. A constructivist might well be interested in other axiom systems within finite mathematics, but finite set theory is the most interesting case because those constructivists who accept it have accepted a counterpart of virtually all of classical mathematics.

The fact that finite mathematics has epistemic notions at its core should seem congenial to those constructivists who base their constructivism on broadly epistemic considerations. One can, for example, provide a rather direct account of how we can have knowledge of indefinitely large collections in any of several ways. Finite set theory, unlike ordinary infinitary set theory, is compatible with the Dummettian "meaning is use" thesis [Dum75, pp. 98ff.], and so it seems to me that finite set theory should be acceptable to those who are constructivists for Dummettian reasons.

Some intuitionists, for example Brouwer, have already accepted fundamental ideas on which the notion of an indefinitely large set can be based. (I have in mind the idea of some kind of epistemic availability, and the idea that mathematics has a temporal aspect.) Such intuitionists have been willing to accept intuitionistic Peano arithmetic on the basis of intuitive considerations. I believe that they should be equally willing to accept Fin(ZFC) on the basis of the similar, and similarly clear, intuitive considerations that have been presented here. Admittedly, such considerations will not sway those—like Hilbert—who are unwilling to accept even Peano arithmetic without further justification.

IX

Extrapolations

§1. Natural Models

When we say that Gödel's sentence that is independent of Peano arithmetic is true, we mean that it is true of the natural numbers. When a question of the theory-independent truth of a sentence arises, the normal case is that we are interested in whether or not the sentence is true in some intended natural model of the theory—the natural numbers, or the real numbers, or the universe of sets. But the old familiar natural models are no part of finite mathematics. What is to take their place?

Are we to conclude, for example, that the Gödel sentence for Fin(PA) (by which I mean a regular relativization of the Gödel sentence for PA) is true in some of the finite models of Fin(PA), false in others, and that's the end of it? Certainly not. In the infinitary case, we discover that the Gödel sentence is true by proving that it is true of the natural numbers. The finite-mathematics version of that proof must show the finite-mathematics counterpart of the sentence to be true in some sense or other. But true where, and in what sense? That is the question we answer here.

Natural models of theories of finite mathematics—natural models of the sort introduced in Chapter VIII—can play the required role in the case of arithmetic. For example, it will turn out that the Gödel sentence for Fin(PA)— while it is indeed true in some models of Fin(PA) and false in others—is true in every *natural* model of Fin(PA). I shall discuss arithmetic in some detail, and turn to the more difficult case of natural models of set theory at the end of the section.

The basic idea is simple enough. Let me begin with the following example. It is immediate from the work of Martin Davis, Yuri Matijacevič, and Julia Robinson (see, for example, [DMR76]) that there is an equation between two polynomials with natural-number coefficients that has no solution in the natural numbers and yet which is such that it cannot be proved in Peano arithmetic that there is no solution. It cannot be proved in Peano arithmetic that there *is* a solution either (assuming that Peano arithmetic is ω-consistent)—since then there would be a solution, contrary to hypothesis. When I say that there is no solution I mean that there is no solution in the natural numbers. Since the theorem that there is no solution is independent of Peano arithmetic, there will be solutions in some other models of Peano arithmetic. Likewise, there will be models of Fin(PA) that include solutions.

Let us require that a *natural model* of Fin(PA) be one with domain a set of natural numbers in which the operations of addition and multiplication are ordinary correct addition and multiplication on the members of the Ωs. We do not need to go beyond finite mathematics to make sense of "ordinary correct addition and multiplication" since PRA suffices to characterize those operations. With that definition of natural model, which certainly seems a reasonable one, there will be no natural model of Fin(PA) in which our polynomial has a solution, since such a solution—given our condition on natural models—could be proved to be a solution to the polynomial in PRA, and hence in PA, contrary to the construction of the polynomial.

The familiar use of the intended model of Peano arithmetic transfers to the natural models of finite arithmetic in the way we wanted: The fact that the polynomial has no solution, which cannot be proved in PA, cannot be proved in Fin(PA) either. It is a "fact" because the polynomial has no solution in the appropriate natural model or models. The natural models provide the appropriate theory-transcendent notion of truth. Moreover, a finite-mathematics version of the usual proof that the polynomial has no solution in the natural numbers shows that it has no solution in any natural model of Fin(PA). (That is not surprising, since the usual proof essentially takes place in PRA already.)

It turns out, as a result of the work of Davis, Matijacevič, and Robinson, that the Gödel sentence is essentially equivalent to one that asserts the nonexistence of a solution to a polynomial, and so the above considerations handle not only the case of polynomials but also the Gödel sentence. But the subtle and sophisticated nature of the argument may make it look like it is nearly an accident that the natural models of Peano arithmetic do what we wanted. In particular, it provides no reason to believe that, for example, natural models of Fin(ZFC) could play a similar role.[1] It is therefore worthwhile to see

---

1. Natural models of Fin(ZFC) would be as required if one stipulated that a natural

how to obtain the result that the Gödel sentence is true in all natural models of Fin(PA) (assuming Fin(PA) is consistent) without reducing the sentence to a polynomial.

We could require that addition and multiplication be correct in a natural model precisely because addition and multiplication are primitive recursively definable—it was PRA that gave us the required antecedent standard of correctness. Addition and multiplication are not the only primitive recursively definable functions on the natural numbers that are definable in Peano arithmetic. Indeed, all primitive recursively definable functions on the natural numbers are definable in Peano arithmetic, as follows from work of Gödel.

Augment the language of Peano arithmetic by new function symbols for every primitive recursive function. Let PA^+ be an axiom system in the augmented language that consists of the axioms of Peano arithmetic plus explicit definitions in Peano arithmetic of all of the primitive recursive functions. Since PA^+ is an extension of Peano arithmetic by definitions, we consider it to be little more than a notational variant of Peano arithmetic. (See, for example, [Sho67, §4.6] for a careful discussion of extensions by definitions.) Moreover, the obvious way to axiomatize PA^+ will be to use schemas, and so there will be an obvious corresponding theory $Fin(PA^+)$, which I take to be little more than a notational variant of Fin(PA). We may now define a natural model of a finite set of axioms of $Fin(PA^+)$ to be a model of those axioms with domain a set of natural numbers in which all of the function symbols that appear in the set of axioms—no longer just addition and multiplication— receive their ordinary interpretations on all of the members of the Ωs (that is, as before, their ordinary interpretations as given by PRA). Note that I have not defined a "natural model" but a "natural model of a set of axioms": the condition that a model must satisfy in order to be natural depends on the set of axioms.

The Gödel sentence—and indeed any Π_1 sentence—is readily seen in PA^+, and hence in $Fin(PA^+)$, to be equivalent to a sentence that is just the universal closure of an equation, since the matrix of the Gödel sentence is primitive recursive [Smo77] and therefore equivalent to an equation involving an appropriate primitive recursive function. The Gödel sentence will therefore be true in any natural model of any set of axioms of $Fin(PA^+)$ that includes

---

model was one in which the addition and multiplication operations on the Mirimanoff–von Neumann natural numbers of the model (in the sense of the model) are standard. But that seems too *ad hoc*.

the definition of that primitive recursive function—by the same reasoning we used above for solutions of polynomials. That is what we needed: any natural model of any set of axioms that includes the definition is a model in which the Gödel sentence is true. The same argument works for any Π_1 sentence.

What about sentences independent of Peano arithmetic that are more complicated than Π_1? Let me consider the case of a true independent Π_2 sentence for concreteness and then comment on the general case. Such a sentence is equivalent to one of the form $(\forall x)(\exists y) f(x, y) = 0$, where f is one of the function symbols of PA^+.[2] There will be natural models of finite sets of axioms of $Fin(PA^+)$ including the definition of f in which the sentence is false. Consider, however, what a *bad* natural model—one in which the sentence is false—must look like. There must be some x_0 in the domain of the model such that there is no y in the model for which $f(x, y) = 0$ holds—call x_0 a witness that the model is bad. We know, however, that a suitable y exists, say, y_0, and that we can prove that $f(x_0, y_0) = 0$ in $Fin(PA^+)$. Thus for every x_0 there is a finite set S of sentences of $Fin(PA^+)$ such that x_0 is not a witness that any natural model of any set of sentences of $Fin(PA^+)$ that includes S is bad. Say that x_0 is a *robust* witness if it serves as a witness that every natural model of every finite set of sentences of $Fin(PA^+)$ is bad.[3] Then we have shown that there are no robust witnesses. We would require a robust witness to conclude that our sentence is false, and we know we cannot get one. Thus, the sentence is true. Now, if we know that our independent sentence is true of the natural numbers, in the normal case that will be because we have proved that it is, though in some theory stronger than Peano arithmetic. The finite-mathematics version of the same proof will show that no natural number is a robust witness. We can thus carry out essentially the argument I have just given within finite mathematics.

The reasoning in the Π_2 case is more recondite than it was in the Π_1 case, and as we go up the hierarchy things may sound even worse. In fact, however, there is a uniform condition for a sentence of finite mathematics to be a truth of a system of finite natural models, a condition that formally resembles the

---

2. See [Gol90, §3] for an example of such a sentence. The example is pretty nearly $(\forall x)(\exists y)(\text{"}x \text{ is a set of sentences of } Fin(PA)\text{"} \rightarrow \text{"}y \text{ is a model of } x\text{"})$.

3. The notion of a robust witness defined in the text goes beyond finite mathematics by quantifying over every natural model and finite set of sentences of $Fin(PA^+)$. But the reader already knows how to give an appropriate counterpart of the notion in finite mathematics. I omit the details.

Putnam–Hellman condition for being true of Zermelo's open totality of all normal domains. The considerations of this section suffice to show that the uniform condition holds in the cases at hand. The uniform condition plays a role in finite mathematics much like the familiar role that truth in a single natural model does in ordinary mathematics. I give it in §2.

The usual proof that the Gödel sentence is true is often phrased so as to rely on the assumption that Peano arithmetic holds of the natural numbers. In fact, all that is used is that the Σ_1 theorems of Peano arithmetic hold of the natural numbers. In a like manner, the proof that the specific example of an independent Π_2 sentence mentioned above is true uses that the Σ_2 theorems of Peano arithmetic hold of the natural numbers. (See [Gol90].) The differences between the arguments concerning the two independent sentences discussed in this section are not a phenomenon unique to finite mathematics. They parallel the finer analysis just mentioned, which is more or less familiar from ordinary mathematics.

How do the above remarks on Peano arithmetic apply to natural models of set theory or any other theory? To obtain a definition of a kind of natural model, one considers extensions of a theory by definitions and then defines natural models to be those in which the defined predicates and functions are correct, when there is an antecedent sense of correctness. That is about all I can say in the general case, but in the case of set theory it is possible to be more precise.

For the theory of numbers we have PRA as a natural background theory. For set theory, it might seem that PRS could play a comparable role, but the situation is more complicated: We cannot simply require that a natural model of a finite set of axioms of Fin(ZFC) give finitely many set-theoretically primitive recursive functions their intended interpretation. For example, in ordinary infinitary set theory we have that $\bigcup \omega = \omega$, but there is no hfpset w such that $\bigcup w = w$, and so the union operation in a hereditarily finite model of Fin(ZFC) cannot be the usual one on whatever set plays the role of ω. That is not a defect in the theory Fin(ZFC) or in its models. After all, the whole point of the counterpart of ω in a model of Fin(ZFC) is that not all of its members are available, while all the definitions in PRS, including that of union, are from below, with the presumption that everything needed is available.

It is not at all clear how to give a general characterization of what natural constraints we can place on models of Fin(ZFC) beyond the ones of §VIII.4. That is, I claim, partially why we are genuinely unclear about various questions of set theory. There are, nonetheless, at least a few plausible constraints:

models with bigger gaps between a limit ordinal and its predecessors are more natural. In particular, models with a bigger gap between the counterpart of ω and the natural numbers that precede it are more natural. That consideration does not yield a single naturalness constraint but a succession of them. We should, for example, feel free to require that the counterpart of ω be at least double any of the natural numbers that precede it in a natural model should that be useful for some purpose, and to impose the stronger requirement that it be at least the factorial of any natural number that precedes it, if and when that turns out to be useful. We might also produce an extension by the definition of the power-set operation, and require that its interpretation be correct in the sense of PRS. That would be a natural way of requiring that, in a certain sense, no subsets of any set are omitted. Those are, I believe, the sorts of considerations that will be relevant to arguing for new axioms of set theory.

§2. Many Models

In this section we discuss the formal relationship between Putnam–Hellman modal set theory and finite set theory. Putnam–Hellman modal set theory provides a formal means of making sense of the notion of truth in Zermelo's open progression of normal domains, as distinct from the notion of truth in a single model of set theory. In finite set theory, like infinitary set theory on Zermelo's account, there is no distinguished model that is of particular interest, and so when we claim that something is true of the sets, we mean something analogous to what Zermelo did. Putnam and Hellman made the notion precise on Zermelo's behalf, and the formal analog within finite mathematics is the one we need there. In finite mathematics not just set theory but every theory involves an open progression of models, and so the picture developed in this section will apply to every theory in finite mathematics, not just finite set theory.

In ordinary mathematics we deal with the models of a fixed theory in a fixed language, while in finite mathematics we are often simultaneously concerned with many theories (various finite sets of axioms of Fin(T)) that may be in different languages. It will therefore be convenient to use a nonstandard definition of a model and of various model-theoretic notions. We reserve the word *structure* for the standard notion of a model. From now on a *model* is an ordered pair with first element a set of sentences and second element a structure for the language of that set of sentences in which all of the sentences are true. One model is *(isomorphic to a) submodel* of another if the set of sentences of the first is a subset of the set of sentences of the second and the structure of the first is (isomorphic to) a substructure of the structure that results from the structure of the second when it is restricted to the language of the first. A model of Fin(T), for any given T, is just a model in which the set of sentences is a set of sentences of Fin(T).

We can now set up our notion of *truth in a progression of models.* Fix a type of model: for finite set theory or number theory, we would use some kind of natural model of the theory; for some other theory within finite mathematics we might just use all the models of the theory. To set up a modal frame say that a model $\mathfrak{A}$ is accessible from a model $\mathfrak{B}$ just if $\mathfrak{B}$ is isomorphic to a substructure of $\mathfrak{A}$ (and let the counterparts of members of $\mathfrak{B}$ be determined by the isomorphism). We must always exclude any *bad* models of a theory from a frame, where a bad model $\mathfrak{A}$ is one such that there is a finite set S of axioms of the theory such that no model of S is accessible from $\mathfrak{A}$.

Observe that the frame for Putnam semantics is essentially a special case of our modal framework: We take the progression of models to consist of models in which the first member of the pair is the set of all the axioms for ordinary set theory and the second is a normal domain. Since we always use all the axioms, there are no bad models.

In Putnam semantics, we transformed a sentence ϕ to a sentence ϕ^* by replacing $\forall$ by $\Box\forall$ and $\exists$ by $\Diamond\exists$. We shall do the same thing here, so that $\forall x \in \Omega_p$ becomes $\Box\forall x \in \Omega_p$ and $\exists x \in \Omega_p$ becomes $\Diamond\exists x \in \Omega_p$. In Putnam semantics, one said that ϕ is true of the progression of normal domains just if ϕ^* is true in the modal frame just specified. For finite mathematics a slight alteration is made necessary by the fact that the different models may be models for different theories: we want to ensure that for whatever finite theory we start out with there is a finite extension of that theory such that the sentence under consideration is true of the progression of models of every extension of that extension. We shall therefore say that a sentence ϕ is true of a progression of models just if $\Box\Diamond\Box\phi^*$ is true in the appropriate modal frame.

It is a routine exercise to check that for the modal frame associated with Putnam semantics the new condition is equivalent to the old one, that is, that for that frame and for all ϕ, $\Box\Diamond\Box\phi^*$ is true if and only if ϕ^* is true.[4] Thus, the modified condition does not represent a departure from Putnam seman-

---

4. The claim follows directly from the fact that either ϕ^* is true on a final segment or false on a final segment, as is established by a straightforward induction on the formation of ϕ.

tics. The condition just given for truth in a progression of models is the one mentioned in §1: it is easily seen that the sentences claimed to be true there for various notions of natural model are true of the corresponding progressions of models in the sense defined here. It turns out that every sentence is true or false in Putnam semantics. With our comparable semantics that need not be the case, since our accessibility relation may have many branches. It is therefore possible that a given notion of natural model for a theory may settle some questions that are independent of the theory without settling them all.

In giving the definition of truth in a progression of models, I have used infinitary notions quite freely. The official definition should be the finite-mathematics counterpart of the one given here. It can be read off the definition I have in fact given in the canonical way.

§3. One Model or Many? Sets and Classes

We extrapolate from Fin(ZFC) to ZFC by setting the bounds on our quantifiers—the Ωs—equal to one another to form a single domain of quantification V that is so large that it can never require enlargement in any context. Thus, extrapolation brings with it the idea of a single maximal domain of all sets. (The choice of ZFC is just for convenience. Similar remarks apply *mutatis mutandis* to other standard theories of sets.) On the other hand, as we have seen in some detail in §2, the model theory of Fin(ZFC) is intimately associated with Zermelo's idea of an unfolding progression of domains. Thus, the idea that our intuitions about infinitary set theory have their source in extrapolations from finite set theory explains the source of the conflict we actually find between two different conceptions of infinitary set theory—the idea of a single maximal domain arises by extrapolation from finite set theory, while the idea of a progression of domains arises by extrapolation from the model theory of finite set theory. The fact that the extrapolation view provides such an explanation counts as strong evidence in its favor: the more of the actual phenomena it explains, the more likely it is to be correct. Let us use the extrapolation view to get clearer about the two sides of the conflict.

Extrapolation from the natural models of Fin(ZFC) yields that there is a single maximal natural model for ZFC, but it leaves that model ontologically remote: The extrapolation enables us to determine many properties of the model, but it is neutral about the metaphysical nature of the things in the domain of the model just because the theory in finite mathematics with which the extrapolation begins is similarly neutral.

In the finite case, it is not hard to provide a plausible metaphysical account; in the infinite case, if the extrapolation view is correct, we should not expect to be able to say much more than that the infinitary objects are extrapolated ("idealized") versions or counterparts of the finite ones. That, as a matter of fact, describes the actual situation rather well. Once one accepts the extrapolation view, that state of affairs does not seem unsatisfactory, since one can describe the nature of the idealization so precisely. The state of affairs is not much worse than that of, say, our knowledge of what it is like in the core of a neutron star. Without extrapolation, however, understanding the nature of infinite sets has been one of the main mysteries of the philosophy of mathematics: How could we know anything about such remote entities—even have any idea whether or not they exist—since they are so unlike anything we experience?

Naturally enough some mysteries remain, even given the extrapolation view. The chief mystery among them is that the universe of sets must go on for unimaginably long, since any length we can imagine could be prolonged in some larger context. We can therefore have no grasp on just how long unimaginably long is. That suggests that our notion of the universe of all sets is irreducibly vague or ambiguous—in Parsons's phrase [Par80], systematically ambiguous. Vagueness or ambiguity of a notion usually suggests the possibility of making the notion more precise or of disambiguating it in various ways. Parsons suggested that any universe we chose for set theory would inevitably become a set in some larger universe, falsifying the idea that the original universe was the universe of all sets, and concluded [Par80, p. 91] that "the 'totality' of sets is irreducibly potential." That suggests a modal view, and Parsons has indeed suggested such a view, but he also said that it is unclear whether that can be made to work and speculated that [Par80, p. 91], "Perhaps a certain 'dialectical' character is essential to the axioms of set theory."

The fundamental problem with supplying a modal or dialectical interpretation of infinitary set theory is that either kind of interpretation seems to require a possibility that the ontological status of at least some sets can change (that possible sets can become actual, or some such). As Parsons himself has suggested, the dimension in which sets change cannot be time. (Compare §V.5.) But there does not seem to be anything that can take time's place either. Proponents of modal views have therefore tended simply to take the modality as a primitive, without attempting to explain it further. But that is just to rename the mysterious systematically ambiguous character of the universe of all sets. It does not represent progress toward understanding it.

I hope it will have been obvious to the reader where the above discussion has been headed: Finite set theory is at its base "dialectical" in the requisite sense, though I would not like to have to argue that it is dialectical in any very precise sense of the term. Finite set theory makes essential use of the idea of context sensitivity, which is often enough sensitivity to a purpose, and I take it that that is close enough. The mysterious systematically ambiguous character of the extrapolated universe V of all sets is inherited from the not-at-all-mysterious systematically ambiguous character of our reasoning about the Ωs, which give rise to V. Extrapolation yields that there is a single universe, but it does nothing to resolve the systematic ambiguity of the Ωs. It is the systematic ambiguity of the Ωs that gives rise to the progression of models of finite set theory, and that progression extrapolates to yield the idea of a progression of models of infinitary set theory.

There are many Ωs, and they are to some extent interchangeable. So long as we persist in identifying them, we will be faced with the mystery of a single maximal universe of sets that should, it seems, itself be a set. That does not show that we should abandon infinitary set theory as incoherent: It is perfectly coherent to quantify over a single V_α (or its counterpart in a theory without Foundation—I have been talking only about ZFC for convenience) conceived as embedded in a larger one, and for many purposes we can leave it somewhat vague just which V_α it is that we are using. What I am advocating is nothing more than Zermelo's idea, but now supplemented with a diagnosis of what is problematic about its rival: The totality of all sets arises by extrapolation, but the idea that it is fixed is a mere artifact of the extrapolation. There is nothing wrong with that *per se,* but it requires a theory of the relevant sort of fixed totality. But the context-sensitive way in which the Ωs that lead to the totality can vary is certainly of no help, and there does not seem to be any way to provide an adequate substitute.

We now turn to the distinction between sets and classes. The model theory of finite set theory is inevitably and naturally associated with Zermelo's idea of a progression of domains for set theory. Indeed, the idea of a fixed universal domain for finite set theory is not even slightly appealing and it is clear that any class in a model of finite set theory is itself a set that may turn out to be a member of some larger model. Thus, within finite set theory even more than within Zermelo's picture of infinitary set theory, it seems natural and inevitable that there should be a distinction between sets and classes— since no model is a final one—and that the distinction is a thoroughly relative one—relative to a model or epistemic context. The distinction between sets and classes arises not because of some need to avoid paradox but out of a need to distinguish the hfpsets in a given natural model from the hfpsets outside it. "The class of all sets is not a set" becomes, within finite set theory, a way to express the fact that there is a canonical choice of a hfpset that is not in the present finite model, namely, the domain of that very model.

So far we have been discussing ZFC but, since the distinction between sets and classes is such a natural and inevitable part of finite set theory, it seems worth considering a theory of sets that admits both sets and classes to see what the distinction looks like within finite set theory. To avoid misunderstanding, let me remind you that the classes we are considering are combinatorial classes—collections that are too big to be counted (counted within the present universe, we might now add)—not Russellian extensions with their attendant paradoxes. Within set theory, the notion of a proper class—that is, a class that is not a set—has a very simple definition: a proper class is a class that is not a member of any other class, a class that is maximal with respect to the membership relation, briefly, an $\in$-maximal class.

It may seem absolutely obvious in light of the considerations just above that any model within finite mathematics will contain classes, since any finite model will of necessity contain some $\in$-maximal members. Though that is more or less correct, a slightly more careful analysis is required: we need to look not at the definition of proper class just considered but at its counterpart within finite mathematics. The canonical counterpart runs as follows: if x is in Ω_0, then x is a proper class if for all y in Ω_1, x is not in y. By indiscernibility, it follows that if x is in Ω_p, then x is a proper class if for all y in Ω_q x is not in y, where Ω_q is the largest Ω under consideration and Ω_p is the next to largest. If we consider a language without function symbols then nothing outside of Ω_q makes any difference,[5] and so we do in fact essentially recover the basic idea that a class is an $\in$-maximal collection.

Note the connection between the idea of a proper class as a collection maximal merely with respect to the present universe and Parsons's idea of systematic ambiguity [Par80, pp. 219, 240]: "if I take your quantifiers to range over 'all' sets, this may only show (from a 'higher' perspective) my lack of a more comprehensive conception of set than yours. But then it seems that a perspective is always possible according to which your classes are really sets." "Leaving aside the possibility of an 'absolute' interpretation, the

---

5. If function symbols are allowed, then we must also take account of objects that result from applying the functions to members of Ω_q, objects that need not be in Ω_q. Though I do not believe that that causes any essential difficulty, it adds significantly to the complexity of the necessary formulations.

generality which such a discourse . . . has which transcends any particular set as range of its quantifiers must lie in some sort of systematic ambiguity, in that indefinitely many such sets will do.'

§4. Natural Axioms

In §V.5 I proposed an axiomatization of set theory that seemed to me to be more natural and conceptually based than the usual ones—though I doubt it is the last word on the subject. Here I show that the finite-mathematics counterpart of that axiomatization can be motivated within finite mathematics. That is not absolutely automatic from the discussion of Fin(ZFC) in §VIII.4 because ZFC is a theory exclusively of sets, while the theory of §V.5 also has classes.

Two of the five axioms can be disposed of immediately: The motivation for Axiom V.5.1, the Axiom of Extensionality for classes, is the same as that for the Axiom of Extensionality for sets, Axiom (18) of §VIII.4. More precisely, the two axioms are similar to each other in form, and as a consequence their counterparts in finite mathematics are similar to each other in form and can be seen to be true in pretty much the same way, a way that was discussed in §VIII.4. The counterpart within finite mathematics of Axiom V.5.3, the Axiom of Infinity, is an immediate consequence of the Axiom of a Zillion, Axiom (24) of §VIII.4, and we have already discussed that axiom in detail. Limitation of Comprehensiveness, Axiom V.5.4, can also be seen to be handled by the considerations of §VIII.4: It is a biconditional. The forward direction is a union axiom, and the finite-mathematics counterpart will be a consequence of Weak Union, Axiom (24) of §VIII.4, and of Limitation of Size, which we shall discuss below, since Limitation of Size yields the requisite separation principle. Similarly, the reverse direction is a consequence of Power Set, Axiom (22) of §VIII.4, and Limitation of Size, once more as a result of a separation principle.

Two axioms remain, the full schematic version of Axiom V.5.2, which I shall call Comprehension, and Limitation of Size, Axiom V.3.1. After discussing those, we shall discuss how Limitation of Comprehensiveness and Limitation of Size fit together.

The finite-mathematics counterpart of Comprehension (which is nothing more than a semiregular relativization of Comprehension) reads as follows:

$$(\exists y \in \Omega_0)(\forall x \in \Omega_1)((\exists z \in \Omega_2)\, x \in z \to (x \in y \leftrightarrow \phi(x))),$$

where ϕ is any semiregular formula in any language for set theory we come to accept in which y does not occur free and in which there are no occurrences of any Ω_p such that $p \le 1$.

Note that the fact that we use a given instance of Comprehension means that the relevant ϕ is available to us. Thus, Comprehension in effect requires that the class of sets that satisfy ϕ be available for any available ϕ. The restriction to classes in Ω_1 that are sets with respect to Ω_2 is exactly the restriction to things in Ω_1 that are ever in any class at all since, by indiscernibility, we may as well take Ω_2 to be the Ω with the highest subscript under consideration. That restriction to things that are ever in any class at all is a very natural one to impose on things to be considered for membership in a particular class. The finite-mathematics counterpart of Comprehension is thus a codification of the same principle that makes, say, the class of cats available to us in virtue of the fact that we know what cats are—despite the fact that it may never be the case that each individual cat is available to us. Once one abandons the idea of transitive models, that seems nearly inevitable.

We now turn to Limitation of Size, von Neumann's version of the Cantorian idea that sets are the collections that can be counted. The forward direction of the axiom says that no available function has absolutely all sets in its range on an available set. It is just the idea that there is no set of all sets: If there were, it would be the range of a function with domain an ordinal (since that is Cantor's definition of a set), hence the range of a function with domain a set. If there were a function from a set onto the class of all sets, then, since there is a function from an ordinal onto the set, composing yields a function from an ordinal onto the class of all sets, which would therefore be a set.

The converse direction is, as von Neumann said explicitly, a strengthening of an intuitively clear idea, not itself directly self-evident. It says that every proper class can be mapped onto the class of all sets, that is, that every proper class is as large as possible. When we consider the finite classes of finite mathematics, that seems plainly impossible since all proper classes are finite subclasses of the finite class of all sets. What we really need to look at is not von Neumann's axiom but its counterpart within finite mathematics. That reads as follows, where I have used Ω_p for the largest Ω under consideration:

$$(\forall x \in \Omega_0)((\forall y \in \Omega_p)x \notin y \rightarrow$$
$$(\exists f \in \Omega_1)((\forall uvw \in \Omega_p)(f(u) = v \wedge f(u) = w \rightarrow v = w) \wedge$$
$$(\forall y \in \Omega_2)((\exists z \in \Omega_p)y \in z \rightarrow (\exists z \in \Omega_p)(z \in x \wedge f(z) = y)))).$$

I did not spell out "$f(x) = y$" because that would have made the resulting formula even harder to read without changing the aspects that are relevant to the present discussion. Thus, what is required is not that all proper classes be the size of the class of all sets, but that every proper class has at least as many members in Ω_p as there are sets in Ω_2, whatever is in Ω_2. (That last clause represents the force of the fact that f is in Ω_1. The function is available independently of what happens to be in Ω_2.) That is a new requirement that goes beyond what is obviously necessary, but it seems natural enough to impose the requirement that, however many sets are available, at least that many members of any proper class are available in virtue of that availability. That is obviously reasonable in the case of the class of all sets. In the case of the class of ordinals, it ensures that there are enough ordinals to count off the sets. It then ensures that any class in Ω_0 that has fewer members than there are sets in Ω_2—any small class—is a set.

According to the Limitation of Size Axiom a proper class is a collection such that as more sets become available at higher degrees of availability, more members of the collection become available in virtue of that fact. Since there is no upper bound on how many sets may become available, there can be no upper bound on the number of members of a proper class. But that justifies the Limitation of Comprehensiveness Axiom: A given set has a fixed upper bound on the number of members in its union, and so the union of a set cannot be a proper class—it is a set. Conversely, a given union of a class yields a fixed upper bound on the number of available members of the class, and so if the union of a class is a set, so must be the class itself. When we work within finite mathematics, the known properties (known from PRS) of finite sets enable us to see how Limitation of Comprehensiveness fits within the picture given by Limitation of Size.

§5. Second Thoughts

Although we are not making any use of second-order logic in our final account of set theory, second-order logic remains a perfectly legitimate formalism, and the idea of full second-order logic and the doubts concerning it seem worth understanding better. As Skolem taught us, every second-order theory has a first-order version. We use that fact to lift finite mathematics from the first-order case to second order.

In investigating second-order logic, I shall consider only second-order quantification over subclasses of a domain, omitting quantification over n-

ary relations and quantification over functions. The method for obtaining the general formalism for second-order logic within finite mathematics is only notationally more complicated than that special case, and when the classes under consideration are classes of sets, the special case is equivalent to the more general one. It is easy to convert from a second-order formalism to a first-order one: We add a unary relation symbol C, to be interpreted "is a class," and we add a binary relation symbol $\in$, to be interpreted "is a member of the class." We need an axiom of extensionality, and one usually assumes some form of comprehension. It is then possible to convert a sentence of second-order logic to a sentence of first-order logic in an obvious way: we define a transformation * that takes $(\forall x)\phi$ to $(\forall x)((\exists y)(C(y) \wedge x \in y) \to \phi^*)$, $(\forall X)\phi(X)$ to $(\forall x')(C(x') \to \phi(x')^*)$, $X(x)$ to $x \in x'$, and passes through everything else. Note that we have taken the domain of the first-order quantifiers to be the class of objects that are members of any class in the model. A *natural* model of the resulting first-order formalism is one in which the interpretation of C is a collection of classes of objects in the domain of the model and the interpretation of $\in$ is the membership relation. A *full* model is a natural model in which the interpretation of C contains every subclass of the domain of the first-order quantifiers. It is virtually immediate that every model (full or not) for a theory in second-order logic is canonically associated with a natural first-order model for the corresponding first-order theory and *vice versa,* and that the first-order model canonically associated with a full second-order model is full and *vice versa.*

If we put the notion of fullness aside for a moment, the above results let us obtain finite second-order mathematics from finite first-order mathematics: If T is a second-order theory, then there is a first-order theory T^* associated with it as just described, and a theory $\mathrm{Fin}(T^*)$ within finite mathematics associated with that. We now let $\mathrm{Fin}(T)$ (a notion that has not yet been defined, since T is second order) be a theory such that $\mathrm{Fin}(T)^*$ is logically equivalent to $\mathrm{Fin}(T^*)$.[6] The second-order models canonically associated with natural models of $\mathrm{Fin}(T^*)$ will be models of $\mathrm{Fin}(T)$, as is straightforward to check. The virtue of extending finite mathematics to the second-order case in the way we just did is that it is now clear that all of the theorems of §VIII.3.1

---

6. The two theories are required to be logically equivalent instead of identical because one of them will have quantifiers with variables bounded by Ωs then Cs, while that order will be reversed in the other.

go over to the second-order case in an automatic way—but so far we have neglected a key idea, that of fullness.

Here, as in earlier chapters, I am discussing fullness under the presumption that the idea makes sense. I shall feel free to talk about all subclasses of a domain as if that makes sense, since it must for the idea of fullness to make sense. After describing the situation under that presumption, I shall come back and evaluate the presumption.

From my perspective, the best way to characterize a full second-order model is as a model of *every* instance of full schematic comprehension (that is, the full schematic version of Axiom V.5.2, which allows formulas in an expanded language). (I do in fact believe that *any* instance is true—see §VII.4— but that is not at all the same thing.) Suppose T is a second-order theory that includes the full schematic comprehension schema. Then $\mathrm{Fin}(T)$ will be a theory within finite mathematics of the sort introduced above. However, when we extrapolate from $\mathrm{Fin}(T)$, something subtle but important happens: Within finite mathematics, we are interested in finite sets of axioms of a theory and their models; when we extrapolate, we are interested in the set of all axioms of a theory and its models. Thus, when we extrapolate from $\mathrm{Fin}(T)$ we do not recover T, with its comprehension schema. We end up with a theory T' in which the comprehension schema has been replaced by all of its instances.

In the case of ordinary schemas (which allow substitution of formulas only from a language that has been fixed in advance), extrapolation is warranted for the reasons discussed in §VIII.2, but the case of full schemas requires separate consideration. Since each instance of a full schema is an ordinary first-order formula, each instance will extrapolate for the reasons just mentioned. It follows that we can extrapolate a full schema within finite mathematics to a full schema within infinitary mathematics. That legitimates the move from $\mathrm{Fin}(T)$ to T.

The additional move being contemplated here, the one that requires additional discussion, is the extrapolation from a full schema to the set of all of its instances, which takes us from $\mathrm{Fin}(T)$ to T' and hence to full second-order logic. Extrapolation yields the idea that there is a single maximum set of all the instances of full schematic comprehension, but it does not, so far as I can see, provide us with any nontrivial information (that is, information going beyond what follows from the first-order theory) about the structure of that set. That situation is, I believe, important confirming evidence for the extrapolation view: the view accounts for the situation in which we actually find ourselves, one in which there is a clear intuition in favor of the idea of full second-order models and yet there is no apparent way to spell that idea out.

The thesis that extrapolation is the source of our intuitions about the infinite does not entail that we should accept those intuitions. Once we see the mechanism that is the source of our intuitions, we are in a position to reevaluate them. We did so in §VIII.2 and concluded that we should accept extrapolated set theory for two reasons: it accounts for the fact that the only theories of interest in finite mathematics are those that have extrapolations, and it accounts for the idea that many physical quantities have objective values. But neither of those reasons provides any justification for the additional extrapolation from full schemas to full second-order logic. That is certainly not a decisive objection to full second-order logic, but, coupled with the fact that full schemas do practically all of the work that has been proposed for full second-order logic and the fact that full second-order logic poses additional philosophical problems, it seems to me, on balance, that we should reject the idea that we have a clear conception of all the subclasses of a domain in virtue of having a clear conception of the domain. We should therefore reject any foundational role for full second-order logic. That does not prevent us from using full second-order logic as a convenient formalism when, for example, a background set theory provides an independent idea of the set of all subsets of a domain. Therefore, as a practical matter, these considerations only suggest avoiding full second-order set theory when taken as a basic theory and leave, for example, full second-order number theory intact, at least so long as it is granted that the notion of all subsets of the set of numbers is provided by a theory of sets.

§6. Schematic and Generalizable Variables

As discussed in §VI.4, there are two theories that are often nearly interchangeably referred to as PRA. In the first, universal PRA (uPRA), the free variables of theorems of PRA are read as if they were implicitly universally quantified. The theorems themselves therefore represent true assertions of PRA. When one employs uPRA, the theorem $x < Sx$ is read as stating $(\forall x)\, x < Sx$, which, I take it, is one of the theorems that might lead one to think that PRA embodies a commitment to potential infinity.

In the second theory, schematic PRA (sPRA), theorems of PRA in which there are free variables are not themselves assertions of the theory: they are schemas with the property that any instance that does not have any variables in it is a true assertion of the theory. When one employs sPRA, the theorem $X < SX$ is read as licensing the assertion of $0 < S0$, $S0 < SS0$, $SS0 < SSS0$, and a host of similar theorems. As discussed in detail in §VI.3, sPRA does not commit one to any form of infinity, potential or actual: it commits one only to the variable-free instances of theorems of sPRA that one comes to use, and of course any such set of instances has a finite model. Moreover, other ways of taking sPRA are question begging. If, for example, one takes a schema to be an abbreviation for the set of all of its infinitely many instances, then sPRA is probably committed to some form of infinity. But that way of taking schemas already employs the notion of actual infinity.

Since for any theory T we can construct Fin(T) in a canonical manner, we immediately have Fin(uPRA) and Fin(sPRA) at our disposal. Fin(uPRA) has theorems like $(\forall x \in \Omega_0)\, x < Sx$. It is not a fragment of classical mathematics, and so it does not facilitate the comparison of Fin(ZFC) and ZFC. The situation for Fin(sPRA) is rather different. It has theorems like $0 < S0$, $S0 < SS0$, and so forth. Since the theorems of sPRA do not contain any variables, the Ωs just go along for the ride. When I say that PRA is a part of finite mathematics, that is true in the straightforward sense that sPRA is a part of Fin(sPRA). Of course, Fin(sPRA) is not literally sPRA, since it contains superfluous Ωs: it is a conservative extension of sPRA that can be proved to be conservative within sPRW.

The situation has not yet been shown to be fully satisfactory, for I prove claims about one part of finite mathematics, Fin(ZFC), using another, sPRA. I shall now show that Fin(sPRA) is, in a relevant sense, a part of Fin(ZFC). Thus, we can in a straightforward manner take sPRA to be a common neutral ground between ZFC and Fin(ZFC): it is a part of both theories.

In what sense is PRA a part of ZFC? It is, unfortunately, a bit awkward to state, because the domains of the two theories are different. It is much easier to state the sense in which PRA is a part of PA, namely, that PRA is a subtheory of a definitional extension of PA. To avoid irrelevant awkwardness in what follows, I shall discuss PA and Fin(PA) instead of ZFC and Fin(ZFC).

It follows from the fact that uPRA is a subtheory of a definitional extension of PA that Fin(uPRA) is a subtheory of a corresponding finite-mathematics definitional extension of Fin(PA). (If PA* is a definitional extension of PA, then Fin(PA*) is the *corresponding finite-mathematics definitional extension* of Fin(PA).) One can also prove that Fin(uPRA) is a subtheory of a finite-mathematics definitional extension of Fin(PA) directly in a familiar way, making only obvious modifications to the usual proofs.

Schematic PRA is a subtheory of uPRA, in the sense that every truth of sPRA—every variable-free theorem—is a theorem of uPRA, and hence Fin(sPRA) is a subtheory of Fin(uPRA). Since, as shown above, sPRA is a subtheory of Fin(sPRA), we have that sPRA is a subtheory of a finite-mathematics definitional extension of Fin(PA), as required.

The above proves all that is required, so far as I can see. However, since I am trading so heavily on the notion of a schema, one might want to require that a finite-mathematics definitional extension of Fin(PA) has all of the *schemas* of sPRA as theorems, in a suitable sense, in addition to having their variable-free instances as theorems. That can be accomplished as follows: The sentence $x + 0 = x$ is not an axiom of sPRA and cannot be used in finite mathematics, since it has $(\forall x)\, x + 0 = x$ as a logical consequence, which is an infinitary statement. That is why we introduced a new kind of—"schematic"—variable, X, Y, Z, ... in addition to the usual—"generalizable"—variables x, y, z, ..., in §VI.3. (The affinity to the difference between substitutional and referential variables is clear.) We allow either kind of variable indiscriminately in terms and atomic formulas, but we do not otherwise alter our syntax. In particular, we do not introduce quantifiers over the new variables. Our logic is standard except that we introduce two new rules of inference: For example, start with Enderton's axiomatization of logic [End72, pp. 102–108]. The first new rule—*schematic instantiation*—is this: from ϕ infer $\phi\binom{X}{t}$ for any formula ϕ, schematic variable X, and term t that does not have any occurrences of generalizable variables. Let me postpone introducing the second rule for a moment. The theory uPRA is PRA written with generalizable variables, and the theory sPRA is PRA written with schematic variables.

Sentences—formulas with no free variables—cannot include any schematic variables. Only sentences are truth bearers. Thus, any finite set of truths of sPRA has a model, as I argued above. Moreover, every truth of sPRA is also a consequence of uPRA—as one can easily prove in sPRW. The schemas derivable in sPRA that include occurrences of schematic variables are naturally not derivable in uPRA—since that theory makes no use whatever of schematic variables. However, those schemas would in fact become derivable in uPRA if one added the following rule—*schematic specification*—as a rule of logic: from $\forall x \phi(x)$ infer $\phi(X)$. The rule is equivalent to allowing terms with schematic variables in the substitution axioms, as one can show in sPRW.[7] Even with the new rule, the schemas of sPRA are still not quite derivable from Fin(uPRA). For example, from $(\forall x \in \Omega_0)\, x + 0 = x$ one can

---

7. It is more convenient to employ an axiomatization of logic with all the substitution axioms than the axiomatization described here with the rule of schematic specification.

only infer $\Omega_0(X) \to X + 0 = X$. However, as one can show in sPRW, sPRA *is* derivable from $\text{Fin(uPRA)} \cup \{\Omega_0(X)\}$, and that theory is easily seen (in sPRW) to be a conservative extension of Fin(uPRA). Thus, the full sPRA, schemas and all, is a subtheory of a finite-mathematics definitional extension of $\text{Fin(PA)} \cup \{\Omega_0(X)\}$, which is a conservative extension of Fin(PA). All of that can be shown in sPRW. A fully parallel remark, let me emphasize, applies to $\text{Fin(ZFC)}^+ = \text{Fin(ZFC)} \cup \{\Omega_0(X)\}$. That is the theory I take to be the counterpart of ZFC in finite mathematics formalized with schematic (and generalizable) variables. I take all that to show that in the relevant sense PRA is not only a part of finite mathematics but even a part of Fin(ZFC)^+, the counterpart in finite mathematics of ZFC.

The axiomatization described here has a property that is convenient for expository purposes: any theory, like uPRA, that has no schematic variables in its axioms will have no schematic variables in those theorems proved without the use of schematic specification.

Bibliography

[Ack37] W. Ackermann. Die Widerspruchsfreiheit der allgemeinen Mengenlehre. *Mathematische Annalen*, 114:305–315, 1937.

[Acz88] Peter Aczel. *Non-Well-Founded Sets*. Number 14 in CSLI Lecture Notes. Center for the Study of Language and Information, Leland Stanford Junior University, Palo Alto, California, 1988.

[Ass60] Günter Asser. Rekursive Wortfunktionen. *Zeitschrift für mathematische Logik und Grundlagen der Mathematik*, 6:258–278, 1960.

[BB85] Errett Bishop and Douglas Bridges. *Constructive Analysis*. Number 279 in Grundlehren der mathematischen Wissenschaften. Springer-Verlag, New York, 1985.

[BBHL05] René Baire, Emile Borel, Jacques Hadamard, and Henri Lebesgue. Cinq lettres sur la théorie des ensembles. *Bulletin de la Société Mathémathique de France*, 33:261–273, 1905. All page references are to the translation by G. H. Moore, pp. 311–320 [Moo82].

[Bee85] Michael J. Beeson. *Foundations of Constructive Mathematics*. Number 6 in the 3rd series in Ergebnisse der Mathematik und ihrer Grenzgebiete. Springer-Verlag, New York, 1985.

[Bel37] E. T. Bell. *Men of Mathematics*. Simon and Schuster, New York, 1937.

[Ben73] Paul Benacerraf. Mathematical truth. *Journal of Philosophy*, 70:661–679, 1973. All page references are to the reprinting as pp. 403–420 [BP83].

[Ben85] Paul Benacerraf. Skolem and the skeptic. *Proceedings of the Aristotelian Society*, supplementary volume 59:85–115, 1985.

[Ber34] George Berkeley. *The analyst*. London and Dublin, 1734. All page references are to the reprinting as vol. 4, pp. 65–102 [Ber51].

[Ber35a] George Berkeley. *A defence of free-thinking in mathematics. In answer to a pamphlet of Philalethes Cantabrigiensis, intituled,* Geometry no friend to Infidelity, or a defence of Sir Isaac Newton, and the British Mathematicians. *Also an appendix concerning Mr. Walton's* Vindication of the principles of fluxions against the objections contained in *the Analyst; wherein it is attempted to put this controversy in such a light as that every reader may be able to judge thereof*. Printed by M. Rhames, for R. Gunne, Dublin, 1735. All page references are to the reprinting as vol. 4, pp. 109–141 [Ber51].

[Ber35b] Paul Bernays. Sur le platonisme dans les mathématiques. *L'Enseignement mathématique*, 1st ser., 34:52–69, 1935. All page references are to the translation by Charles D. Parsons, pp. 258–271 [BP83].

[Ber51] George Berkeley. *The Works of George Berkeley Bishop of Cloyne*. Edited by A. A. Luce. Thomas Nelson and Sons, London, 1951. Reprinted by Kraus Reprint, Nendeln, Liechtenstein, 1979.

[Ber76] Paul Bernays. A system of axiomatic set theory. In Gert H. Müller, editor, *Sets and Classes; On the Work by Paul Bernays*, number 84 in Studies in Logic and the Foundations of Mathematics, pp. 1–119. North-Holland Publishing Company, New York, 1976. Originally published in the *Journal of Symbolic Logic,* 2:65–77, 1937; 6: 1–17, 1941; 7:65–89, 1942; 7:133–145, 1942; 8:89–106, 1943; 13:65–79, 1948; 19:81–96, 1954.

[Boo71] George Boolos. The iterative conception of set. *Journal of Philosophy*, 68:215–231, 1971. All page references are to the reprinting as pp. 486–502 [BP83].

[Bos74] H. J. M. Bos. Differentials, higher-order differentials and the derivative in the Leibnizian calculus. *Archive for History of Exact Sciences*, 14:1–90, 1974.

[Bos80] H. J. M. Bos. Newton, Leibniz, and the Leibnizian tradition. In I. Grattan-Guiness, editor, *From the Calculus to Set Theory, 1630–1910*, pp. 49–93. Gerald Duckworth & Co., London, 1980.

[Bot86] Umberto Bottazzini. *The Higher Calculus: A History of Real and Complex Analysis from Euler to Weierstrass*. Springer-Verlag, New York, 1986. Translation by Warren Van Egmond, with substantial corrections and revisions, of *Il Calcolo sublime: storia dell'analisi matematica de Euler a Weierstrass*. Editore Boringhieri, Turin, 1981.

[Boy49] Carl B. Boyer. *The Concepts of the Calculus, A Critical and Historical Discussion of the Derivative and the Integral*. Hafner Publishing Company, 1949. All page references are to the reprinting as *The History of the Calculus and Its Conceptual Development* (Dover Publications, New York, 1959).

[BP83] Paul Benacerraf and Hilary Putnam, editors. *Philosophy of Mathematics*, 2d ed. Cambridge University Press, Cambridge, England, 1983.

[Bro52] Luitzen Egbertus Jan Brouwer. Historical background, principles and methods of intuitionism. *South African Journal of Science*, 49:139–146, 1952. All page references are to the reprinting as vol. 1, pp. 508–515 [Bro75].

[Bro75] Luitzen Egbertus Jan Brouwer. *Collected Works*. Edited by Arend Heyting. North-Holland Publishing Company, Amsterdam, 1975.

[Bur83] John P. Burgess. Why I am not a nominalist. *Notre Dame Journal of Formal Logic*, 24:93–105, 1983.

[Caj85] Florian Cajori. *A History of Mathematics*, 4th ed. Chelsea, New York, 1985.

[Can82] Georg Cantor. Ueber unendliche, lineare Punktmannigfaltigkeiten, 3. *Mathematische Annalen*, 20:113–121, 1882. All page references are to the reprinting as pp. 149–157 [Can32b].

[Can83] Georg Cantor. *Grundlagen einer allgemeinen Mannigfaltigkeitslehre. Ein mathematische-philosophischer Versuch in der Lehre des Unendlichen*. B. G. Teubner, Leipzig, 1883. All page references are to the reprinting as pp. 165–208 [Can32b]. Translated as [Can76].

[Can84] Georg Cantor. Ueber unendliche, lineare Punktmannigfaltigkeiten, 6. *Mathematische Annalen*, 23:453–488, 1884. All page references are to the reprinting as pp. 210–244 [Can32b].

[Can87] Georg Cantor. Mitteilungen zur Lehre vom Transfiniten. *Zeitschrift für Philosophie und philosophische Kritik*, 91:81–125, 1887. All page references are to the reprinting as pp. 378 *et seq.* [Can32b].

[Can91] Georg Cantor. Über eine elementare Frage der Mannigfaltigkeitslehre. *Jahresbericht der Deutschen Mathematiker-Vereinigung*, 1:75–78, 1890–1891. Reprinted as pp. 278–280 [Can32b] and translated as Appendix B to Chapter IV of this book.

[Can95] Georg Cantor. Beiträge zur Begründung der transfiniten Mengenlehre, I. *Mathematische Annalen*, 46:481–512, 1895. Reprinted as pp. 282–311 [Can32b] and translated in [Can15].

[Can97] Georg Cantor. Beiträge zur Begründung der transfiniten Mengenlehre, II. *Mathematische Annalen*, 49:207–246, 1897. Reprinted as pp. 312–351 [Can32b] and translated in [Can15].

[Can15] Georg Cantor. *Contributions to the Founding of the Theory of Transfinite Numbers*. Open Court Publishing Company, Chicago, 1915. Translation of [Can95] and [Can97] by Philip E. B. Jourdain. With an introduction by the translator. Reprinted by Dover Publications (New York, 1955).

[Can32a] Georg Cantor. Cantor an Dedekind. In Ernst Zermelo, editor, *Gesammelte Abhandlungen*, pp. 443–447, 451. Verlag von Julius Springer, Berlin, 1932. Includes Zermelo's editorial comments. All page references are to the translation by Stefan Bauer-Mengelberg and Jean van Heijenoort, pp. 113–117 [vH67].

[Can32b] Georg Cantor. *Gesammelte Abhandlungen*. Edited by Ernst Zermelo. Verlag von Julius Springer, Berlin, 1932.

[Can76] Georg Cantor. Foundations of a general theory of manifolds. *Campaigner, Journal of the National Caucus of Labor Committees*, 9:69–96, 1976. Translation by Uwe Parpart of [Can83].

[Chu56] Alonzo Church. *Introduction to Mathematical Logic*. Princeton University Press, Princeton, New Jersey, 1956.

[Cof79] J. Alberto Coffa. The humble origins of Russell's paradox. *Russell*, 33–34:31–37, 1979.

[Con87] Francisco Rodríguez Consuegra. Russell's logicist definitions of numbers, 1898–1913: Chronology and significance. *History and Philosophy of Logic*, 8:141–169, 1987.

[Cor80] John Corcoran. Categoricity. *History and Philosophy of Logic*, 1:187–207, 1980.

[Cur51] Haskell B. Curry. *Outlines of a Formalist Philosophy of Mathematics*. North-Holland Publishing Company, New York, 1951.

[Dau71] Joseph Warren Dauben. The trigonometric background to Georg Cantor's theory of sets. *Archive for History of Exact Sciences*, 7:181–216, 1971.

[Dau79] Joseph Warren Dauben. *Georg Cantor: His Mathematics and Philosophy of the Infinite*. Princeton University Press, Princeton, New Jersey, 1979.

[Dau80] Joseph Warren Dauben. The development of Cantorian set theory. In I. Grattan-Guiness, editor, *From the Calculus to Set Theory, 1630–1910*, pp. 181–219. Gerald Duckworth & Co., London, 1980.

[Dau88] Joseph W. Dauben. Abraham Robinson and nonstandard analysis: History, philosophy, and foundations of mathematics. In William Aspray and Philip Kitcher, editors, *History and Philosophy of Modern Mathematics*, number 11 in Minnesota Studies in the Philosophy of Science, pp. 177–200. University of Minnesota Press, Minneapolis, Minnesota, 1988.

[Ded72] Richard Dedekind. *Stetigkeit und irrationale Zahlen*. Viewig, Braun-schweig, 1872. All page references are to the translation by Wooster Woodruff Beman, pp. 1–27, in Richard Dedekind, *Essays on the Theory of Numbers* (Dover Publications, New York, 1963), a reprinting of the 1901 Open Court edition.

[Ded88] Richard Dedekind. *Was sind und was sollen die Zahlen?* Viewig, Braunschweig, 1888. (2d ed., 1893; 3d ed., 1911). All page references are to the translation of the second edition by Wooster Woodruff Beman, pp. 28–115, in Richard Dedekind, *Essays on the Theory of Numbers* (Dover Publications, New York, 1963), a reprinting of the 1901 Open Court edition.

[Des54] René Descartes. *The Geometry*. Translated by David Eugene Smith and Marcia L. Latham. Dover Publications, New York, 1954. With a facsimile of the first (1637) edition. Reprinting of the 1925 Open Court edition.

[Det86] Michael Detlefsen. *Hilbert's Program*. Number 182 in Synthese Library. Reidel, Boston, 1986.

[DMR76] Martin Davis, Yuri Matijacevič, and Julia Robinson. Hilbert's tenth problem. Diophantine equations: Positive aspects of a negative solution. *Proceedings of Symposia in Pure Mathematics*, 28:223–378, 1976.

[Dum67] Michael A. E. Dummett. Frege, Gottlob. In Paul Edwards, editor, *The Encyclopedia of Philosophy*, vol. 3, pp. 225–237. Macmillan and The Free Press, New York, 1967.

[Dum75] Michael A. E. Dummett. The philosophical basis of intuitionistic logic. In H. E. Rose and J. C. Shepherdson, editors, *Logic Colloquium '73. Proceedings of the Logic Colloquium (Bristol, 1973)*, number 80 in Studies in Logic and the Foundations of Mathematics, pp. 5–40. North-Holland Publishing Company, New York, 1975. All page references are to the reprinting as pp. 97–129 [BP83]. Also reprinted as pp. 215–247 [Dum78].

[Dum77] Michael A. E. Dummett. *Elements of Intuitionism*. Oxford Logic Guides. Clarendon Press, Oxford, 1977.

[Dum78] Michael A. E. Dummett. *Truth and Other Enigmas*. Harvard University Press, Cambridge, Massachusetts, 1978.

[Edw88] Harold Edwards. Kronecker's place in history. In William Aspray and Philip Kitcher, editors, *History and Philosophy of Modern Mathematics*, number 11 in Minnesota Studies in the Philosophy of Science, pp. 139–144. University of Minnesota Press, Minneapolis, Minnesota, 1988.

[End72] Herbert B. Enderton. *A Mathematical Introduction to Logic*. Academic Press, Orlando, Florida, 1972.

[Eva83] Diane Wilkinson Evans. *Understanding Zero and Infinity in the Early School Years*. Ph.D. thesis, University of Pennsylvania, 1983.

[FBHL73] A. Fraenkel, Yehoshua Bar-Hillel, and Azriel Levy. *Foundations of Set Theory*, 2d rev. ed. Number 67 in Studies in Logic and the Foundations of Mathematics. North-Holland Publishing Company, Amsterdam, 1973.

[Fef77] Solomon Feferman. Theories of finite type related to mathematical practice. In Jon Barwise, editor, *Handbook of Mathematical Logic*, number 90 in Studies in Logic and the Foundations of Mathematics, pp. 913–971. North-Holland Publishing Company, Amsterdam, 1977.

[Fef88] Solomon Feferman. Weyl vindicated: "Dass Kontinuum" 70 years later. In *Temi e prospettive della logica e della filosofia contemporanee: Atti del congresso, Cesena 7–10 gennaio, 1987*, vol. 1, pp. 59–93. CLUEB, Bologna, 1988.

[Fef91] Solomon Feferman. Reflecting on incompleteness. *Journal of Symbolic Logic*, 56:1–49, 1991.

[Fie89] Hartry Field. *Realism, Mathematics and Modality*. Basil Blackwell, Oxford, 1989.

[Fra21] A. Fraenkel. Über die Zermelosche Begründung der Mengenlehre. *Jahresbericht der Deutschen Mathematiker-Vereinigung (Angelegenheiten)*, 30:97–98, 1921.

[Fra22a] A. Fraenkel. Der Begriff "definit" und die Unabhängigkeit des Auswahlsaxioms. In *Sitzungsberichte der Preussischen Akademie der Wissenschaften, Physikalisch-mathematische Klasse*, pp. 253–257. 1922. Translation by Beverly Woodward, pp. 285–289 [vH67].

[Fra22b] A. Fraenkel. Zu den Grundlagen der Cantor-Zermeloschen Mengenlehre. *Mathematische Annalen*, 86:230–237, 1922.

[Fra23] A. Fraenkel. *Einleitung in die Mengenlehre*, 2d ed. Springer-Verlag, Berlin, 1923.

[Fra25] A. Fraenkel. Unterzuchen über die Grundlagen der Mengenlehre. *Mathematische Zeitschrift*, 22:250–273, 1925.

[Fra26] A. Fraenkel. Axiomatische theorie der geordneten Mengen (unterzuchen über die Grundlagen der Mengenlehre II). *Journal für die reine und angewandte Mathematik*, 155:129–158, 1926.

[Fra27] A. Fraenkel. *Zehn Vorlesungen über die Grundlegung der Mengenlehre*. B. G. Teubner, Leipzig and Berlin, 1927.

[Fra28] A. Fraenkel. *Einleitung in die Mengenlehre*, 3d ed. Springer-Verlag, Berlin, 1928.

[Fra32] A. Fraenkel. Axiomatische theorie der Wohlordnung (unterzuchen über die Grundlagen der Mengenlehre III). *Journal für die reine und angewandte Mathematik*, 167:1–11, 1932.

[Fre95] Gottlob Frege. Kritische Beleuchtung einiger Punkte in E. Schröder's Vorlesungen über die Algebra der Logik. *Archiv für systematische Philosophie*, 1:433–456, 1895. All page references are to the translation by Peter Geach, pp. 210–228 [Fre84].

[Fre67] Gottlob Frege. Letter to Russell. In Jean van Heijenoort, editor, *From Frege to Gödel: A Source Book in Mathematical Logic, 1879–1931*, pp. 127–128. Translated by Beverly Woodward. Harvard University Press, Cambridge, Massachusetts, 1967.

[Fre80] Gottlob Frege. *Philosophical and Mathematical Correspondence*. Abridged from the German edition by Brian McGuinness, translated by Hans Kaal, and edited by Gottfried Gabriel *et al.* University of Chicago Press, Chicago, 1980.

[Fre84] Gottlob Frege. *Collected Papers on Mathematics, Logic, and Philosophy*. Edited by Brian McGuinness. Basil Blackwell, New York, 1984.

[Fri71] Harvey Friedman. Higher set theory and mathematical practice. *Annals of Mathematical Logic*, 2:326–357, 1971.

[Geo85] Alexander George. Skolem and the Löwenheim-Skolem theorem: A case study of the philosophical significance of mathematical results. *History and Philosophy of Logic*, 6:75–89, 1985.

[Geo91] Alexander George. Goldbach's conjecture can be decided in one minute: On an alleged problem for intuitionism. *Proceedings of the Aristotelian Society*, 91, Part 2:187–189, 1990–1991.

[GG71] I. Grattan-Guiness. The correspondence between Georg Cantor and Philip Jourdain. *Jahresbericht der Deutschen Mathematiker-Vereinigung*, 73:111–130, 1971.

[GG74] I. Grattan-Guiness. The rediscovery of the Cantor-Dedekind correspondence. *Jahresbericht der Deutschen Mathematiker-Vereinigung*, 76:104–139, 1974.

[GG77] I. Grattan-Guiness. *Dear Russell–Dear Jourdain*. Columbia University Press, New York, 1977.

[GG80a] I. Grattan-Guiness. The emergence of mathematical analysis and its foundational progress, 1780–1880. In I. Grattan-Guiness, editor, *From the Calculus to Set Theory, 1630–1910*, pp. 94–148. Gerald Duckworth & Co., London, 1980.

[GG80b] I. Grattan-Guiness, editor. *From the Calculus to Set Theory, 1630–1910*. Gerald Duckworth & Co., London, 1980.

[GGR72] I. Grattan-Guiness and J. R. Ravetz. *Joseph Fourier 1768–1830*. MIT Press, Cambridge, Massachusetts, 1972.

[GJ79] Michael R. Garey and David S. Johnson. *Computers and Intractability: A Guide to the Theory of NP-Completeness*. W. H. Freeman, San Francisco, 1979.

[Göd30] Kurt Gödel. Einige metamathematische Resultate über Entscheidungsdefinitheit und Widerspruchsfreiheit. *Anzeiger der Akademie der Wissenschaften in Wien*, 67:214–215, 1930. Translated by Stefan Bauer-Mengelberg as pp. 595–597 [vH67].

[Göd34] Kurt Gödel. Review of [Sko33]. *Zentralblatt für Mathematik und ihre Grenzgebiete*, 7:193–194, 1934. Translated as vol. 1, pp. 379, 381 [Göd90].

[Göd40] Kurt Gödel. *The Consistency of the Axiom of Choice and of the Generalized Continuum Hypothesis with the Axioms of Set Theory*. Number 3 in Annals of Mathematics Studies. Princeton University Press, Princeton, New Jersey, 1940. Lecture notes by George W. Brown. Reprinted with additions in 1951 and 1966. All page references are to the reprinting as vol. 2, pp. 33–101 [Göd90].

[Göd44] Kurt Gödel. Russell's mathematical logic. In Paul Arthur Schilpp, editor, *The Philosophy of Bertrand Russell*, number 5 in Library of Living

Philosophers, pp. 123–153. Northwestern University Press, Evanston, Illinois, 1944. All page references are to the reprinting as pp. 447–469 [BP83].

[Göd47] Kurt Gödel. What is Cantor's continuum problem? *American Mathematical Monthly*, 54:515–525, 1947. Errata, vol. 55, p. 151. Revised and expanded for the first (1964) edition of [BP83]. All page references are either to the reprinting of the revised and expanded version as pp. 470–485 [BP83] or to the reprinting of the original version as vol. 2, pp. 176–187 [Göd90].

[Göd90] Kurt Gödel. *Collected Works*. Edited by Solomon Feferman, John W. Dawson, Jr., Stephen C. Kleene, Gregory H. Moore, Robert M. Solovay, and Jean van Heijenoort. Oxford University Press, New York, vol. 1, 1986; vol. 2, 1990.

[Gol90] Warren Goldfarb. Herbrand's theorem and the incompleteness of arithmetic. *Iyyun, A Jerusalem Philosophical Quarterly*, 39:45–64, 1990.

[Goo57] Reuben Louis Goodstein. *Recursive Number Theory*. Studies in Logic and the Foundations of Mathematics. North-Holland Publishing Company, Amsterdam, 1957.

[Gos75] J. C. B. Gosling. *Plato: Philebus*. Clarendon Press, Oxford, 1975.

[Gri88] Nicholas Griffin. The Tiergarten programme. *Russell*, n.s. 8:19–34, 1988.

[Gro80] E. Grosholz. Descartes' unification of algebra and geometry. In Stephen Gaukroger, editor, *Descartes, Mathematics and Physics*. Harvester, Hassocks, England, 1980.

[Hal84] Michael Hallett. *Cantorian Set Theory and Limitation of Size*. Clarendon Press, Oxford, 1984.

[Hal90] Michael Hallett. Physicalism, reductionism & Hilbert. In A. D. Irvine, editor, *Physicalism in Mathematics*, number 45 in The University of Western Ontario Series in Philosophy of Science, pp. 183–257. Kluwer Academic Publishers, Dordrecht, The Netherlands, 1990.

[Haw75] Thomas Hawkins. *Lebesgue's Theory of Integration: Its Origins and Development*, 2d ed. Chelsea Publishing, New York, 1975.

[Haw80] Thomas Hawkins. The origins of modern theories of integration. In I. Grattan-Guiness, editor, *From the Calculus to Set Theory, 1630–1910*, pp. 149–180. Gerald Duckworth & Co., London, 1980.

[HB70] David Hilbert and Paul Bernays. *Grundlagen der Mathematik*, 2d ed. 2 vols. Number 40 and Number 50 in Die Grundlehren der mathematischen Wissenschaften in Einzeldarstellungen. Springer-Verlag, New York, 1968 and 1970. First edition, 1934 and 1939.

[HCV52] David Hilbert and Stephan Cohn-Vossen. *Geometry and the Imagination*. Chelsea, New York, 1952. This work is a translation by P. Nemenyi of *Anschauliche Geometrie* (Verlag von Julius Springer, Berlin, 1932).

[Hea56] Sir Thomas Heath, editor. *Euclid; The Thirteen Books of The Elements*. Dover Publications, New York, 1956. Reprinting of the second edition (Cambridge University Press, Cambridge, England, 1925).

[Hea81] Sir Thomas Heath. *A History of Greek Mathematics*. Dover Publications, New York, 1981. Reprinting with corrections of the original Clarendon Press edition (Oxford, 1921).

[Hel89] Geoffrey Hellman. *Mathematics without Numbers: Towards a Modal-Structural Interpretation*. Clarendon Press, Oxford, 1989.

[Hen47] Leon Henkin. *The Completeness of Formal Systems*. Ph.D. thesis, Princeton University, Princeton, New Jersey, 1947.

[Her30] Jacques Herbrand. *Recherches sur la théorie de la démonstration*. Ph.D. thesis, University of Paris, 1930. All page references are to the translation of chapter 5 by Burton Dreben and Jean van Heijenoort, pp. 529–579 [vH67].

[Her31] Jacques Herbrand. Unsigned note on Herbrand's thesis. *Annales de l'Université de Paris*, 6:186–189, 1931. All page references are to the translation as pp. 272–276 [Her71].

[Her71] Jacques Herbrand. *Jacques Herbrand: Logical Writings*. Edited and translated by Warren D. Goldfarb. Harvard University Press, Cambridge, Massachusetts, 1971.

[Hil26] David Hilbert. Über das Unendliche. *Mathematische Annalen*, 95:161–190, 1926. Translation by Stefan Bauer-Mengelberg, pp. 369–392 [vH67]. Partial translation by Erna Putnam and Gerald J. Massey, 1964. All page references are to the second edition of the Putnam and Massey translation, pp. 183–201 [BP83].

[Hil28] David Hilbert. Die Grundlagen der mathematik. *Abhandlungen aus dem mathematischen Seminar der Hamburgischen Universität*, 6:65–85, 1928. All references are to the translation by Stefan Bauer-Mengelberg and Dagfinn Føllesdal, pp. 464–479 [vH67].

[Hin78] Peter G. Hinman. *Recursion-Theoretic Hierarchies*. Perspectives in Mathematical Logic. Springer-Verlag, New York, 1978.

[HMŠS85] Leo A. Harrington, Michael D. Morley, A. Ščedrov, and Stephen G. Simpson, editors. *Harvey Friedman's Research on the Foundations of Mathematics*. North-Holland Publishing Company, Amsterdam, 1985.

[Hyl90] Peter Hylton. Logic in Russell's logicism. In David Bell and Neil Cooper, editors, *The Analytic Tradition: Meaning, Thought and Knowledge*, number 1 in Philosophical Quarterly Monographs, pp. 137–172. Basil Blackwell, Oxford, 1990.

[IK80] Shôkichi Iyanaga and Yukiyosi Kawada, editors. *Encyclopedic Dictionary of Mathematics*. Translated by the Mathematical Society of Japan with the cooperation of the American Mathematical Society. MIT Press, Cambridge, Massachusetts, 1980.

[Jec78] Thomas Jech. *Set Theory*. Pure and Applied Mathematics. Academic Press, New York, 1978.

[JK71] Ronald B. Jensen and Carol Karp. Primitive recursive set functions. In Dana Scott, editor, *Axiomatic Set Theory*, number 13, part 1 in Proceedings of Symposia in Pure Mathematics, pp. 143–176. American Mathematical Society, Providence, Rhode Island, 1971.

[Jou15] Philip E. B. Jourdain. Introduction to [Can15], 1915.

[Ken80] Hubert C. Kennedy. *Peano, Life and Works of Guiseppe Peano*. Reidel, Boston, 1980.

[Kit83] Philip Kitcher. *The Nature of Mathematical Knowledge*. Oxford University Press, New York, 1983.

[Kit88] Philip Kitcher. Mathematical naturalism. In William Aspray and Philip Kitcher, editors, *History and Philosophy of Modern Mathematics*, number 11 in Minnesota Studies in the Philosophy of Science, pp. 293–325. University of Minnesota Press, Minneapolis, Minnesota, 1988.

[Kle76] Virginia Klenk. Intended models and the Löwenheim-Skolem theorem. *Journal of Philosophical Logic*, 5:475–489, 1976.

[Kle88] Stephen C. Kleene. The work of Kurt Gödel. In Stuart G. Shanker, editor, *Gödel's Theorem in Focus*, pp. 48–71. Routledge, New York, 1988.

[Kli72] Morris Kline. *Mathematical Thought from Ancient to Modern Times*. Oxford University Press, Oxford, 1972.

[Knu74] Donald Ervin Knuth. *Surreal Numbers: How Two Ex-Students Turned On to Pure Mathematics and Found Total Happiness: A Mathematical Novelette.* Addison-Wesley Publishing Company, Reading, Massachusetts, 1974.

[Kre71] Georg Kreisel. Observations on popular discussions of foundations. In Dana S. Scott, editor, *Axiomatic Set Theory*, number 13, part 1 in Proceedings of Symposia in Pure Mathematics, pp. 189–198. American Mathematical Society, Providence, Rhode Island, 1971.

[Lak78] Imre Lakatos. Cauchy and the continuum: The significance of nonstandard analysis for the history and philosophy of mathematics (edited by J. P. Cleave). In John Worrall and Gregory Currie, editors, *Mathematics, Science, and Epistemology*, pp. 43–60. Cambridge University Press, Cambridge, England, 1978. Volume 2 of Lakatos's unpublished papers.

[Lan87] Gregory Landini. Russell's substitutional theory of classes and relations. *History and Philosophy of Logic*, 8:171–200, 1987.

[Lan89] Gregory Landini. New evidence concerning Russell's substitutional theory of classes. *Russell*, n.s. 9:26–42, 1989.

[Lav91] Shaughan Lavine. Is quantum mechanics an atomistic theory? *Synthese*, 89:253–271, 1991.

[Lav92] Shaughan Lavine. Penelope Maddy: Realism in mathematics. *Journal of Philosophy*, 89:321–326, 1992. Book review.

[Lav95] Shaughan Lavine. Finite mathematics. Forthcoming (1995) in *Synthese*.

[Lea77] Jonathan Lear. Sets and semantics. *Journal of Philosophy*, 74:86–102, 1977.

[Lev68] Azriel Lévy (Levy). On von Neumann's axiom system for set theory. *American Mathematical Monthly*, 75:762–763, 1968.

[Lev69] Azriel Lévy (Levy). The definability of cardinal numbers. In Jack J. Bulloff, Thomas C. Holyoke, and S. W. Hahn, editors, *Foundations of Mathematics: Symposium Papers Commemorating the Sixtieth Birthday of Kurt Gödel*, pp. 15–38. Springer-Verlag, New York, 1969.

[Lev79] Azriel Levy. *Basic Set Theory*. Perspectives in Mathematical Logic. Springer-Verlag, New York, 1979.

[Löw15] Leopold Löwenheim. Über Möglichkeiten im Relativkalkül. *Mathematische Annalen*, 76:447–470, 1915. All page references are to the translation by Stefan Bauer-Mengelberg, pp. 232–251 [vH67].

[Mad88a] Penelope Maddy. Believing the axioms. I. *Journal of Symbolic Logic*, 53:481–511, 1988.

[Mad88b] Penelope Maddy. Believing the axioms. II. *Journal of Symbolic Logic*, 53:736–764, 1988.

[Mad89] Penelope Maddy. The roots of contemporary Platonism. *Journal of Symbolic Logic*, 54:1121–1144, 1989.

[Mad90] Penelope Maddy. *Realism in Mathematics*. Clarendon Press of Oxford University Press, Oxford, 1990.

[Mad92] Penelope Maddy. Indispensability and practice. *Journal of Philosophy*, 89:275–289, 1992.

[Mah67] F.-K. Mahn. Zu den primitiv-rekursive Funktionen über einem Bereich endlicher Mengen. *Archiv für mathematische Logik und Grundlagenforschung*, 10:30–33, 1967.

[Mah73] M. S. Mahoney. *The Mathematical Career of Pierre de Fermat*. Princeton University Press, Princeton, New Jersey, 1973.

[Mak77] M. Makkai. Admissible sets and infinitary logic. In Jon Barwise, editor, *Handbook of Mathematical Logic*, number 90 in Studies in Logic and the Foundations of Mathematics, pp. 233–281. North-Holland Publishing Company, Amsterdam, 1977.

[Mal36] Anatolii Ivanovich Maltsev. Untersuchungen aus dem Gebiete der mathematischen Logik. *Matematicheskii sbornik*, 1:323–336, 1936. Translated as pp. 1–14 of [Mal71]. The author's name is also transliterated Mal'cev.

[Mal41] Anatolii Ivanovich Maltsev. On a general method for obtaining local theorems in group theory. *Ivanovskii Gosudarstvennii Pedagogicheskii Institut im. D. A. Furmanova. Ivanovskoye matematicheskoye obshchestvo. Ucheniye zapiski*, 1:3–9, 1941. In Russian. Translated as pp. 15–21 of [Mal71]. The author's name is also transliterated Mal'cev.

[Mal71] Anatolii Ivanovich Maltsev. *The Metamathematics of Algebraic Systems: Collected Papers 1936–1967*. Edited and translated by Benjamin F. Wells III. North-Holland Publishing Company, Amsterdam, 1971. The author's name is also transliterated Mal'cev.

[Man89] Paolo Mancosu. The metaphysics of the calculus: A foundational debate in the Paris Academy of Sciences 1700–1706. *Historia Mathematica*, 16:224–248, 1989.

[Mar75] Donald A. Martin. Borel determinacy. *Annals of Mathematics*, 102:363–371, 1975.

[Mar91] Norman L. Martin. "The Song That Doesn't End." Norman Martin Music (BMI), 1989–1991.

[McC83] Charles McCarty. Intuitionism: An introduction to a seminar. *Journal of Philosophical Logic*, 12:105–149, 1983.

[Med91] Fyodor A. Medvedev. *Scenes from the History of Real Functions*. Number 7 in Science Networks—Historical Studies. Birkhäuser, Boston, 1991. Translated from the Russian work of 1975 by Roger Cooke, with corrections and new references supplied by the author.

[MG81] Gregory H. Moore and Alejandro Garciadiego. Burali-Forti's paradox: A reappraisal of its origins. *Historia Mathematica*, 8:319–350, 1981.

[Mir17a] D. Mirimanoff. Les antinomies de Russell et de Burali-Forti et le problème fondamental de la théorie des ensembles. *L'Enseignement Mathématique*, 19:37–52, 1917.

[Mir17b] D. Mirimanoff. Remarques sur la théorie des ensembles et les antinomies Cantoriennes. I. *L'Enseignement Mathématique*, 19:209–217, 1917.

[Mon65] Richard Montague. Set theory and higher-order logic. In J. N. Crossley and Michael A. E. Dummett, editors, *Formal Systems and Recursive Functions*, Studies in Logic and the Foundations of Mathematics, pp. 131–148. North-Holland Publishing Company, Amsterdam, 1965.

[Moo78] Gregory H. Moore. The origins of Zermelo's axiomatization of set theory. *Journal of Philosophical Logic*, 7:305–329, 1978.

[Moo80] Gregory H. Moore. Beyond first-order logic: The historical interplay between mathematical logic and axiomatic set theory. *History and Philosophy of Logic*, 1:95–137, 1980.

[Moo82] Gregory H. Moore. *Zermelo's Axiom of Choice, Its Origins, Development, and Influence*. Number 8 in Studies in the History of Mathematics and Physical Sciences. Springer-Verlag, New York, 1982.

[Moo86] Gregory H. Moore. Note to 1947 and 1964. In Solomon Feferman, John W. Dawson, Jr., Stephen C. Kleene, Gregory H. Moore, Robert M. Solovay, and Jean van Heijenoort, editors, *Kurt Gödel, Collected Works*, vol. 2, pp. 154–175. Oxford University Press, New York, 1986.

[Moo88a] Gregory H. Moore. The emergence of first-order logic. In William Aspray and Philip Kitcher, editors, *History and Philosophy of Modern Mathematics*, number 11 in Minnesota Studies in the Philosophy of Science, pp. 95–135. University of Minnesota Press, Minneapolis, Minnesota, 1988.

[Moo88b] Gregory H. Moore. The roots of Russell's paradox. *Russell*, n.s. 8:46–56, 1988.

[Myc81a] Jan Mycielski. Finitistic real analysis. *Real Analysis Exchange*, 6:127–130, 1980–1981.

[Myc81b] Jan Mycielski. Analysis without actual infinity. *Journal of Symbolic Logic*, 46:625–633, 1981.

[Myc86] Jan Mycielski. Locally finite theories. *Journal of Symbolic Logic*, 51:59–62, 1986.

[Myc89] Jan Mycielski. The meaning of pure mathematics. *Journal of Philosophical Logic*, 18:315–320, 1989.

[Myc92] Jan Mycielski. Quantifier-free versions of first-order logic and their psychological significance. *Journal of Philosophical Logic*, 21:125–147, 1992.

[Nel86] Edward Nelson. *Predicative Arithmetic*. Number 32 in Mathematical Notes. Princeton University Press, Princeton, New Jersey, 1986.

[Par71] Charles Parsons. Ontology and mathematics. *Philosophical Review*, 80:151–176, 1971. All page references are to the reprinting as pp. 37–62 [Par83a].

[Par74] Charles Parsons. Sets and classes. *Noûs*, 8:1–12, 1974. All page references are to the reprinting as pp. 209–220 [Par83a].

[Par76] Charles Parsons. Some remarks on Frege's conception of extension. In Matthias Schirn, editor, *Studien zu Frege*, vol. 1, pp. 265–277. Friedrich Frommann, Stuttgart, 1976.

[Par77] Charles Parsons. What is the iterative conception of set? In R. E. Butts and J. Hintikka, editors, *Logic, Foundations of Mathematics, and Computability Theory. Proceedings of the Fifth International Congress of Logic, Methodology, and the Philosophy of Science (London, Ontario, 1975)*. Reidel, Dordrecht, The Netherlands, 1977. All page references are to the reprinting as pp. 503–529 [Par83a].

[Par80] Charles Parsons. Mathematical intuition. *Proceedings of the Aristotelian Society*, n.s. 80:145–168, 1979–1980.

[Par83a] Charles Parsons. *Mathematics in Philosophy; Selected Essays*. Cornell University Press, Ithaca, New York, 1983.

[Par83b] Charles Parsons. Quine on the philosophy of mathematics. In *Mathematics in Philosophy; Selected Essays*, pp. 176–205. Cornell University Press, Ithaca, New York, 1983.

[Par83c] Charles Parsons. Sets and modality. In *Mathematics in Philosophy; Selected Essays*, pp. 298–341. Cornell University Press, Ithaca, New York, 1983.

[Par86] Charles Parsons. Introductory note to 1944. In Solomon Feferman, John W. Dawson, Jr., Stephen C. Kleene, Gregory H. Moore, Robert M. Solovay, and Jean van Heijenoort, editors, *Kurt Gödel: Collected Works*, vol. 2, pp. 102–118. Oxford University Press, New York, 1986.

[Par90] Charles Parsons. The uniqueness of the natural numbers. *Iyyun, A Jerusalem Philosophical Quarterly*, 39:13–44, 1990.

[Paw89] Janusz Pawlikowski. Remark on locally finite theories. *Abstracts of Papers Presented to the American Mathematical Society*, 10:172, 1989. Abstract.

[Pea89] Guiseppe Peano. Arithmetices principia, nova methodo exposita (Fratres Bocca, Turin, 1889). Reprinted in *Opera scelte,* vol. 2, pp. 20–55 (Edizioni cremonese, Rome, 1958). All page references are to the excerpt translated by Jean van Heijenoort, pp. 85–97 [vH67].

[Poz71] Lawrence Pozsgay. Liberal intuitionism as a basis for set theory. In Dana S. Scott, editor, *Axiomatic Set Theory*, number 13, part 1 in Proceedings of Symposia in Pure Mathematics, pp. 321–330. American Mathematical Society, Providence, Rhode Island, 1971.

[Put67] Hilary Putnam. Mathematics without foundations. *Journal of Philosophy*, pp. 5–22, 1967. All page references are to the reprinting as pp. 295–311 [BP83].

[Put79a] Hilary Putnam. *Mathematics, Matter and Method*, 2d ed. Cambridge University Press, Cambridge, England, 1979.

[Put79b] Hilary Putnam. What is mathematical truth? In *Mathematics, Matter and Method*, 2d ed., pp. 60–78. Cambridge University Press, Cambridge, England, 1979. All page references are to the reprinting as pp. 49–65 of [Tym86].

[QC90] Willard Van Orman Quine and Rudolf Carnap. *Dear Carnap, Dear Van*. Edited and with an Introduction by Richard Creath. University of California Press, Berkeley, 1990.

[Qui61] Willard Van Orman Quine. *From a Logical Point of View*, 2d ed. Harvard University Press, Cambridge, Massachusetts, 1961.

[Qui70] Willard Van Orman Quine. *Philosophy of Logic*. Foundations of Philosophy Series. Prentice-Hall, Englewood Cliffs, New Jersey, 1970.

[Qui86] Willard Van Orman Quine. Reply to Charles Parsons. In Lewis Edward Hahn and Paul Arthur Schilpp, editors, *The Philosophy of W. V. Quine*, number 18 in Library of Living Philosophers, pp. 396–403. Open Court, La Salle, Illinois, 1986.

[Res80] Michael D. Resnik. *Frege and the Philosophy of Mathematics*. Cornell University Press, Ithaca, 1980.

[Röd67] D. Rödding. Primitiv-rekursive Funktionen über einem Bereich endlicher Mengen. *Archiv für mathematische Logik und Grundlagenforschung*, 10:13–29, 1967.

[RT81] B. Rang and W. Thomas. Zermelo's discovery of the "Russell paradox." *Historia Mathematica*, 8:15–22, 1981.

[Rus03] Bertrand Russell. *Principles of Mathematics*. George Allen and Unwin, London, 1903. Second edition (1938) reprinted by W. W. Norton and Company (New York, no date).

[Rus05] Bertrand Russell. On some difficulties in the theory of transfinite numbers and order types. *Proceedings of the London Mathematical Society*, 2d ser. 4:29–53, 1905. All page references are to the reprinting as pp. 135–164 [Rus73].

[Rus08] Bertrand Russell. Mathematical logic as based on the theory of types. *American Journal of Mathematics*, 30:222–262, 1908. All page references are to the reprinting as pp. 152–182 [vH67].

[Rus19] Bertrand Russell. *Introduction to Mathematical Philosophy*. Macmillan, New York, 1919.

[Rus67a] Bertrand Russell. *The Autobiography of Bertrand Russell*. Little, Brown and Company, Boston, 1967.

[Rus67b] Bertrand Russell. Letter to Frege. Translated by Beverly Woodward. In Jean van Heijenoort, editor, *From Frege to Gödel: A Source Book in Mathematical Logic, 1879–1931*, pp. 124–125. Harvard University Press, Cambridge, Massachusetts, 1967.

[Rus73] Bertrand Russell. *Essays in Analysis*. Edited by Douglas Lackey. George Braziller, New York, 1973.

[Sco74] Dana Scott. Axiomatizing set theory. In Thomas J. Jech, editor, *Axiomatic Set Theory*, number 13, part 2 in Proceedings of Symposia in Pure Mathematics, pp. 207–214. American Mathematical Society, Providence, Rhode Island, 1974.

[Sha85] Stewart Shapiro. Second-order languages and mathematical practice. *Journal of Symbolic Logic*, 50:714–742, 1985.

[Sha90] Stewart Shapiro. Second-order logic, foundations, and rules. *Journal of Philosophy*, 87:234–261, 1990.

[Sha91] Stewart Shapiro. *Foundations without Foundationalism*. Clarendon Press, Oxford, 1991.

[Sho67] Joseph R. Shoenfield. *Mathematical Logic*. Addison-Wesley Series in Logic. Addison-Wesley Publishing Company, Reading, Massachusetts, 1967.

[Sko23a] Thoralf Skolem. Begründung der elementaren Arithmetik durch die rekurrierende Denkweise ohne Anwendung scheinbarer Veränderlichen mit unendlichem Ausdehnungsbereich. *Videnskapsselskapets skrifter, I. Matematisk-naturvidenskabelig klasse*, (6), 1923. All page references are to the translation by Stefan Bauer-Mengelberg, pp. 303–333 [vH67].

[Sko23b] Thoralf Skolem. Einige Bemerkungen zur axiomatischen Begründung der Mengenlehre. In *Matematikerkongressen i Helsingfors den 4–7 Juli 1922, Den femte skandinaviska matematikerkongressen, Redogörelse*, pp. 217–232. Akademiska Bokhandeln, Helsinki, 1923. All page references are to the translation by Stefan Bauer-Mengelberg, pp. 291–301 [vH67].

[Sko30] Thoralf Skolem. Einige Bemerkungen zu der Abhandlung von E. Zermelo: "Über die Definheit in der Axiomatik." *Fundamenta Mathematicae*, 15:337–341, 1930.

[Sko33] Thoralf Skolem. Über die Unmöglichkeit einer vollständigen Charakterisierung der Zahlenreihe mittels eines endlichen Axiomensystems. *Norsk Matematisk Forenings, Skrifter (2)*, (10):73–82, 1933.

[Sko34] Thoralf Skolem. Über die Nicht-charakterisierbarkeit der Zahlenreihe mittels endlich oder abzählbar unendlich vieler Aussagen mit ausschliesslich Zahlenvariablen. *Fundamenta Mathematicae*, 23:150–161, 1934.

[Smo77] C. Smorynski. The incompleteness theorems. In Jon Barwise, editor, *Handbook of Mathematical Logic*, number 90 in Studies in Logic and the Foundations of Mathematics, pp. 821–865. North-Holland Publishing Company, Amsterdam, 1977.

[Ste75] Mark Steiner. *Mathematical Knowledge*. Cornell University Press, Ithaca, New York, 1975.

[Ste88] Howard Stein. *Logos,* logic, and *logistiké:* Some philosophical remarks on nineteenth century transformation of mathematics. In William Aspray and Philip Kitcher, editors, *History and Philosophy of Modern Mathematics*, number 11 in Minnesota Studies in the Philosophy of Science, pp. 238–259. University of Minnesota Press, Minneapolis, Minnesota, 1988.

[Str87] Dirk J. Struik. *A Concise History of Mathematics*, 4th ed. Dover Publications, New York, 1987.

[Tai81] William W. Tait. Finitism. *Journal of Philosophy*, 78:524–546, 1981.

[Tai90] William W. Tait. The iterative hierarchy of sets. *Iyyun, A Jerusalem Philosophical Quarterly*, 39:65–79, 1990.

[Tap93] Jamie Tappenden. The liar and Sorites paradoxes: Toward a unified treatment. *Journal of Philosophy*, 90:551–577, 1993.

[Tar38] Alfred Tarski. Über unerreichbare Kardinalzahlen. *Fundamenta Mathematicae*, 30:68–89, 1938.

[TD88] Anne Sjerp Troelstra and Dirk van Dalen. *Constructivism in Mathematics: An Introduction*. 2 vols. Number 121 in Studies in Logic and the Foundations of Mathematics. North-Holland Publishing Company, New York, 1988.

[Tym86] Thomas Tymoczko, editor. *New Directions in the Philosophy of Mathematics*. Birkhäuser, Boston, 1986.

[Urq88] Alasdair Urquhart. Russell's zigzag path to the ramified theory of types. *Russell*, n.s. 8:82–91, 1988.

[Vau67] Robert L. Vaught. Axiomatizability by a schema. *Journal of Symbolic Logic*, 32:473–479, 1967.

[Vau85] Robert L. Vaught. *Set Theory, An Introduction*. Birkhäuser, Boston, 1985.

[Vau86] Robert Vaught. Introductory note to 1934c and 1935. In Solomon Feferman, John W. Dawson, Jr., Stephen C. Kleene, Gregory H. Moore, Robert M. Solovay, and Jean van Heijenoort, editors, *Kurt Gödel, Collected Works*, vol. 1, pp. 376–379. Oxford University Press, New York, 1986.

[VB89] Jean Paul Van Bendegem. Foundations of mathematics or mathematical practice: Is one forced to choose? *Philosophica*, 43:197–213, 1989.

[Vel93] Daniel J. Velleman. Constructivism liberalized. *Philosophical Review*, 102:59–84, 1993.

[vH67] Jean van Heijenoort, editor. *From Frege to Gödel: A Source Book in Mathematical Logic, 1879–1931*. Harvard University Press, Cambridge, Massachusetts, 1967.

[vN23] John von Neumann. Zur Einführung der transfiniten Zahlen. *Acta litterarum ac scientiarum Regiae Universitatis Hungaricae Francisco-Josephinae, Sectio scientiarum mathematicarum*, 1:199–208, 1923. Reprinted as pp. 24–33 [vN61]. All page references are to the translation by Jean van Heijenoort, pp. 347–354 [vH67].

[vN25] John von Neumann. Eine Axiomatisierung der Mengenlehre. *Journal für die reine und angewandte Mathematik*, 154:219–240, 1925. Errata 155:128. Reprinted as pp. 34–56 [vN61]. All page references are to the translation by Stefan Bauer-Mengelberg and Dagfin Føllesdal, pp. 394–413 [vH67].

[vN28] John von Neumann. Über die Definition durch transfinite Induktion und verwandte Fragen der allgemeinen Mengenlehre. *Mathematische Annalen*, 99:373–391, 1928. All page references are to the reprinting as vol. 1, pp. 320–338 [vN61].

[vN29] John von Neumann. Über eine Widerspruchfreiheitsfrage in der axiomatischen Mengenlehre. *Journal für die reine und angewandte Mathematik*,

160:227–241, 1929. All page references are to the reprinting as vol. 1, pp. 494–508 [vN61].

[vN61] John von Neumann. *Collected Works*. Edited by A. H. Taub. Pergamon Press, New York, 1961.

[Vol86] Klaus Volkert. Die Geschichte der pathologischen Funktionen—Ein Beitrag zur Entstehung der mathematischen Methodologie. *Archive for History of Exact Sciences*, 37:193–232, 1986.

[Vop79] Petr Vopěnka. *Mathematics in the Alternative Set Theory*. Teubner Verlagsgesellschaft, Leipzig, 1979.

[Wan74] Hao Wang. *From Mathematics to Philosophy*. Routledge and Kegan Paul, London, 1974. Chapter 6, "The concept of set," is reprinted as pp. 530–570 [BP83].

[Wan86] Hao Wang. Theory and practice in mathematics. In Thomas Tymoczko, editor, *New Directions in the Philosophy of Mathematics*, pp. 131–152. Birkhäuser, Boston, 1986.

[Wey10] Hermann Weyl. Über die Definitionen der mathematischen Grundbegriffe. *Mathematish-naturwissenschaftliche Blätter*, 7:93–95, 109–113, 1910. Reprinted as vol. 1 , pp. 298–304 [Wey68].

[Wey18] Hermann Weyl. *Das Kontinuum*. Veit, Leipzig, 1918.

[Wey46] Hermann Weyl. Mathematics and logic. A brief survey serving as a preface to a review of "The Philosophy of Bertrand Russell." *American Mathematical Monthly*, 53:2–13, 1946. Reprinted as vol. 4, pp. 268–279 [Wey68].

[Wey68] Hermann Weyl. *Gesammelte Abhandlungen*. Edited by K. Chandrasekharan. Springer-Verlag, New York, 1968.

[Whi67] Derek T. Whiteside, editor. *The Mathematical Works of Isaac Newton*. Johnson Reprint, New York, 1967.

[WR57] Alfred North Whitehead and Bertrand Russell. *Principia Mathematica*, 2d ed., vol. 1. Cambridge University Press, Cambridge, England, 1957.

[You76] A. P. Youschkevitch. The concept of function up to the middle of the 19th century. *Archive for History of Exact Sciences*, 16:37–85, 1976.

[You91] Palle Yourgrau. *The Disappearance of Time—Kurt Gödel and the Idealistic Tradition in Philosophy*. Cambridge University Press, Cambridge, England, 1991.

[YV70] A. S. Yessenin-Volpin. The ultra-intuitionistic criticism and the antitraditional program for foundations of mathematics. In A. Kino, John Myhill, and Richard Eugene Vesley, editors, *Intuitionism and Proof Theory. Proceedings of the Summer Conference at Buffalo, New York, 1968*, pp. 3–45. North-Holland Publishing Company, Amsterdam, 1970. The author's name is also transliterated Esenin-Vol′pin.

[Zar82] Andrzej Zarach. Unions of ZF⁻-models which are themselves ZF⁻-models. In D. van Dalen, D. Lascar, and T. J. Smiley, editors, *Logic Colloquium '80*, number 108 in Studies in Logic and the Foundations of Mathematics, pp. 315–342. North-Holland Publishing Company, Amsterdam, 1982.

[Zer04] Ernst Zermelo. Beweis, daß jede Menge wohlgeordnet werden kann. *Mathematische Annalen*, 59:514–516, 1904. All page references are to the translation by Stefan Bauer-Mengelberg, pp. 139–141 [vH67].

[Zer08a] Ernst Zermelo. Neuer Beweis für die Möglichkeit einer Wohlordnung. *Mathematische Annalen*, 65:107–128, 1908. All page references are to the translation by Stefan Bauer-Mengelberg, pp. 183–198 [vH67].

[Zer08b] Ernst Zermelo. Untersuchungen über die Grundlagen der Mengenlehre I. *Mathematische Annalen*, 65:261–281, 1908. All page references are to the translation by Stefan Bauer-Mengelberg, pp. 200–215 [vH67].

[Zer29] Ernst Zermelo. Über den Begriff der Definheit in der Axiomatik. *Fundamenta Mathematicae*, 14:339–344, 1929.

[Zer30] Ernst Zermelo. Über Grenzzahlen und Mengenbereiche. *Fundamenta Mathematicae*, 16:29–47, 1930.

[Zer32] Ernst Zermelo. Über Stufen der Quantifikation und die Logik des Unendlichen. *Jahresbericht der Deutschen Mathematiker-Vereinigung (Angelegenheiten)*, 41:85–88, 1932.

[Zer35] Ernst Zermelo. Grundlagen einer allgemeinen Theorie der mathematischen Satzsysteme. *Fundamenta Mathematicae*, 25:136–146, 1935.

Index

ℵ, 8n, 97, 209
α-recursion theory, 96
Abel, N., 34, 35
abstract structure, 2n, 46–47
abstractness, 163, 165, 221; *See also*
    remoteness
  attitudes toward, 246
  finite and infinite mathematical objects,
      contrasted, 2, 162, 164
  finite mathematics, 254
Academy of Science of Paris, 29
Ackermann, W., [Ack37], 210
Aczel, P., [Acz88], 140, 140n
AD, *See* determinacy, axiom of
adequalities, in finite analysis, 280
algebra, 13, 21, 22, 112, 187, 190–191, 192
algebraic numbers, 42, 90, 91
analytic expression, *See* function, rule-
    governed
ancestral, 224n
anti-foundation, axiom of, 131, 140, 145–146,
    227n, 230n, 239
antifoundationalism, *See* foundationalism
any vs. every, 75–76, 187, 190, 192, 207, 208;
    See also schemas
Archimedes, 15
artists, 166
Asser, G., [Ass60], 209–210
available objects, *See* availability, semiformal
    account
availability, *See also* indefinitely large, size,
    axioms of
  degrees of, 261–262
    not temporal, 262, 299

examples, 255, 261, 262–263
finite set theory, 293, 294–295, 295–296,
    297, 298–299, 299–300, 313, 321,
    322
how specified in finite mathematics, 258,
    264
imposes a lower bound, 261
indefinitely large size, relation to, 262
intuitionism, 308
as a nontemporal modality, 177–178
schemas not required for formalizing,
    260
semiformal account, 260–261, 260–261n,
    261n, 261–262
availability functions, *See* availability,
    semiformal account
axiom systems, 65; *See also* finite
    mathematics of arbitrary theories;
    indefinitely large, sets, theory of;
    primitive recursive
  choice of, 183, 185, 187
  consistency of, *See* consistency
  for intended domains, 215
    Cantor did not use, 47, 142
    first-order, *See* first-order set theory
    natural, 151–153, 320–322
    not categorical, *See* Skolem, T.,
        paradox of
    quasi-categorical, 238
    require clarification, 9, 106, 239–240,
        243–245
    second-order, *See* second-order set
        theory
    Skolem dismissed, 126, 127–128

349

countable, *See* denumerability; 54n
counting
  ℵ theorem, 97, 98
  Cantor's conception of sets, 3, 54–55, 81, 98, 249
  choice, axiom of, 4
  examples, 166
  experience and mathematical knowledge, 163–164, 166
  finite set theory, 298–300, 321, 322
  finitism, 167
  Frege, G., 55
  idealization, 55, 167–168, 176–177
  infinite too many to be subject to, 247–248, 248–249
  law, 85
  limitation of size, 98
  more than one, of the same infinite set, 45, 54, 81
  not all collections subject to, 54, 76, 319
  ordinal numbers, 3, 54, 81, 242, 300
  power sets, 4
  power sets, axiom of, 95
  predecessor and, 300
  real numbers, 95
  well-ordering, 53–54, 300
  witnessing function, 83
Curry, H., 183–184, 188
  [Cur51], 183
curves
  Bernoulli, Johann, 22
  bounding region of integration, 49, 50
  characteristic triangle, 20
  Descartes, R., took polynomials to determine, 15
  Eudoxus and Archimedes, studied geometry of, 15
  Euler, L., 24
  freely drawn, 24
  Leibniz, G.
    assimilated to sequences of numbers, 18, 19, 20, 285
    suggested functions, 21
  motion, 15, 16, 18
  Newton, I., 21
  Peano, 48, 49
  study of, led to notion of function, 7
  vibrating-string problem, 26–27, 28, 29

cuts, *See* Dedekind cuts
cycloid, 15

d'Alembert, J., 26, 27, 28
Darwin, C., 248n
Dauben, J., 94
  [Dau71], 41
  [Dau79], 38, 44, 47, 49, 92n, 94
  [Dau80], 43, 46
  [Dau88], 288
Davis, M., 309, 310
  [DMR76], 309
Dedekind, R., 35, 47
  [Ded72], 37
  [Ded88], 46, 47, 48
  abstract structure, 46
  dimension, 43
  irrational numbers, theory of, 37, 38
  natural numbers, 47, 48
Dedekind cuts, 11, 38, 92, 94n
deductivism, 185–189
definiteness, 152, 238; *See also* Fraenkel functions
  Fraenkel, A., 129, 132, 133, 142; *See also* Fraenkel functions
  Skolem, T., 120, 125, 132, 133, 134, 139, 142
  von Neumann, J., 130n, 131–132, 142; *See also* classes, von Neumann's conception of
  Weyl, H., 124–125, 133, 142
  Zermelo, E., 117–118, 120, 133, 136, 139, 141, 142, 152
definitional extension, 311, 326–327, 328
denumerability, 50n, 90n; *See also* models, denumerable
  actual, 168, 176–177
  choice, axiom of, 88, 109
  measure, 50
derivative, *See* differentiation
derived set, 40–41, 44–45, 122n, 158, 242, 300
Descartes, R., 15, 287
  [Des54], 15
determinacy
  axiom of, 293n
  Borel, 121n, 178
Detlefsen, M., 198, 201n